D0764231

DISCARDED

A COLD WELCOME

A COLD WELCOME

The Little Ice Age and Europe's Encounter with North America

Sam White

Harvard University Press

Cambridge, Massachusetts

London, England

2017

Copyright © 2017 by the President and Fellows of Harvard College
All rights reserved
Printed in the United States of America

Maps and illustrations copyright © 2017 by Sam White.

First Printing

Library of Congress Cataloging-in-Publication Data

Names: White, Sam, 1980– author.
Title: A cold welcome : the Little Ice Age and Europe's encounter with
    North America / Sam White.
Description: Cambridge, Massachusetts : Harvard University Press, 2017. |
    Includes bibliographical references and index.
Identifiers: LCCN 2017016539 | ISBN 9780674971929 (alk. paper)
Subjects: LCSH: Europeans—North America—History. | Indians of North America—
    First contact with Europeans. | Human beings—Effect of climate on—North America. |
    Archaeology and history—North America. | North America—History—Colonial period,
    ca. 1600–1775. | North America—Discovery and exploration.
Classification: LCC E46 .W48 2017 | DDC 970.01—dc23
LC record available at https://lccn.loc.gov/2017016539

To my parents, and the memory of my grandparents

Contents

Maps

Author's Note

The Atlantic world of the sixteenth and seventeenth centuries can be a confusing place. This book endeavors to make it as approachable as possible without trying to hide its wonderful idiosyncrasies. I have tried to give a sense of the original spelling used by writers of the time while modernizing some of the more difficult early modern orthography in order to make it readable for a modern audience. I made use of existing translations of primary sources when they appeared trustworthy. When quoting or citing significant passages, however, I have almost always checked the original text in languages I can (more or less) read myself: Spanish, French, German, Italian, Latin, Dutch, and Turkish. On occasion, I made adjustments or corrections that seemed appropriate. The one significant exception has been for documents originally gathered from the Simancas archives, which I have not had the opportunity to access. I use the terms *Native Americans, Amerindians, Indians,* and (in Canadian examples) *First Nations* all interchangeably simply to avoid repetition. I have capitalized *Native(s)* when it stands in for *Native American(s)* but left it lowercase when paraphrasing contemporary European descriptions of indigenous populations.

This book breaks with convention in one important respect: all dates have been converted to the Gregorian calendar and never left in the original Julian calendar, which was still in use in Protestant countries. This way, the reader can see the simultaneity of events in different French, Spanish, and English colonies. It also clarifies the time of year in which events took place, reminding us, for example, that many English and early Spanish expeditions got started rather late in the spring or summer and consequently wound up at sea during the Atlantic hurricane season.

Finally, I have tried to keep discussions of climate reconstruction and climate history as simple as possible, avoiding unnecessary jargon and technicalities. Good climate history is rarely as easy as finding a single tree-ring study or one description in a historical source. Many seemingly straightforward statements in this book about temperature, drought, or sea ice represent weeks of research, weighing dozens of relevant sources. I encourage readers interested in learning more about this subject to consult Raymond S. Bradley, *Paleoclimatology: Reconstructing Climates of the Quaternary*, 3rd ed. (Amsterdam: Elsevier, 2015); Sam White, Christian Pfister, and Franz Mauelshagen, eds., *The Palgrave Handbook of Climate History* (London: Palgrave Macmillan, 2018); and the resources of the Climate History Network (http://climatehistory.net).

A COLD WELCOME

Introduction

From 1605 to 1607, three expeditions set out from Denmark in one of the strangest and least remembered episodes of Europe's encounter with the New World. These were not voyages of trade or colonization, but a mission to find the long-lost Viking colonies of Greenland. There had been no record of contact with these Vikings since the mid-fifteenth century, but Denmark's king Christian IV (r. 1588–1648) had never lost faith in their survival. He personally authored the expedition's instructions. He had studied, as he wrote, "old documents, both Norwegian and Icelandic," from the time of the Vikings, and he was convinced he knew where to find their descendants. As for communication, he "did not doubt, that they either understand Icelandic or the old Norwegian language."[1]

The first expedition began with three ships, but after some disagreement among the captains, the flagship separated from the others. Within a month, the two remaining vessels found "a very high ragged lande," as the expedition's English navigator, James Hall, described the southeast coast of Greenland. As they approached within 9 miles (15 kilometers) of shore, the ships found their way blocked, "compassed about with ilandes of ice." They sailed around Cape Farewell at the southern tip of Greenland, in continuous danger from icebergs. Proceeding north at a distance from the coast, they eventually entered a fjord where they encountered Inuit people for the first time. Hall's narrative recounts only a few days of peaceful barter before those Inuit inexplicably turned on the Europeans, slinging stones at their ships. But from other sources we know that the expedition had kidnapped four of the Natives to bring back to Denmark, and that the sailors of the separated flagship had taken another two Inuit men from a nearby fjord. In any case, the frozen, treeless land was not what the sailors had expected. The captain aboard Hall's ship feared mutiny from the crew, and in late June he decided to turn back. By August, all three ships had returned to Denmark.[2]

The second expedition followed one year later. Its five ships caught sight of Greenland in July 1606 and spent a month navigating through icy waters, "the aire very cold, as with us in the moneth of Januarie." The sailors encountered more Inuit settlements and managed some modest exchange of goods. They returned three of the Natives seized on the previous expedition but kidnapped five more. The sailors brought back some minerals from Greenland to assay for precious metals, but they turned out to be nothing more than worthless rocks.[3]

The third and final expedition, in May 1607, proved even more disappointing than the previous two. One of its two ships was driven off course by a storm. The other came close enough to spy the high glaciers of western Greenland but never made land. Even in midsummer the Greenland ice "extended far out to sea," as one contemporary chronicler described; "there was ice piled upon ice so high that it resembled great cliffs." Unable to find an approach, the captain was forced to turn his ship for home, where the "the king of Denmark received his excuses, and the impossibility [of the task]."[4]

The Danish expeditions never found the Viking colonies, of course, because those colonies had vanished more than a century before, victims of a climatic episode known as the Little Ice Age. When the Vikings had first settled Greenland in the tenth century, the world in general and the North Atlantic in particular had been a slightly warmer place.[5] Greenland was never truly green, but fjords in the southeast and southwest of the island provided enough shelter for two modest colonies with their sheep, goats, and few cattle. Relying on these livestock, seal and walrus hunting, and occasional commerce with the Scandinavian mainland, the Greenland Vikings managed to hold on for more than 300 years. Then, apparently suddenly and violently, their western colony disappeared around 1350. Within a century the larger, eastern colony had vanished as well.[6]

In hindsight, the colonies were clearly vulnerable. Their land was remote and cold, bare of trees and fuel, unable to support crops, and even in good times scarcely able to sustain enough pasture for their livestock. They faced competition from the Inuit, who had arrived in Greenland around the same time as the Vikings but with better-adapted technologies, including kayaks and harpoons for fishing and whaling.[7] Meanwhile, the Vikings struggled to uphold a religion, a social hierarchy, and a way of life brought from a distant homeland and disparate environment.

Yet for more than three centuries they did survive. They supported families, churches, and communities on Greenland and even left traces of a settlement on Newfoundland. They adapted their way of life enough to keep alive, while not losing their sense of identity and community. Had the Greenland expeditions actually rediscovered their descendants in 1607, those descendants just might have welcomed the Scandinavian sailors as fellow countrymen.

They did not survive longer because climatic change made a difficult way of life all but impossible. It had always been challenging to keep livestock alive through the long Greenland winters and to find enough food for people and animals during the critical weeks of late winter and early spring. Analysis of ice cores drilled from Greenland glaciers tell a grim tale of colder years with shorter growing seasons during the fourteenth and fifteenth centuries, implying scarce pasture, dying animals, and desperate humans. There are many reasons the Greenland colonies might have died out or moved out eventually. However, the close fit between some of the coldest decades in the ice core records and the most likely dates for the expiration of the colonies offers the best explanation for when and how Viking Greenland really did come to an end.[8]

King Christian IV's confidence in their survival seems all the more remarkable given the poor record of European colonization in northern North America to that date. By the time of the Danish Greenland expeditions of 1605–1607, more than a century had passed since Columbus reached the New World. Dozens of Spanish, French, and English expeditions had explored or tried to settle in the lands of the present United States and Canada. Yet there was still no enduring European colony on the mainland north of Spain's tenuous outpost at St. Augustine, Florida.

Nor had the world stopped cooling since the end of the Viking settlements. During the late 1500s global temperatures fell to new lows, and around 1600 they reached the coldest point for centuries or even millennia. In neighboring Iceland, where Viking settlements had held on through the onset of the Little Ice Age by turning to fishing, the situation was often dire. A long run of harsh seasons had brought a

famine during the years 1602–1604 in which 9,000 people are reported to have died of hunger and disease. Historical records describe
heavy sea ice around the island in 1602, 1604, 1605, 1608, and 1610,
sometimes reaching the southern coast of Iceland even in June. As far
as climate went, it was one of the worst times in centuries to sail the
North Atlantic.[9]

Nevertheless, during the same years that Denmark searched in vain
for the lost Viking colonies, England, France, and Spain launched four
major colonizing enterprises in North America. At first all four seemed
as vulnerable to the Little Ice Age as the Viking settlements on Greenland. One, in the present-day state of Maine, would be abandoned in
less than a year, its "hopes . . . frozen to death," in the words of one
contemporary. Remarkably, the other three would survive, if only just.
Jamestown, Quebec, and Santa Fe—all established within a year of
each other in 1607 and 1608—would stake the territorial claims of
rival empires and shape the cultural heritage of America for centuries
to come. All three were the outcome of a century of trial and error in
the European struggle to come to grips with the new climates and environments of an unfamiliar continent.

This book tells the parallel story of these colonizing enterprises, and
through them the story of how climate, weather, and the Little Ice Age
shaped Europe's first exploration and settlement in what would become the United States and Canada. Along the way, it examines the
fate of many other European expeditions and colonies that failed or
never launched at all. It is a story of extreme seasons and untimely
weather, misfortune and perseverance, and how groups and individuals succeeded or failed to survive. It is the story of mistakes and inge-
nuity, improvisations and prayers, violence and desperation, and all
too often disappearance and death. It is a history of North America's
first colonies written from the vantage point of global warming, produced with the help of new tools to reconstruct the climates of the
past, and conscious of the challenges posed by climate change.

The period covered in this book has been described as a "forgotten
century" of American history.[10] Over the past generation, a gulf has
opened between advances in historical scholarship and what most
Americans and Canadians actually know of early colonial history, often
from primary school lessons. In the United States, these lessons fre-

quently skip over the century between the first voyage of Columbus, in 1492, and the landing of the Plymouth Pilgrims, in 1620, because the story in between remains complicated and unfamiliar. Moreover, it is hardly a story meant for schoolchildren. As historians have revealed, this first century of colonial history is a true horror story: a tale filled with violence, starvation, disease, and death. These are "the barbarous years" of North American history, as one historian has described them— or the "creation story from hell," in the words of another.[11]

Yet understanding this period has become more urgent than ever. It is not only that our picture of North American history remains incomplete without it. The early colonial period is also an era that addresses concerns of the present. It represents another age when America spoke many languages and when its future, its environment, and its place in the world were all uncertain. It was another age when climatic change and extremes threatened lives and settlements. And, as this book aims to demonstrate, the real horror story of North America's colonial beginnings also holds much more fascination than the primary school version.

At the same time, new research and new methods have made it possible to understand this period in new ways. Better historical scholarship and diligent examination of the written records—such as the work of historian Karen Kupperman on Jamestown—have opened new windows onto the early colonial past.[12] In the last decade alone, many primary materials in English, Spanish, and French have been published for the first time or in superior editions.[13] Pathbreaking archaeological discoveries and excavations of early European colonies and important Native American sites have turned up enlightening and unexpected finds.[14] The new research not only fills in missing details of this story but also challenges the history we thought we knew.

Beyond better historical scholarship, we also have better science. In this regard, the case of the Greenland Vikings offers us another lesson. Modern scholars uncovered their story by carefully assembling evidence from physical remains and drawing out data from an impressive range of disciplines. These included zooarchaeology (examining animal bones to see what those settlers ate), palynology (examining buried pollen to understand how their environment changed), and bioarchaeology (examining human skeletons to reconstruct their demography, health, and nutrition). Historians can, and must, embrace this science to retell the story of America's early colonies, too. As often as not, the written record provides only half the truth, with the other half literally buried in human and animal bones, sediment, and pollen. These sources can provide answers to questions that until recently could only be speculated about. When early Spanish and English explorers told tales of North American Indians asking them to pray for rain, they were probably reporting the truth: tree-ring records confirm that those Indians faced major droughts and possibly famine (see Chapter 6). There really was a sudden environmental change in Quebec between the time of Jacques Cartier and that of Samuel de Champlain, which possibly forced Iroquois villages out of the St. Lawrence Valley (see Chapter 9). Settlers really did turn to cannibalism during the "starving time" at Jamestown—and archaeologists have uncovered the butchered skeleton of an English girl to prove it (see Chapter 10).

Above all, we can now discover far more about climate and what it meant for this period of Atlantic history. This new knowledge comes from two kinds of sources. On the one hand, paleoclimatologists study climate "proxies," or physical remains that record some element of past climates—tree rings and ice cores being only two of the best-known examples. On the other hand, historical climatologists study human records, from diaries to logbooks to tax records, evaluating them for direct or indirect evidence of past weather and climates. Combining these two kinds of sources in order to understand how past climate influenced the course of human events, we arrive at the interdisciplinary field known as climate history. Although the work of climate history began in Europe and China decades ago, the field has grown rapidly around the world, spurred by advances in climatology and concern over global warming. Recently it has begun to shed new light on North America's history, especially in the colonial period.[15]

Building on this new research—historical, archaeological, and scientific—this book revisits North America's colonial beginnings. It makes the case that climate and climatic change—both as ideas and as physical realities—had a profound impact on the way Europeans encountered the land and people of North America, and shaped the course of the first North American expeditions and colonies. By understanding the role of climate and the Little Ice Age, we can better understand what made early colonial history a "creation story from hell" and what that might mean for us as we confront climate change in the twenty-first century.

The book tells this story both chronologically and geographically, starting with Spain's first expeditions. Chapter 1 examines Europeans' fundamental misunderstandings about climate, and the ways that Spanish writers labored to make sense of the weather and seasons they encountered in the Americas, even as the Little Ice Age brought climatic instability and change. The impacts of climate and weather on early Spanish expeditions in the Southeast, on French and Spanish Florida, and then the English settlement at Roanoke follow in Chapters 2 and 3. Chapter 4 makes the case that climatic events on both sides of the Atlantic at the turn of the seventeenth century undermined Spanish imperial resolve to defend its exclusive claims on North America, opening a window for the next English and French expeditions. However, as described in Chapters 5 and 6, English plans for Jamestown suffered from many of the same geographical mistakes and meteorological misfortunes that had undercut previous Spanish settlement attempts. Exceptional cold and drought in Virginia aggravated early problems of hunger, disease, and conflict that decimated the first settlers. Chapter 7 tells the story of the Sagadahoc colony during the "great winter" of 1608. Chapters 8 through 10 examine the trials and survival of settlements at Santa Fe and Quebec, and finally the "starving time" and rescue of Jamestown.

Throughout this story, four themes stand out. First is the complexity of early modern climate change and the phenomenon we call the Little Ice Age. As this book explains, the Little Ice Age was much more than centuries of cold weather. It brought changes that varied over time and space, affecting each phase and region of colonial history in particular ways. At each step, this book will examine exactly what we know of past weather and climate in particular regions and how it related to the large-scale changes of the Little Ice Age.

The second and related theme is the many and often indirect ways that climate and weather influenced colonial history. In some cases their impacts are obvious. Droughts withered crops and brought famine. Freezing winters left settlers dead or demoralized. Storms wrecked ships or blew them off course. Yet just as often, climate and weather shifted the course of events in more subtle and pervasive ways. Misguided expectations about seasons led settlers to plant the wrong crops. Unusual

extremes left enduring misimpressions about the suitability of potential colonies. Climatic changes shifted storm tracks and sea ice patterns.
Drought affected the health of water supplies, and freezing weather
affected the viability of foraging.

Above all, climate and weather had complicated and consequential
impacts on early contacts between Europeans and Native Americans,
and these constitute a third theme in our story. The focus of this book
is on Europe's encounter with North America in the Little Ice Age: the
ways that people from one continent tried to make sense of another
continent with new and unfamiliar environments during an era of climate variability and change. Nevertheless, the Little Ice Age brought
significant challenges for Native Americans as well. Through the historical and archaeological record, we can glimpse how different populations
responded to these challenges. Moreover, we can see how environmental
pressures interacted with cultural differences and conflicting political
aims to provoke conflict between Natives and newcomers in so many
early European expeditions.

Fourth and finally, climate history highlights the role of chance and
contingency in North America's early colonial history. These historic
expeditions often depended on so few people in such fraught conditions that small accidents took on an outsized role in human affairs. It
may be true, in the widest sense, that European diseases and technologies would determine the eventual outcome. Yet in the short term
there was nothing easy or inevitable about the European invasion and
colonization of North America, and nothing predetermined about
which of the first settlements would survive or disappear.

For all that this book will try to frame the larger historical forces at
work in this period, it remains in no small part a story of bizarre coinci-
dences and accidents. Time and again expeditions happened to arrive
during record droughts or abnormally strong winters. Critical pieces
of information appeared or disappeared at decisive moments. Supply
ships were lost just when they were needed most, or else they arrived
only just in time to save the day. Some of the drama in this history
surely owes to its dramatic telling in original historical sources—but by
no means all. Climatology, archaeology, and corroborating historical
evidence often confirm some of the most incredible accounts from contemporary narratives. The history of early colonial America can be truly
strange and exciting, and it would be misleading to tell it any other way.

1

Where Everything Must Be Burning

The moment that Christopher Columbus reached the Americas in 1492, Europeans faced a geographical conundrum. There was not only the question of whether Columbus had reached Asia or some new continent, or whether Europeans might go on to sail around the world. There was also the more immediate question of why these lands discovered in the tropics—a region supposed to have nothing but dry, burning deserts— were habitable at all.

Columbus had based his decision to head south across the ocean on idiosyncratic ideas about world geography that few of his contemporaries shared or understood.[1] For most observers at the time, the habitability of the tropics was one of the greatest mysteries of what was soon recognized as a New World. The Italian Pietro Martire d'Anghiera (known to the English as Peter Martyr) was one of the first Europeans to gather information about the new discoveries and one of the first to marvel at the healthful climate of the American tropics, so unexpected and so different from African lands at the same parallel: "What the causes of these differences may be, I do not know. They are due rather to the conditions of the earth than to those of the sky; for we know perfectly well that snow falls and lies on the mountains of the torrid zone, while in northern countries far different from that zone the inhabitants are overcome by great heat."[2]

As historian Karen Kupperman has put it, Europeans crossing the Atlantic faced a "puzzle of the American climate." It was not simply that the climate of the New World was different from that of Europe, or that European settlers had first to come to grips with a novel environment. It was that the climate of North America defied European preconceptions handed down across centuries from classical Greece and Rome.[3]

In the work of the geographer Ptolemy (ca. 90–168 CE), still influential in early modern Europe, climate *was* latitude. The words were more or less synonymous.[4] No separate term for *climate* as we use it today yet existed in European languages, although expressions such as *airs* or *seasons* might be used in similar fashion.[5] In the Ptolemaic vision, as it was simplified and popularized in Renaissance geography, the world was divided by parallel concentric bands: from a "frozen zone" at the poles to a dry, burning "torrid zone" in the tropics, with "temperate zones" in between. It was a view of climates taken from ancient Greek and Roman experiences of northern Europe, the Sahara, and the Mediterranean, but one that proved entirely misleading when extended across the Atlantic to continents whose existence Ptolemy had never imagined.[6]

As Europeans acquired experience over a wider area of the New World, the puzzle of American climates only deepened. The conquest of Aztec Mexico, led by Hernán Cortés in 1519–1521, and that of Inca Peru, led by Francisco Pizarro in 1532–1533, left Spain a vast empire in mainland North and South America. Educated migrants to the New World—officials, priests, and missionaries—tried to make sense of American climates, combining their classical learning with empirical observation and reasoning. They found that prevailing winds in the tropics blew east to west, rather than west to east, as in Europe. Some regions near the equator were found to be temperate. Weather in the tropics was cooler and rainier in the summer months than in the winter months. Patterns of seasons, winds, and precipitation could change dramatically over short distances.

Yet the equation between climate and latitude would continue to sow confusion about the New World for more than a century, shaping nearly all Spanish, English, and French designs on North America. Three main factors lay behind its persistence: the key role of latitude in efforts to explore and colonize the Americas, the challenges of explaining American climates based on the science of the times, and the climatic changes and extremes brought by the Little Ice Age.

It would take generations of European explorers after Columbus to determine the outlines of the new continents, much less to understand the land that lay within. Throughout the first century of Spanish, French, and English settlement efforts in North America, confusion reigned regarding the relative locations of Florida, Mexico, and New Mexico, for instance, or the distance between America's east coast and the Pacific Ocean. Geographical features were stretched and shrunk, and whole rivers and inland seas would appear and disappear from maps of the New World.[7]

The fundamental problem was that people of the time had no way to measure longitude. Before the development of the marine chronometer in the 1700s, ships in unfamiliar waters could only estimate the distances traversed, leading to misdirected voyages and fatal accidents.[8] Without familiar landmarks along well-traveled routes, sailors could do little more than guess their geographical position.

In the meantime, the European art of navigation continued to make progress on other fronts. Whereas Columbus relied on his feel for winds and currents and a talent for dead reckoning, sailors of the following generations acquired more solid skills in mathematics, cosmography, and the use of instruments such as the compass, astrolabe, and backstaff.[9] Even when explorers and other sailors might have only the faintest sense of distances or locations from east to west, they could determine their location from north to south with increasing confidence and precision. In this way, latitudes took on an oversized role in colonial planning. Often the only sure information about a proposed site for settlement was its latitude, and therefore how the location lined up with lands across the ocean in Europe.

Even the briefest glance at a map of the Atlantic reveals the dangers in this approach. Today it is common to forget just how far north most of Europe lies compared to the populous parts of America and Canada. Britain is well above the continental United States. Even Paris lies north of Quebec City. The Mediterranean, from southern France to northern Tunisia, lines up with the coast from Maine to North Carolina. Nevertheless, given prevailing ideas about latitude and climate, it was only natural for educated Europeans to assume that today's eastern United States would grow the crops of Italy or Israel and that Canada might have the mild winters of France.

In time, the real experiences of sailors and settlers exposed the shortcomings in classical views about climate. The Jesuit priest and naturalist José de Acosta (1539–1600) captured the sentiment of many educated Spaniards arriving in Mexico or Peru:

> As I had read the exaggerations of the philosophers and poets, I was convinced that when I reached the equator I would not be able to bear the dreadful heat; but the reality

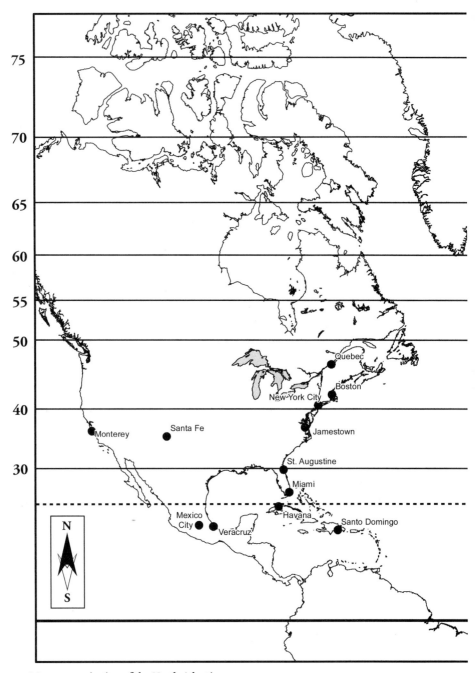

1.1 Mercator projection of the North Atlantic

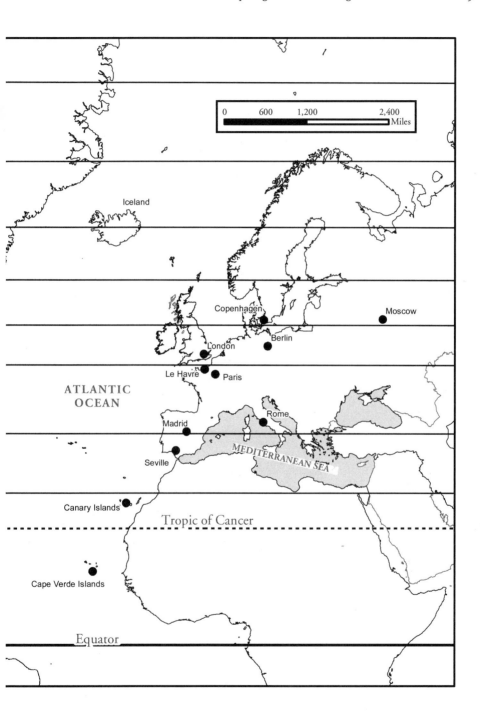

was so different that at the very time I was crossing it I
felt such cold that at times I went out into the sun to keep
warm. . . . I will confess here that I laughed and jeered
at Aristotle's meteorological theories and his philosophy,
seeing that in the very place where, according to his rules,
everything must be burning and on fire, I and all my companions were cold.[10]

Or as his fellow Jesuit Bernabé Cobo insisted: "As for this result that
seemed obvious to the ancients, guided by inference from general
causes, we who live in this land now see and experience that it is so
false, that there is no need of other argument to refute it than well-
known experience."[11]

However, it proved easier for Spanish writers to dismiss classical
theories about climates than to come up with new ways to explain them.
The first efforts to make sense of the Americas' seasons and weather
demonstrated a blend of old and new thinking characteristic of Europe's first generations of encounter with the New World: the power of
novel experiences in the Age of Discovery to shake up received wisdom,
combined with a persistent influence of classical learning and a biblical worldview. Their reasoning reflected the intellectual atmosphere
of an era perched between medieval conceptions handed down over
centuries and new ideas brought about by exploration and the expansion of printing and literacy. This was the era of Nicolaus Copernicus,
Galileo Galilei, William Shakespeare, and Francis Bacon, but also the
era of witch hunts, astrology, pseudoscience, and an abiding dread of
divine wrath and end times.[12]

To appreciate the puzzles facing naturalists such as Acosta, and to
appreciate how far they came in solving those mysteries, it helps first
to understand our modern scientific model of atmospheric circulation.
To simplify somewhat, the direct solar radiation reaching the tropics
drives convection: that is, it warms and raises air near the surface of
the earth. That convection creates a band of rain and thunderstorms
that was known to early modern sailors as the doldrums but is now
called the Intertropical Convergence Zone, or ITCZ. The ITCZ moves
north in the Northern Hemisphere summer months and south in the
winter months, bringing seasonal rains to the northern and southern
tropics. The air raised by convection at the ITCZ slides down toward

higher latitudes, moving from west to east because it carries the momentum of the equator's rotation, a phenomenon known as the Coriolis effect. This air descends at the edge of the tropics, leaving zones of heat and high pressure and creating the deserts of the Sahara, northern Mexico, western Australia, and the southwestern United States. At the same time, the low-pressure band around the tropics draws in air from higher latitudes with less rotational momentum. The tropical trade winds thus blow from east to west (making these "easterly" winds in the language of navigation and meteorology), and the lower atmosphere around the tropics turns over in a sort of loop, called the Hadley cell.

This large-scale circulation pattern shapes climates in ways that proved especially confusing for observers in Spanish Mexico and Peru. The tropics themselves turned out to be milder than the dry desert subtropics, especially during the summer rainy season. In other words, temperatures did not always rise as one got closer to the equator. Moreover, winds in the tropics generally blew in the opposite direction of those

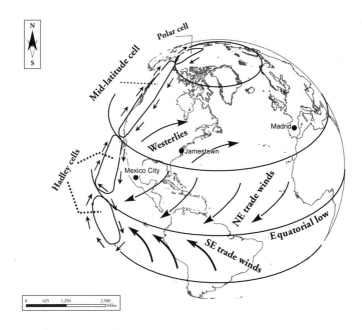

1.2 Simplified schema of atmospheric circulation, illustrating Hadley cells and prevailing winds

in the mid-latitudes—a phenomenon with no explanation in classical geography.

In their efforts to overturn Ptolemaic ideas about climate and to find new explanations, Spanish writers such as Acosta and Cobo fell back on other classical writings, especially Aristotle's *Meteorology*. This work was not only part of a university education but also enjoyed a wide circulation among the educated public through popularized versions.[13] In essence, it worked as follows: The heavenly spheres of the moon, planets, and stars were perfect and unchanging, while the sublunary world was still imperfect and in flux. The heavenly bodies rotated from east to west around the earth and drew wet out vapors and dry exhalations from the ground. These vapors were heated by the sun and raised through the three concentric layers of the atmosphere, which were warm, cold, and hot, respectively. The mixing, movement, expansion, and contraction of these vapors and exhalations produced the weather phenomena—called "meteors"—that people observed on the ground.

Classical meteorology suffered from fundamental problems—not least that it was mostly wrong and very incomplete. The sun and stars do not actually draw "exhalations" from the earth. The whole model assumed a geocentric universe, that is, one with the earth at the center.[14] Nothing in the system explained real patterns of winds. The "meteors" of classical meteorology included all sorts of phenomena, such as comets and earthquakes, that had nothing to do with actual weather. Moreover, classical meteorology was closely related to medical theories left over from ancient writers such as Hippocrates and Galen.[15] For centuries to come, discussions of climate would remain tangled up in notions of disease, race, and supposed influences of New World environments on European bodies.[16]

In spite of these limitations, sixteenth-century Spanish American writers would advance some original and perceptive explanations for the climates they found in the New World. Many of these focused on Spanish America's physical geography: its mountains, deserts, rivers, and lakes. For instance, the Jesuit-trained doctor Juan de Cárdenas proposed that the mountainous lands of Peru reached closer to the second (cold) layer of the atmosphere, and that since South America was full of caverns and rivers, the tropical sunlight drew out more cooling vapors.[17] Some writers proposed that the varied topography of the

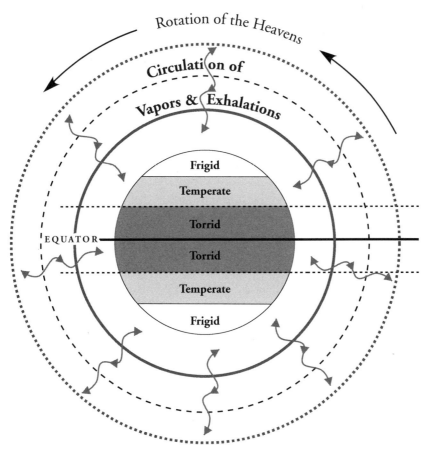

1.3 Simplified schema of classical meteorology and geography, illustrating the zones of the earth, three layers of the atmosphere, and circulation of vapors and exhalations

Americas alternately gathered and dispersed the sun's rays, making certain locations unusually hot and others unusually cold.[18]

Still others proposed what could be called the first theories of atmospheric circulation. Observers in the tropics quickly caught on to the idea that direct sunlight created something like atmospheric convection, leading to rain and storms. As early as 1518, the influential geographer Martín Fernández de Enciso proposed that heat from the sun caused exhalations and vapors to rise together to the second layer of the atmosphere, where the vapors congealed and the hotter exhalations burst through them, causing rains and thunder.[19] By 1590, José de Acosta had worked out a more comprehensive approach toward New World climates. Based on his Aristotelian training, he imagined that winds were vapors and exhalations pulled by celestial bodies. He assumed the earth stood still and the sun and stars rotated east to west, drawing vapors and exhalations with them. Since the earth was widest at the tropics, that meant the heavens rotated most quickly there, creating easterly trade winds that carried cooling ocean vapors. The vapors and exhalations carried by the trade winds must expand under the heat of the tropical sun, Acosta reasoned. Those expanding vapors and exhalations would push air in the mid-latitudes in the opposite direction, creating the westerlies of the mid-latitudes.[20]

In some respects, Acosta's work represented a real breakthrough. He helped turn climates, in the modern sense, into a subject for empirical inquiry. His analysis bridged the previously divided disciplines of geography and meteorology, and it opened new investigations into prevailing winds and their influence on temperatures. By the mid-seventeenth century, Acosta's ideas would give rise to new understandings of winds, weather, and seasons, and in time, these observations and reasoning would help lay the foundations for the modern science of climatology.[21] That intellectual breakthrough came too late, however, to influence the planning and execution of the expeditions described in this book.

In the meantime, Europeans faced a still more intractable puzzle when it came to the climates of North America. Even today, many people remain confused about why western Europe has such mild seasons, even at the latitudes of Edinburgh or Stockholm, when compared to most of the United States and Canada. Some assume incorrectly that it is all about the Gulf Stream. Basically the difference is that between maritime and continental climates.

Recall that air rises from the tropics and spreads north and south. As it descends on the mid-latitudes, it carries more west-to-east momentum, driving winds that blow in the same direction. There are often northerly, southerly, and occasionally easterly winds in the mid-latitudes, too—typically the result of cyclones and anticyclones created by the waves in jet streams high up in the atmosphere. However, the prevailing winds in the temperate latitudes are westerly, and so mid-latitude air masses generally move from west to east. In Europe, this means the prevailing winds bring maritime air from over the Atlantic.

In the eastern United States and Canada, the prevailing westerlies bring continental air from the interior. That difference is crucial, because land heats and cools more quickly than large bodies of water. Consequently, regions that lie on the eastern side of an ocean typically enjoy a milder climate, and those that lie in the interior or on the eastern side of a continent have climates with greater contrasts from day to day and season to season. Average lows on the west coast of France stay above freezing even in January. At the same latitude in New Brunswick, Canada, they average around –10°C (about 13°F) and can drop far below that.

This contrast between Europe's maritime climates and North America's continental climates proved even harder for Europeans to identify and explain than the puzzles of climate in the American tropics. Working from Aristotle's meteorology, some Europeans actually expected mild ocean vapors to travel east to west, keeping eastern North America more temperate than Europe. Moreover, the differences between maritime and continental climates are not constant and absolute, but reflected in averages over seasons and years. Weather in America is often mild, and that in Europe sometimes harsh. When severe weather was encountered during brief expeditions, the promoters of colonization could always find ways to write it off as an aberration, insisting that the true climates of North America were similar to those along the same latitudes across the Atlantic. The argument may have been especially persuasive during the Little Ice Age, an era of climate instability and change.

For many readers, the expression "Little Ice Age" suggests an age of perpetual cold and human misery, stretching imprecisely from the late Middle Ages to the dawn of the industrial age. It conjures images of hungry German peasants, Dutchmen skating on frozen canals, or winter landscapes such as Pieter Brueghel's famous *Hunters in the Snow*. These images from popular history are not altogether wrong, but they are not the whole picture, either. Ongoing research by climatologists and historians now provides a more nuanced sense of the period essential to understanding the situation during Europe's encounter with North America.[22]

The expression "Little Ice Age" originally referred to glacial advances in between the great ice ages that have dominated the last 2.5 million years of earth history. Climatologists later limited the term to glacial advances around the Northern Hemisphere circa 1300–1850 CE. The advancing ice suggested that temperatures had cooled since the high Middle Ages, but at first there was no way of telling exactly when, where, and by how much. Since the mid-twentieth century scholars have tackled this problem with new methods, in order to reconstruct past climates at finer scales.

First, historical climatologists have carefully compiled personal weather narratives and public and private records of various climatic indicators, ranging from grape harvest dates to the freezings of lakes and rivers. They have also indexed these findings in long time series and compared the results to modern instrumental measurements of temperature and precipitation. This research allows us to understand climatic variations from season to season and year to year in medieval and early modern Europe, China, and other parts of the world. It reveals particular periods of unusual cold during the Little Ice Age, and sometimes of unusual storminess, floods, drought, or even warmth.[23]

At the same time, paleoclimatologists have found more and better ways to get information out of so-called climate proxies. These are physical records, such as ice cores and tree rings, which can be measured for indications of past temperature or precipitation. The following chapters of this book will describe just some of the many proxy records now available for Europe and North America and what they can (and can't) tell us. When it comes to understanding the Little Ice Age as a whole, it is important to recognize two important breakthroughs that have recently come from this kind of evidence.[24] First, we now have enough proxy climate records to speak meaningfully and with some confidence about regional and global temperatures for each decade or even year over the past two millennia. The data are far from perfect and there are still many gaps. However, the coverage is far ahead of where it used to be just a decade ago, and it improves continuously.[25] Second, we now have techniques that help us identify both long-term changes and year-to-year differences in temperature with much more accuracy and precision. To simplify, it has often been the case that certain types of evidence or measurements identified short-term variations in climate (high-frequency variability) but not gradual movements (low-frequency variability), or vice versa. New studies have developed better ways to find both.[26]

These breakthroughs in climate reconstruction make it clear that the Little Ice Age was in reality more than one phenomenon, with more than one cause. Over several millennia, slight changes in the earth's orbit around the sun very slowly reduced the intensity of solar radiation reaching the Northern Hemisphere, leading to the very slow cooling of summer temperatures. That cooling may have sped up around 1260 CE, as the spread of land and sea ice reflected more sunlight back into space, generating a feedback loop of cooling. Nonetheless, this rate of cooling reached only about 0.1°C per century—a background trend that mainly affected populations in marginal climates.[27]

Medium- and short-term events changed climate more dramatically on the scale of decades or years. First, regional temperatures and precipitation naturally fluctuate according to oscillations in atmospheric and oceanic circulation. This book will visit some of the most important of these, including the North Atlantic Oscillation and the El Niño Southern Oscillation. Second, the Little Ice Age witnessed several solar minima, including the Spörer Minimum of the mid-1400s and the Maunder Minimum of the late 1600s, when there were few or no sunspots. These events might have brought slight cooling by reducing energy coming from the sun. Third, and most important, from 1257 to 1815 there happened to be an unusual number of large tropical volcanic eruptions. These cast dust and sulfates high into the atmosphere, scattering back sunlight and reducing global temperatures for up to a few years, particularly after the very biggest eruptions or when eruptions followed one another in close succession.[28] Finally, during the late 1500s there was a slight, temporary reduction in the concentration of greenhouse gases in the atmosphere, for reasons discussed below.

Such phenomena happened more or less independently of one another. Internal oscillations and external "forcings" (to give them their more technical name) took place at different timescales and different geographical ranges and had different impacts. Added together, they meant that the period from about 1300 to 1850 CE was on average cooler than the centuries that came before or the era of global warming that has followed, especially in the Northern Hemisphere. In that sense, the period deserves the name "Little Ice Age." However, within this period there were notable variations. For some decades, even generations, temperatures in Europe and North America were not much different than in the early twentieth century. At other times, when these oscillations and forcings aligned, the impact was considerable.

Moreover, the climate data are not the whole story. What we call the Little Ice Age was as much a human event as an atmospheric one. If all of the cooling had taken place where no one noticed, or no one was harmed, we probably wouldn't talk of a Little Ice Age at all. It mattered how people experienced and perceived the climate, and how it influenced their history. What made the mid-1500s to early 1600s the epitome of the Little Ice Age—what the celebrated French historian Emmanuel Le Roy Ladurie dubbed the "hyper-LIA"—was not just the strength of its climatic anomalies but the vulnerability of populations.[29] The Little Ice Age played a major role in Europe's encounter with North America not because the climate was unrelentingly cold but because it was variable and unpredictable, and Europe's colonial enterprises were often at their most vulnerable when the climate was most extreme. One of the steepest declines in Northern Hemisphere temperatures in perhaps thousands of years took place in the half century leading up to the foundings of Jamestown, Quebec, and Santa Fe.[30]

Nor did contemporaries have any way to know that they were living through a Little Ice Age. The first, unreliable weather instruments appeared only at the end of this period. Very few observers survived long enough and recorded the weather carefully enough to find meaningful trends or patterns in the climate. One of those few who did, the Swiss scholar Renward Cysat (1545–1614), was astonished: "For already some time now the years have shown themselves to be more rigorous and severe than in the past, and deterioration amongst creatures, not only among mankind and the world of animals but also of the earth's crops and produce, has been noticed, in addition to extraordinary alterations of the elements, stars, and winds," he wrote in 1600.[31]

Most contemporaries did not notice climatic change per se. They felt the greater frequency and intensity of weather extremes and disasters. To judge from contemporary chronicles, the oldest passengers on the first Jamestown voyage of 1607 might have witnessed or remembered any number of these: the English plague of 1563; the flooding of the Thames in 1564, and the freezing of the Thames in 1565; the "drie sommer," "sharpe winter," and "vehement rage and tempest of windes" in 1567–1568; the floods of 1570 ("to the utter undoing of a great number of subjects"); the "great and sharpe frost" of 1572; the dearth of 1574;

the hailstorms of 1575; the great snowstorm of 1579 (in which "some men and women were overwhelmed and lost"); the "blazing starre" of 1582; the cold summer and "general dearth of graine" in 1586–1587; the windstorms of 1589 and 1590; the frosts of 1598; the cold spring and "great tempest of haile" of 1600; the thunderstorms of 1602; and the "great and violent" wind of 1606.[32] Few could have understood the long-term climatic trends and patterns behind these events. Many people during the Little Ice Age perceived the disruption of the seasons as divine warnings or punishments. If the weather was severe, it must have been a consequence of their own sins.[33]

They just might have been right. In a 2005 book, climatologist William Ruddiman made the case that human greenhouse gas emissions have been changing global climate not only for the past two centuries but for millennia. Methane released from agriculture and cattle, and carbon dioxide from deforestation—although tiny compared to industrial fossil fuel emissions—added up over thousands of years of preindustrial history, keeping another ice age at bay. According to Ruddiman, the spectacular invasion of Old World empires and diseases into the New World after 1492 momentarily reversed that trend. As most of the population of the Americas died, land fell into disuse, clearing and burning stopped, and forests grew back. For a time during the late 1500s, this loss of people and livestock and the regrowth of vegetation actually drew down carbon dioxide from the atmosphere, causing global cooling.[34]

When first proposed, Ruddiman's hypothesis was largely speculative. Another decade of research has generated stronger arguments both for and against. It is now clear that preindustrial agriculture and livestock did influence levels of methane in the atmosphere. Glacial and polar ice reveals traces of pre-Columbian Amerindian fire and the decline of population and burning during the 1500s. Bubbles trapped in ice cores indicate that atmospheric carbon dioxide levels dipped during the late 1500s before rebounding after 1610.[35] Nevertheless, the latest research at the time of writing indicates that a key part of Ruddiman's thesis was incorrect. Natural feedbacks rather than human influence should account for most of the change in atmospheric CO_2. As the earth cooled during the Little Ice Age, soils absorbed more carbon, further reducing temperatures. Human activity may have influenced greenhouse gas concentrations, but it remains an open question whether that influence was enough to affect global climates.[36]

Right or wrong, Ruddiman's thesis underlines the magnitude of environmental transformation in the Americas during the 1500s. The Columbian Exchange—one of the central events of world history—has now become well known from the work of Alfred Crosby, as well as Jared Diamond and other popular historians. By crossing the Atlantic, European conquistadors and colonists reunited the biota of continents separated since the breakup of Pangaea. Amerindians lacking experience with Old World diseases died by the millions from smallpox, influenza, measles, and myriad other pathogens. In their place came European weeds, livestock, and other invasive flora and fauna.[37] Friar Bartolomé de las Casas, the conscience of the Spanish conquest, summed it all up in his history of Hispaniola: "We should remember that we found the island full of people, whom we erased from the face of the earth, filling it with dogs and beasts."[38]

Earlier histories of the Columbian Exchange have left an impression that mortality followed instantly and inescapably from contact: that invading pathogens, plants, and animals swept away the Native American presence far ahead of actual European colonization. Recent research tells a different and even sadder story. Death came not at once but in wave after wave of epidemics and human and natural disasters. In New Spain alone, at least fifteen major epidemics afflicted indigenous peoples during the sixteenth century, the worst during the 1520s, 1540s, and 1570s. The drastic decline of Native American populations did not follow inevitably from immunological differences. War, impoverishment, and the upheavals of conquest all drove the terrible mortality of the Columbian Exchange.[39]

Anecdotal evidence indicates that the climatic variability and extremes of the Little Ice Age were significant factors in this mortality as well. Native writings of the sixteenth century, as in these entries from the anonymous Nahua *Annals of Tlaxcala,* illustrate the interplay of climate, natural disasters, and disease outbreaks:

> 1579: 9 Reed year . . . At this time there was flooding in Tlaxcala and Atlihuetzan. An epidemic began in the time of famine.
>
> 1586: 3 Rabbit year . . . The plants froze. Also at this time there was an epidemic.

1595: 12 Reed year . . . At this time the measles broke out.

1601: 5 House year . . . There was an epidemic.

1604: 8 Flint knife year . . . At this time a pox broke out.

1607: 11 Reed year . . . At this time people perished by water in Mexico City . . .

1610: 1 Rabbit year . . . At this time the wind and snow kept coming. All the trees fell and the animals suffered.[40]

Given the limits of the evidence, it remains hard to say just how much climatic extremes exacerbated Amerindian population loss. The pre-Columbian environment had not been pristine and disease free. Death could have come from endemic microbes as well as exotic ones, from floods and droughts, hunger and violence, or any combination of these.[41] For instance, epidemiologists and climatologists have made a circumstantial case that the alternation of intense droughts and rainy periods in sixteenth-century Mexico could have unleashed a hemorrhagic fever responsible for some of the worst mortality of the era.[42]

Native Americans evidently found the impacts of the Little Ice Age, combined with the shock of invasion and colonization, as bewildering as did early modern Europeans. At the turn of the seventeenth century, the celebrated Inca writer Guaman Poma de Ayala captured this experience, and at the same time revealed the synergy among human oppression, climatic extremes, and mortality:

> And in this life we have seen the eruption of volcanoes and the rain of fire from the inferno and solar sand over a city and its district. Also said to be miracles are earthquakes and the death of many people. . . . Also called a miracle is the pestilence that God sent of measles, smallpox, croup and mumps, of which many people died. Also called a miracle is the great mass of snow and hail that fell from the sky and covered the hills to one fathom, in some places two fathoms; a great many people and livestock died. Also called a miracle of God is the punishment and pestilence of mice on the plains and the great damage done by birds in the fields. . . . Also a pestilence, punishment of God, is the freezing of the maize and potatoes, and hail falling on the crops. Another pestilence that God sends is the bad Christians to rob the possessions of the poor and take their wives and daughters and use them.[43]

This process of recurring disasters and epidemics in Latin America would be repeated during the European colonization of North America. The archaeological evidence is now clear: there was no instant pandemic that swept away North American populations before the Spanish settled in Florida and New Mexico, the English in Virginia, and the French in Canada. In fact, studies of every location discussed in this book have found no convincing signs of any major introduced epidemics prior to the first enduring European settlements.[44] Some first contacts with explorers and failed colonies did introduce deadly diseases.[45] However, we can infer that the effects were temporary and localized. The overwhelming mortality of North American Indians followed sustained interaction through missionization, trade, and slavery.[46] For more than a century, European expeditions into North America encountered societies with their populations intact, which would be an important factor in resource pressures and conflicts during the Little Ice Age.

In Spanish Florida, for instance, the waves of disease and death would begin only in the 1610s, just as Franciscan missionaries made significant inroads into the population. Roughly half of the 16,000 or so Christian Indians of Spanish Florida died that decade. Missionization picked up again in the following generation, but then during a serious drought of the 1650s waves of typhus and smallpox broke out and decimated the converts once more. New epidemics followed during the 1690s and 1700s. Next, slave raiding promoted by English colonies in the Carolinas spread new diseases and carried off many of the Florida Natives who remained. When the Spanish left St. Augustine in 1763, only eighty-three Christian Indians were there to follow them. Whole nations, such as the Timucua, had been shattered. The enormity of this injustice has been obscured from American history by its very completeness.[47]

Climate and climate change were not the only challenges for Europeans and Native Americans during the first century of colonial history. Their historic influence makes sense only in the context of confusions and upheavals accompanying the first generations of exploration and contact. For Europeans, geographical misconceptions, the variable continental climate of North America, and the swings and extremes of the Little Ice Age all went hand in hand. Efforts to make sense of the new continent and its novel climates would make real progress by the seventeenth century. Yet that progress was too little, too late to save early expeditions from misguided expectations and poor planning. For Native Americans, climatic change represented one more set of challenges on top of European invasions and epidemics. When it came to the cold and droughts of the Little Ice Age—or the violence and loss of life from colonization and disease—each region and each nation faced particular challenges and developed particular responses. The combined impacts of both kinds of disasters at once often overwhelmed their capacities to adapt, leaving a pattern of devastation across the first centuries of colonial history.

2

Such Great Snows We
Thought We Were Dead Men

Spain's disappointment with North American climates began with its
first expeditions to Florida, led by Juan Ponce de León. A former governor of Puerto Rico, Ponce de León was following reports of a land
called Bimini, assumed to be another island of the Caribbean. He obtained a royal patent for its exploration and settlement and outfitted an
expedition at his own expense in early 1513. The popular legend of his
quest for a mythical fountain of youth has tended to obscure the less
romantic reality of the voyages.

Setting out from Puerto Rico in March, the venture sailed through
the Bahamas and then turned northwest, passing through bad weather
off the east coast of Florida. The ships reached the mainland probably
somewhere near Cape Canaveral, where Ponce de León found a "beautiful view of many and cool woodlands . . . level and uniform." Struggling
to sail back south against the strong Gulf Stream, his ships eventually
rounded the peninsula in mid-May, traveling along the uninhabited
Florida Keys, where they restocked from the teeming bird and marine
life. The voyage then turned north and northeast to meet Florida's
west coast, possibly at San Carlos Bay. Dropping anchor, the Spaniards
tried to barter with Native Americans; but, coming under attack from
their canoes, Ponce de León decided to head back to Puerto Rico. His
ships returned with a description of "Bimini" as "a large island, cool,
with many fountains and trees."[1]

Ponce de León returned to Spain and obtained a royal patent to
conquer the land he had renamed Florida. It would take him another
seven years to outfit his next venture, again at his own expense, and
there is scant surviving information about his 1521 expedition. Two
ships carrying around 250 settlers departed from San Juan, Puerto
Rico, in early March and "suffered many hardships during the voyage,"
according to a later chronicler.[2] They landed at an unknown site, pos-

sibly Charlotte Harbor. Ponce de León ordered his men to march inland, but they soon encountered resistance from Indians. A number of the expedition's soldiers were killed, and their leader was wounded by an arrow. He sailed back to Cuba only to die shortly thereafter.

Apparently Ponce de León had seen enough of Florida to leave disappointed. According to the most detailed account of the expedition, that of the celebrated chronicler Gonzalo Fernández de Oviedo y Valdés, he had come with high hopes of building a colony of ranchers and farmers:

> As a good colonist, he took mares and calves and swine and sheep and goats and all kinds of domestic animals useful in the service of people; and for agriculture and the working of fields, he was supplied with all [kinds of] seeds, as if the business of colonization consisted of nothing more than to arrive and cultivate the land and pasture their livestock. But the climate of the region was very unsuitable and different from what he had imagined, and the natives of the land [were] a very harsh and very savage and bellicose people.[3]

Ponce de León's voyages set a pattern for the next half century of Spanish efforts at colonizing La Florida, as the region of today's southeastern United States would come to be known. Expeditions led by aspiring conquistadors began in optimism and geographical confusion and ended with lost investments, lost lives, bitterness, and disillusion. The lack of suitable farmland or pasture and resistance by Native Americans were the most-cited causes of failure. Yet behind so many disasters and disappointments was the unexpected and unpredictable climate of North America, so "unsuitable and different" from what Spaniards imagined.

Even as they gained enough experience with the tropics to dismiss the ancient idea of a "torrid zone," many explorers and colonizers from the Spanish Empire still came to North America expecting the same climates found at the same parallels in Europe. Many hoped for a "New Andalusia" across the ocean at the latitude of Virginia or the Carolinas: a country matching the best parts of Spain that could supply Spanish America with Mediterranean commodities.[4] That did not necessarily mean they came unprepared for the cold or heat. They knew that

Mediterranean lands could have blazing summers and freezing winters. Nevertheless, America's continental climate presented greater unpredictability and extremes.

There were tropical storms and hurricanes rarely encountered in the Mediterranean. Although reports of these storms reached Europe as early as the 1490s, sailors and explorers coming from Spain could not recognize the signs of impending hurricanes and failed to appreciate their destructive power.[5] Moreover, the seasons in most of North America differ fundamentally from those of southern Spain. In the Mediterranean, from around late May to September, high pressure blocks incoming moisture from the Atlantic and inhibits cyclogenesis, that is to say, the formation of storms and low-pressure fronts that usually bring rainfall. This is the basic reason for the warm, sunny days that bring hordes of tourists to Mediterranean beaches today. But that holiday season comes to a close during the months of October to April, when the high pressure lifts and the region can turn gray and soggy. This peculiar seasonal pattern of hot, dry summers and cool, wet winters gives the Mediterranean environment its distinctive character: its herbs and olive trees, its pastures for goats and sheep, and its fields of winter wheat and barley. However, it is not a pattern found at the same latitudes in most of North America, or for that matter in most of the world. The American Southeast gets more of its rainfall during the summer, and its precipitation patterns are less well defined. Without a rainy season for winter grains to germinate or a dry season to let them ripen, Spanish crops rotted in the fields, leaving colonists to starve and giving an impression that La Florida's land was unmanageable and infertile.[6]

Moreover, most of the men and women in Spanish imperial expeditions to North America did not come from Spain at all. Many were creoles—that is, people of Spanish descent born in the Americas, mostly Mexico and the Caribbean. From their perspective, La Florida was a "northern" country, promising none of the comforts or riches of the tropics, and raising fears of extreme cold. More often than not, most expeditionaries were actually African servants and slaves and so-called *indios amigos*—the Native Mexican allies who provided most of the manpower and troops in the Spanish Empire's conquest of the New World. These members often had little or no prior experience of freezing winters. They were also the least paid, least fed, and most poorly clothed, sheltered, and equipped. Africans, Native Americans, and even creoles barely figure in most Spanish accounts. Yet their suffering and their need for food and warmth remained driving forces behind the conflicts and misfortunes of these expeditions.[7]

Climatic changes and extremes brought by the Little Ice Age made a difficult situation far worse. Several lines of evidence indicate the Southeast cooled as much or more than the rest of North America during the Little Ice Age, creating unaccustomed hardships. Reconstructing past temperatures can pose a challenge in a region where cold does not usually limit the growth of vegetation. One option for climatologists is to examine tree rings at high elevations, where the cold does affect tree growth—although in this case, the nearest samples are cedars in the mountains of West Virginia. Moreover, scientists can draw on other kinds of climate proxies in sediment cores taken from offshore, including shallows and estuaries in the Gulf of Mexico and Chesapeake Bay. These indicators include varieties of buried pollen, ratios of carbon and oxygen isotopes in corals and tiny shells, and ratios of calcium to magnesium in buried plankton. These kinds of proxies have trouble distinguishing year-to-year variations, but they can sometimes illustrate temperature trends from decade to decade. Taken together, such studies indicate that average temperatures in the Southeast fell by 1° to 4°C (1.8–7.2°F) during the Little Ice Age, with the exact figure depending on the period, location, and method of reconstruction. By the standards of the wider Little Ice Age, the sixteenth century was not especially frigid. However, the proxy data point to cold anomalies in the northern Gulf of Mexico during the mid-1500s and in Virginia around the turn of the seventeenth century.[8]

Studies of lakes in the Midwest indicate that this cooling came as part of a larger shift in atmospheric circulation. In some cases, different sources of precipitation can leave different isotopic signatures—that is, different ratios of elements with different numbers of neutrons—that show up in lake bottom sediments. In this case, the evidence reveals that the eastern United States received less warm, wet air from the Gulf of Mexico and more cold, dry air and winter snows from the northwest. This situation, perhaps a positive phase of what climatologists call the Pacific North American pattern, dominated during the Little Ice Age. It let up slightly during the late 1400s to early 1500s but reasserted itself for the rest of the sixteenth and seventeenth centuries, bringing cold northerly winds and freezing weather.[9]

This shift in atmospheric circulation also contributed to recurring spring and summer droughts during the first century of expeditions into La Florida. Scientists have reconstructed drought throughout this region, and across temperate North America, by measuring tree rings in species whose annual growth is limited by growing-season moisture. Calibrating recent tree-ring widths to modern instrumental measurements, they can then use older tree rings as a proxy for past drought. American tree-ring experts have refined this technique to the point that they can map out the year-to-year variation in summer drought for several centuries. These reconstructions demonstrate a number of serious dry spells, especially in the Carolinas and Virginia in the late sixteenth and early seventeenth centuries.[10]

Scientific insights such as these, as well as new historical and archaeological research, justify a fresh look at expeditions into Spanish Florida. The proxy data demonstrate that Spanish descriptions of extreme weather had a real basis in the climate anomalies of the period. When considered together, not just as isolated episodes, eyewitness accounts point to a consistent pattern of harsh northerly winds across the Gulf region, including outbreaks of Arctic air. These episodes are not unique to the sixteenth century. In 1899, an Arctic outbreak brought blizzards and deep freezes to some of the same locations described in this chapter, including Pensacola Bay and Mobile, Alabama. However, Arctic outbreaks in the southern United States have been less frequent and less severe since the twentieth century.[11] During the Little Ice Age, the Southeast challenged invaders with weather rarely, if ever, encountered in the region today.

Moreover, climatic shifts brought by the Little Ice Age had profound consequences for the Indians of America's eastern woodlands. By and large, these communities formed part of the Mississippian culture. They had adopted maize agriculture around the turn of the second millennium CE and, to a greater or lesser degree, they shared artistic motifs and ceremonial architecture, including mound-building. They mostly lived in chiefdoms, or communities with a political and social hierarchy based on tribute and redistribution.[12] The largest Mississippian settlement, Cahokia in western Illinois, flourished during the eleventh and twelfth centuries before its population decreased in the 1200s and then disappeared. The new paleoclimate evidence lends further weight to a hypothesis advanced by several archaeologists in recent de-

cades: that the onset of the Little Ice Age drove the collapse of Cahokia, the breakup of population centers in the middle Mississippi Valley, and migrations toward the Southeast, leaving vacant lands in between.[13]

These researchers have also identified a correlation between periods of cold and drought and growing signs of warfare. More skeletons reveal evidence of injury or homicide. Communities built palisades to defend their villages. Vacant regions between those settlements may have been buffer zones separating warring populations. Yet conflict came from within communities as well as between them. As David Anderson and colleagues have demonstrated, Mississippian chiefdoms tended to disintegrate during periods of drought. The evidence of this phenomenon remains incomplete and circumstantial. The best inference, however, is that drought undermined the production of surplus maize and other resources that chiefs could use to shore up their authority. As chiefdoms dissolved, their populations dispersed or joined other communities.[14]

This situation proved particularly dangerous for Spanish invaders during the sixteenth century. Hoping to follow in the footsteps of Hernán Cortés or Francisco Pizarro, they found not unified empires but rather fortified villages and towns of a few hundred to a few thousand people spaced dozens of miles apart. These were societies large and wealthy enough to raise the conquistadors' hopes. Yet they offered no great conquest like that of the Incas or Aztecs, nothing that would allow expedition leaders to recover the personal fortunes sunk into these ventures. At the same time, many of these chiefdoms were experienced in warfare, and many chiefs resisted Spanish demands for maize and other resources they needed to support their own authority. Freezing winters and droughts only aggravated the need for these resources and Indians' reluctance to part with them.

The next major expedition after Juan Ponce de León's has left further evidence of the environmental challenges and shocking cold that Spanish expeditions encountered in La Florida.[15] In 1521, a Spanish slave raid returned from the region around the Santee River, in present South Carolina, bringing rumors that this land they called Chicora was a "New Andalusia" blessed with fertile soil and the climate of southern Spain.[16] Lucas Vázquez de Ayllón, a prominent official of Hispaniola, secured a contract from Emperor Charles V (r. 1516–1556) to colonize

Chicora, or Santa Elena, as he soon renamed it. The would-be conquistador promised to outfit an expedition at his own expense in return for
rights to the land's commodities. In his reports about Santa Elena,
Ayllón exaggerated its location northward, in order to put it on a parallel with Spain.[17] Based on his royal license, Ayllón believed "that silk
can be grown in the said land, and this is a cultivation without much
work and very well suited to the Indians," who were rumored to be a
docile people, ready for Christian civilization.[18] A contemporary letter
observed the excitement surrounding the venture: "Associates will not
fail him, for the entire Spanish nation is in fact so keen about novelties
that people go eagerly anywhere they are called by a nod or a whistle,
in the hope of bettering their condition, and are ready to sacrifice what
they have for what they hope."[19]

Ayllón's expedition finally set out in 1526, taking six ships loaded
with sailors, settlers, servants, and other passengers. Based on Oviedo's chronicle, the settlement effort was doomed even before it began.
As the fleet approached the shore of Santa Elena that August, its
flagship ran aground and "was lost with all its supplies (except that its
people were saved)."[20] The other five ships managed to enter the bay,
but the Spaniards' Native guides deserted them, and the settlers found
the land disappointing. They soon relocated south to found a settlement they named San Miguel de Gualdape.[21] By then it was early
October, and the weather was beginning to turn. "Since they were
running short of food, and did not find any on the land, and it was
very cold, since the land where they had stopped is thirty-three degrees north and it was flat, many of the people became sick and many
died."[22]

Historian Paul Hoffman has observed that the groundwater in this
area reaches close to the surface and is easily contaminated by human
waste. Dysentery, typhoid, or similar diseases could have spread through
the new settlement within weeks.[23] Ayllón himself soon fell ill and
died on October 28. After the loss of their leader and the death of two
men at the hands of Indians, the discontented settlers began to mutiny. Soon the remaining captains decided to abandon the venture. Of
the roughly 600 who departed, according to Oviedo, "not a hundred
and fifty escaped with their lives, most of those sick and hungry." This
loss moved the chronicler to a brief tirade against ignorant captains
who led their people unprepared into the New World, "especially to

those northern parts, where the [Native] people are more fierce and the land very cold."[24]

According to Oviedo, the fleet encountered incredibly frigid weather on the way back to Santo Domingo: "The cold was such that as they had set out unwell and short of provisions they died of cold; in the caravel *Santa Catalina*, seven men froze to death; on the ship *Choruca* occurred something hardly ever seen before, if at all, that one of those poor souls wanted to take his breeches off and all the flesh came away from both legs from the knees downwards, leaving his bones bare, and he died that night."[25] It is true that the region can experience exceptional cold fronts: since instrumental measurements began, November temperatures have fallen as low as $-9°C$ ($15°F$). Yet this story probably tells us less about real temperatures than about the trauma of the settlement's failure and the shock experienced by creoles and Caribbean Natives unaccustomed to freezing weather. Comparing La Florida to the comforts of Spain's tropical empire, Pietro Martire d'Anghiera drew the obvious conclusion: "It is toward the south, not towards the frozen north that those who seek fortune should bend their way."[26]

The next Spanish invasion of La Florida resulted in complete catastrophe, and climate was only one of many causes. Nevertheless, testimonies of the expedition's few survivors have left us with more evidence of Little Ice Age cold and its human impacts in the sixteenth-century Southeast. For contemporaries, their descriptions of adverse climate and extreme weather added to the growing disillusion with the land and climates of La Florida.

Pánfilo de Narváez had been lieutenant to the governor of Cuba in 1519, when Hernán Cortés led his unauthorized expedition to conquer the Aztec Empire of Mexico.[27] In an effort to steal the glory, Narváez led a rival invasion, but Cortés defeated and imprisoned him in 1520. Released from captivity, Narváez returned to Spain and managed to secure a royal license to conquer all of America north of $31°N$ latitude (roughly the northern Gulf Coast). In 1527, at his own expense, Narváez outfitted an expedition of another 600 soldiers, settlers, and slaves. He intended to begin his conquests farther up the coast of Mexico, just north of Cortés's territory.

Within a year and a half, Narváez and most of his followers would be dead. Their troubles began long before their ships even reached La

Florida. During a stopover in Santo Domingo, nearly a quarter of the passengers deserted, perhaps after learning the fate of the recently returned Ayllón expedition.[28] While they resupplied at Trinidad, Cuba, a hurricane struck the island, blowing down churches and houses and killing sixty men and twenty horses aboard the ships.[29] The storm damage forced the expedition to overwinter in Cuba, in order to gather new provisions and to locate a suitable pilot. The man they found, Diego Miruelo, proved a terrible choice. Departing from Cuba in February 1528, he first ran the ships aground and then steered them into another storm, nearly wrecking the fleet a second time. Hopelessly lost, swept along by the Gulf Stream, with no indication of their longitude, the expedition veered much too far east and ended up in Tampa Bay, Florida.[30]

Soon after they made land in mid- or late April, they had their first contact with Florida's Indians: an unfriendly but not yet violent encounter. Short of provisions and in need of a better harbor to await fresh supplies, Narváez took a fateful decision: he sent the ships away, while he led a company of about 300 men and their horses north and west overland. Disappointed with the region around Tampa, Narváez first followed Indian reports of a richer territory in Apalachee (around modern Tallahassee), which the expedition reached in mid-June. Attacking the first village they encountered, the soldiers captured its chief and seized its supplies of corn.

From then on, word of the Spanish invaders raced ahead of them, and Indian villages were ready. Unimpressed by Apalachee, and chasing rumors of richer lands to the west, Narváez and his men ran into frequent ambushes and hit-and-run raids. They eventually arrived in a land they called Aute, only to find its villages abandoned and burned ahead of their arrival. The invaders began to suffer shortages and disease, probably typhus or typhoid.[31] By then, their ships had already returned from Cuba with fresh supplies and were searching the coast in vain for Narváez's company.

By late summer 1528, the Spanish soldiers in Aute fell into desperation. Facing continuous attacks and growing hunger, they traveled down to the coast. They killed off their remaining horses for food and fashioned five log rafts with which to sail along the coast to Mexico. Losing men to illness, starvation, and Indian canoe raids, the expedition held together as far as the mouth of the Mississippi. The outflow of the river scattered their crafts and a storm drove them west. The rafts and their men washed up separately hundreds of miles apart along the Texas coast that November.

As best as can be reconstructed from the surviving narratives, one raft landed as far south as Mustang Island, where local Indians attacked and killed its passengers to a man. Another two landed somewhere near the mouth of the San Bernard River, where some of its men were blown away to sea and the rest overwintered in conditions of such privation that they resorted to cannibalism, leaving only one man alive by spring. The last two rafts washed up on or around Galveston Island. Only four of the men on those rafts, including Álvar Núñez Cabeza de Vaca, would make it back to New Spain alive, following eight years of captivity and peregrinations.

In a region better known for its suffocating heat and humidity, our surviving narratives agree it was intensely cold. They described Apalachee, which they passed in July 1528, as a land of "great cold," covered with fallen trees from perpetual storms and hurricanes. For the men who washed up on Galveston that November, "the cold was

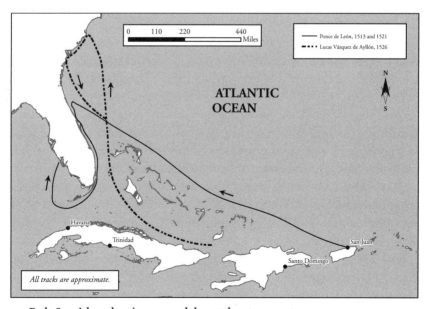

2.1 Early Spanish explorations around the southeastern coast

very great," and that winter they suffered "great cold." Texas Indians also appear to have been caught off guard by the ferocity of the winter: "And those who sought food, in spite of their great labors, could find but very little, because the weather was so severe." Oviedo's narrative, probably taken from the survivor Andrés Dorantes, described northerly winds on the Texas coast so harsh "that even the fish within the sea are frozen from cold." Dorantes "said that he saw it snow and hail together on the same day."[32]

Years later, once news of the survivors' return reached Mexico, their reports raised new enthusiasm for the conquest of La Florida. Hernando de Soto, a veteran of Pizarro's overthrow of the Inca, obtained another royal patent to conquer and govern the territory in 1537. Soto and his captains had the chance to meet Cabeza de Vaca, newly returned to Spain, before they set out on their expedition. Typical of so many early reports of La Florida, Cabeza de Vaca's story left an exaggerated impression of poverty in the regions he actually visited, while hinting at fabulous riches in other lands just beyond. "He described in general the wretchedness of the land and the hardships he had suffered," one of Soto's officers recalled, but somehow also "gave them to understand that it was the richest land in the world."[33]

Climate and weather played a significant and often overlooked role in the Soto expedition and its aftermath. Year after year, unaccustomed cold, rains, and snow demoralized its members and left them vulnerable to hunger, sickness, and attacks. Like reports from the Narváez expedition, accounts of these hardships would over time deflate enthusiasm for further ventures in La Florida. Narratives of the expedition also offer a glimpse at the impacts of the Little Ice Age on Mississippian chiefdoms at the dawn of European contact.[34]

Crossing the Atlantic in 1538, Soto's nine ships and roughly 700 passengers wintered and resupplied in Havana before setting out for the Florida peninsula. Landing in Tampa Bay in early June 1539, the expedition began with an extraordinary stroke of good luck: they found Juan Ortiz, a member of the Narváez expedition who had been captured and enslaved by Florida Indians, and who could now serve the Spanish as an interpreter. Otherwise, Soto's experience of Florida much resembled that of Narváez. Disappointed with the soil and pressed by Indian raids, the Spanish invaders followed rumors of more fertile land to the north. Making a slow crossing through swampy terrain and over several rivers, they faced violent resistance from Native Americans along the way—and delivered even more violence in response.[35] In October, they reached the land of the Apalachee and decided to overwinter around present-day Tallahassee.

During their months of rest, they reconnoitered the coast and gathered information about the interior. Then, much like Narváez, Soto made a fateful decision to split his expedition. About a hundred of the men would take the ships and set up a base for supplies in the promising harbor they had just discovered at Pensacola Bay. Soto, meanwhile, would lead the rest of the expedition northeast into the interior, following rumors of gold, pearls, and other riches in an Indian kingdom called Cofitachequi. He promised to rendezvous with the ships the following year. Instead he launched his company on a four-year odyssey through America that would take the lives of hundreds of his followers, and many thousands more Indians, before the survivors straggled back to New Spain in 1543.

Soto's invasion moved haphazardly, driven by rumors, exaggerations, ambition, and hunger, forced to adapt to each new environment

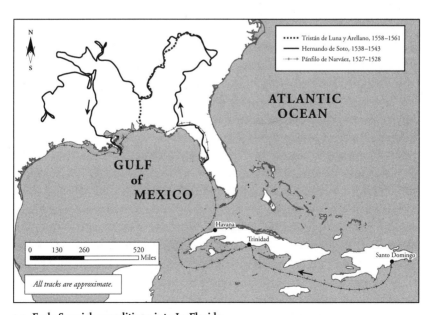

2.2 Early Spanish expeditions into La Florida

and to improvise tactics at each new encounter. Lacking a firm grasp of geography or navigation, he was often at the mercy of Indian guides anxious just to get rid of him and his soldiers.[36] Scholars continue to dispute the sources and details, but extensive research gives a rough indication of their route. In 1540, after wintering in Florida, the expedition traveled first through present-day Georgia and into South Carolina. Disappointed by what they found in Cofitachequi, they turned northwest into the hills of western North Carolina and Tennessee, then southwest along the Coosa River into central Alabama. The invasion had faced opposition along the way, but in October 1540, at Mabila, it provoked the deadliest confrontation yet, losing at least twenty soldiers along with most of the expedition's baggage and clothing.[37]

Sensing the demoralization of his men and fearing the disintegration of his expedition, Soto suppressed information of just how close they already were to the rendezvous planned at Pensacola Bay. He turned his invasion in the opposite direction, heading into the middle of the present-day state of Mississippi, where they passed the winter. In 1541, the expedition continued northwest into Arkansas, striking out in different directions, following rumors of wealthier chiefdoms and in constant pursuit of food and supplies. A foray to the west that autumn took them briefly into the land of bison-hunting Plains Indians before they returned to Arkansas for the winter.

By the spring of 1542, nearly 250 men had died, including the expedition's translator, and more were falling weak and sick from malnutrition and disease. Soto himself finally despaired of making a great conquest and succumbed to an unknown illness in May. His successor, Luis de Moscoso, and the venture's other remaining leaders opted first to head west in search of Mexico. Vastly underestimating the distances involved, lured by vague reports of Spanish settlements to the south, and unable to communicate reliably with indigenous peoples, they took the expedition deep into present-day east Texas. They had gone perhaps as far as the Brazos River when in October they finally gave up and turned back. Arriving at the Mississippi River again in early 1543, the survivors plundered Indian supplies and improvised rafts for the trip downriver and along the coast, finally reaching the town of Pánuco in September.

Throughout the expedition, each winter proved worse than the last. The first suggestions of unusual cold appear in descriptions of the winter of 1539–1540 in Florida, when according to a chronicler "most of the Indians whom they had to serve them, being naked and in chains, died because of the hard life they suffered."[38] Conditions deteriorated each year as supplies ran low, footwear wore out, and clothes and baggage were lost or destroyed. In the winter of 1540–1541, in Chicaza, "it snowed with as much wind as if they were in Burgos and with as much or more cold."[39] The Indians of Chicaza, who had at first fled the invaders, began to harass the Spanish who occupied their town, making regular nightly assaults. Finally in early March, they managed to set fire to the Spanish camp, burning much of their remaining clothing, killing a dozen Spaniards, scores of horses, and hundreds of the expedition's prolific swine. The invaders spent the rest of the winter in "great suffering from the cold . . . The whole night was passed in turning from one side to the other without sleeping, for if they were warmed on the one side they froze on the other."[40]

By November 1541, the expedition was trapped by freezing weather at the village of Utiangue, in central Arkansas, where they endured their worst winter of all. As one chronicler recalled, "There were such great snows and cold weather that we thought we were dead men."[41] In mid-March, they were finally able to decamp and continue their journey. But late that month, according to another account, "such weather occurred that we could not march for four days because of the snow."[42] Little Ice Age weather pursued them even into Texas, where the invaders faced cold rains and snow, and were drenched crossing streams and tributaries. By then, most of the company were barefoot and dressed in makeshift clothes of deerskin and furs, with little protection against the elements. People and horses continued to sicken and die along the way.[43]

In a few places, the narratives and archaeology of Soto's expedition also point to impacts of the Little Ice Age on the chiefdoms they encountered.[44] Towns and villages had been abandoned, some during droughts of the previous decades. At Cofitachequi, Spaniards learned of a disease, probably brought on by famine, that had decimated the population a few years before.[45] The expedition entered Arkansas at the start of a drought in 1541, which apparently hurt the maize harvest and made the invaders even more unwelcome to the region's people.[46]

As news from the returning expedition reached Mexico, it provoked a mixed reaction. At first, their bitter experience of La Florida and failure to seize new wealth or territory only confirmed the negative impressions of past ventures.[47] At the same time, some survivors began to stir up support for a new expedition into the region, spreading rumors of gold, pearls, and fertile soils.[48]

By the 1550s, Spanish officials took a growing interest in La Florida. French and English privateers were taking a toll on shipping in the Caribbean, and Spanish officials feared that a rival colony on the Gulf of Mexico would become a nest for pirates. Spanish sailors shipwrecked along the coast needed a safe harbor. Religious orders pressed the Crown to support their missionaries, especially after the dramatic martyrdom of a Franciscan friar at the hands of Florida Indians in 1549. At the same time, many Spaniards still drastically underestimated the overland distance from Mexico to Florida, and new publications spread the myth of an inland sea or a passage to the Pacific through America's east coast.[49] The notion of a "New Andalusia" clung on in spite of all evidence to the contrary. As a 1554 publication from Mexico City put it, "It is, of course, that Florida, lying only an easy and very short voyage by sea, and by land neither a long nor a difficult journey, should be conquered by the Spaniards . . . for whatever is produced by the other Spain in the Old World, from which merchandise is imported into ours with so much delay and difficulty, all this would be supplied by Florida."[50]

Colonial promoters concocted an elaborate scheme to link Mexico, the Gulf Coast, and the Atlantic through the most promising sites discovered in past expeditions: Ochuse (Pensacola Bay), Coosa (in present-day Georgia), and Santa Elena.[51] In 1557 that ambitious plan won approval in the Council of Indies, the key imperial decision-making body in Spain. Early the following year the new king, Philip II (r. 1556–1598), and the viceroy of New Spain, Don Luis de Velasco, entrusted the mission to Tristán de Luna y Arellano, a younger son of an aristocratic Castilian family.

The expedition began with high hopes and promise. Viceroy Velasco bought into the most optimistic depictions of La Florida from survivors of the Soto expedition, "specifically, a number who had been kept by the Indians in the province of Coosa," as a later chronicler recalled. "They had described this country as being very rich, and told marvelous tales of things which were later not found. Such reports, and those on other provinces, remembered ever since they had been disseminated by the soldiers of Pánfilo de Narváez and Hernando de Soto, imbued everyone with a great desire to make the expedition."[52] All the same, Velasco and Luna spent hundreds of thousands of pesos to equip the voyage with clothing, arms and armor, basic supplies, and enough food to last for months without depending on Native Americans. Although among the least well known today, it was probably the largest and most expensive European expedition to North America during the entire sixteenth century.

An initial reconnaissance voyage missed Pensacola Bay but discovered nearby Mobile Bay in late 1558. Its captain praised the region very highly, claiming that it had "the climate of Spain, both in respect to rain and in occurrence of cold."[53] The main expedition of thirteen ships with around 550 soldiers and 1,000 colonists departed Mexico in June 1559. After first passing up their destination and then doubling back, the entire expedition eventually reached Ochuse in late August.[54] Luna's first report praised Pensacola in superlative phrases. "Seamen say it is the best port in the Indies, and the site taken for the founding of the town is no less good," he wrote to Velasco. The viceroy in turn reported to the king that "the port is so secure that no wind can do them any damage at all."[55]

Colonial promoters in the sixteenth century would make many misjudgments about the geography of North America, but few had such poor timing and tragic irony. Just as the viceroy sent his report to the king, a hurricane swept into the Caribbean, reaching Puerto Rico on September 22, 1559. The island's governor described "a storm that carried off everything [the people] had in the countryside . . . [such that] nothing remained to eat."[56] A week later, it smashed into Pensacola Bay. Luna wrote to the viceroy of "a fierce tempest, which, blowing for twenty-four hours from all directions . . . without stopping but increasing continuously, did irreparable damage to the ships of the fleet." A more dramatic description of the hurricane had the wind snapping cables and smashing boats to pieces. Archaeologists have now rediscovered and partially excavated two sunken ships of Luna's fleet, revealing damage from the storm.[57]

Luna had put such confidence in his harbor that he had left all the expedition's most important supplies aboard the fleet. Arms, armor, shoes, clothing, and, most important, food all went to the bottom of the sea with the largest vessels. Only a caravel and two small barques survived to bring the disastrous news back to the viceroy. "This has reduced us to such extremity," Luna wrote, "I do not know how I can maintain the people."[58]

The area around Ochuse proved thinly populated and unsuitable for farming. Two small parties sent to reconnoiter the region for supplies came back "without discovering anything more than the great hunger they felt after their food ran out," a chronicler recalled.[59] A larger detachment ventured inland in search of a better site, returning a month and a half later with reports of an Indian town at Nanipacana, probably on the lower Alabama River. However, Luna had fallen sick with a fever and refused to move, generating accusations that his illness left him delusional and unbalanced. While he recovered, the expedition stalled at Ochuse, reduced to near starvation. When they finally set out for Nanipacana in late spring, Native Americans retreated before them, taking away everything edible.[60]

News of the expedition's disaster must have reached Viceroy Velasco by early October, and he rushed to organize relief. However, the shortage of available supplies and the bad roads between the capital and the port of Veracruz created weeks of delay. The first, inadequate resupply fleet did not reach Ochuse until the end of 1559. Still vastly underestimating the distances involved, Velasco at first tried to resupply Luna's colony overland through Texas before finally sending provisions by ship.[61] Learning of the deteriorating situation, the viceroy organized a second supply fleet in early 1560, but this one faced even longer delays and failed to reach Ochuse until June, by which time Luna and most of the settlers had decamped to Nanipacana. Even then, Velasco made a disastrous miscalculation, not sending enough food for the whole expedition: "I thought from what you wrote that you and the other people of the camp who went inland would not be likely to lack food; for unless those who were in that country with Soto lied, they never suffered want."[62]

Velasco's optimism was cold comfort for the starving colonists. Despite the first resupply, provisions in Ochuse were running out by March 1560.[63] The move to Nanipacana barely helped. A small Native American village, it could scarcely supply the colonists in the best of times, and during spring and early summer the crops and nuts were not yet ripe.[64] In midsummer, Nanipacana's Indians turned to scorched-earth tactics, evidently trying to starve out the Spanish. Abandoning any pious hopes to convert the Natives, Spanish soldiers resorted to violent seizures of food. Viceroy Velasco specifically ordered the Franciscan friars accompanying the expedition not to get in their way.[65]

The expedition resorted to eating acorns and other famine foods, and sickness spread through the camp. Throughout the summer, the Spanish soldiers and Native Mexican allies petitioned Luna to abandon the enterprise and return to Spain, citing starvation, mortality, and Indian resistance. "We have no prospect of food from any quarter, either from up the river or from down it, or much less from the interior, for the Indians have the whole [country] in revolt and burned over," read one complaint.[66]

Luna, however, was coming under pressure from King Philip and the viceroy not only to stay in La Florida but also to press on with their elaborate plan to link up Pensacola Bay with South Carolina. Spain's European conflicts and a recent French effort to plant a colony in Brazil raised new fears that foreign enemies would encroach on Spanish claims and attack its treasure fleets.[67] Still trusting in the optimistic reports from veterans of the Soto expedition, and probably fearing desertion and the breakup of the expedition if they returned to the coast, Luna opted to continue the exploration of Coosa as an overland bridge to Santa Elena. He dispatched a party of 140 soldiers accompanied by two Franciscan friars and some settlers in late April 1560.[68] For nearly two months, as they journeyed more than 200 miles northeast, they passed only small Indian villages whose people fled before them, hiding their food. Hungry and tired, the detachment finally found sustenance at the larger Indian towns around Coosa in midsummer, just as the corn in the fields and the chestnuts on the trees began to ripen.

Even Coosa was a disappointment. Despite its good soil, it was thickly wooded, and its people grew little surplus food. Some historians have suggested that Coosa had fallen victim to some epidemic contracted from Soto's expedition two decades earlier, but in all likelihood its supposed richness and fertility had always been an exaggeration.[69] The expedition lived for a few weeks on the bare minimum they could ask from Indians, for fear of driving them to resistance or flight. By the time the soldiers and friars sent back their discouraging reports to Nanipacana in June 1560, they had run out of clothing, footwear, horseshoes, and other supplies. "The country is so poor and with such scant opportunities for gainful pursuits that we think it would be difficult to maintain ourselves in it," reported one friar. "The people are very discontented with the country, and it kills them to tell them they will have to stay in it," wrote the other.[70]

Luna tried to hide these reports from his officers and continued with plans to move to Coosa. However, his authority with the soldiers was falling apart, along with living conditions in Nanipacana. By September, once his officers found out about the real situation in Coosa, they openly defied their commander. Led by the quartermaster (*maestre de campo*), they declared Luna mentally unbalanced and unfit for command, and took the whole expedition back to Pensacola Bay to await evacuation.[71]

In the meantime, prodded by Viceroy Velasco, Luna had tried sending a small expedition to Santa Elena by sea. The viceroy heartily approved, but this time his reaction was more circumspect: "I fear that the weather when they went was not favorable . . . for that coast is, they say, rough in winter," he warned in September.[72] In fact, by the time Velasco wrote that letter, the three ships had already been scattered in a storm off Cuba and never completed their mission.[73]

The detachment in Coosa eventually received news of the Nanipa-cana colony's collapse, and they returned to Ochuse in November 1560. By then, it seems, no more than 600 of the original 1,550 members of the expedition remained in La Florida. Some of the sick men, women, and children had gone back with the second resupply fleet in June, but many more must have died of hunger and disease.[74] Throughout the late summer and fall, continual rains, flooding, and storms in Mexico delayed the procurement and delivery of new supplies, leaving the remnants of the expedition at Ochuse to starve. A fleet from New Spain under the captain Ángel Villafañe finally evacuated the colony to Cuba in early 1561.[75]

Late that spring, Villafañe selected ninety of the remaining men to retry the mission to Santa Elena. The captain and other officers later testified that they sailed to Santa Elena but found no suitable place to moor their ships, and so they continued up the coast. About a week later, probably somewhere south of Cape Fear, another storm scattered their fleet, and they sailed back to Havana. All the expedition's leaders testified to the region's unsuitability, pointing to the difficulties of navigation, the lack of good ports, and the bad character of the soil. King Philip was soon apprised of "the poor outcome" of these ventures in La Florida, "given how much was spent in this to so little effect, and the poverty they say there is in that land, so that even if we could settle it we wouldn't gain much from it."[76]

Reports of the region's "bad climate" added to the overwhelming frustration. Luna's poorly clothed and fed colonists had suffered considerably during the winter of 1559–1560, leaving the impression that Florida was a "cold and humid country." The problem of climate had surfaced again when Luna tried to force his colony to continue to Coosa. His officers objected that "the rains might come and the winter set in, so that we should all die." They later complained that "every eight or ten days there are hard rains, cold, and great heat, in such intemperate succession" that their clothes wore out in just weeks. The soldiers and friars who traveled to Coosa that summer were equally disappointed not to find the Mediterranean seasons they needed to raise Spanish grains, vines, and olive trees: "The climate of this country is unequal, with extremes of heat and cold. The rains do not come at fixed times, for it rains now as well as in winter."[77]

The promise of a "New Andalusia"—which had somehow survived the repeated disasters of the Ponce de León, Ayllón, Narváez, and Soto expeditions—now evaporated, leaving in its place an exaggerated impression of cold and sterility. Juan López de Velasco, the official chronicler-cosmographer of the Indies during the 1570s, compiled all the official accounts of New Spain's geography and came to some disappointing conclusions about La Florida. Below 30°N latitude (that is, the Florida peninsula) it might support Mediterranean crops, he concluded. North of that, the native inhabitants were "miserable and very poor," "the land all very cold and uncultivated," and the coast prone to "rough northerlies." The Gulf of Mexico, where the Narváez expedition had traveled, presented "miserable and sterile lands" and "the poorest and most unfortunate peoples." It could be "plagued by hurricanes and north winds," the latter blowing often in winter. López de Velasco acknowledged rumors of rich lands in the interior, but with evident skepticism.[78]

By the time the Luna and Villafañe expeditions ended in failure, the Spanish Empire had lost half a century in its struggle to settle North America, with little to show for its efforts beyond wasted men and money. These expeditions had suffered from chronic problems of geographical ignorance, erratic leadership, outsized ambitions, and aggression and insensitivity toward Native Americans. Yet when they are examined in detail, it becomes clear how many of the Spanish Empire's worst disasters in La Florida came from weather and climate. Storms disrupted the Narváez, Luna, and Villafañe expeditions at critical moments. Freezing winters aggravated the suffering, demoralization, and high mortality of the Ayllón, Narváez, Soto, and probably Luna expeditions. And from the first, the variable continental climate of North America proved a serious disappointment to Spanish conquistadors and colonists in search of a "New Andalusia" in the New World.

Yet climate and weather were more than just accidental and occasional factors in Spanish imperial failures. The Little Ice Age revealed vulnerabilities built into Spain's whole approach to imperial expansion into North America. This expansion did not take place across an open frontier, where small groups of settlers could gradually test the land and its soils and climates. Nor was it a concerted imperial enterprise, with a coordinated strategy and continuous financial and logistical support. Instead it relied primarily on private initiative and investment, compensated by exclusive rights to exploit the territory and its inhabitants. This was a system designed to draw on the private wealth and crusading spirit of the Spanish aristocracy, at minimal cost to the overstretched Castilian treasury. But in retrospect it had serious shortcomings when it came to colonizing an unfamiliar continent with an uncertain climate.

In practice it meant that each expedition represented a high-stakes throw of the dice. For several years, until a conquistador's monopoly expired or his expedition collapsed, Spanish imperial expansion in La Florida remained committed to a single venture, which could be wrecked by an untimely storm or outbreak of disease, or the failure to find food and warmth during freezing winters. In this way, half a century slipped by in one failure after another. Moreover, the system evidently attracted the wrong sort of commanders. These were not cautious leaders willing or able to wait years or decades to turn a profit from local resources. They were mostly bold, impetuous men, drawn by the examples of Cortés in Mexico and Pizarro in Peru—men who were willing to gamble lives and fortunes on the uncertain promise of quick returns from conquering populous Indian kingdoms or plundering silver and gold.

There were lessons to be learned from their failures. These lessons included the need to seek out fertile sites for settlement beyond the sandy coast, and the need to prepare years of continuous logistical support until a new colony could become self-sufficient. However, there are few signs that successive ventures into La Florida learned from mistakes of the past. The Spanish Empire's secrecy, censorship, and failure to publish on new American discoveries formed part of the problem. Rumors, hearsay, and self-serving justifications and accusations left by expedition survivors must have filled that gap in public information.

The lack of a permanent presence in La Florida or of continuity among the expeditions added to the difficulties, making it that much harder to really understand the country's climate and environment. Without that understanding, unusually fair or foul weather could leave a lasting impression of an entire regional climate, giving credit to the hyperbole of promoters or disparagement of detractors. Rather than constructing a more accurate and useful picture of La Florida, each venture created a mix of exaggerated fears and false hopes—warnings of unendurably harsh seasons and sterile soils, mixed with empty promises of great fertility and precious commodities in a few inland provinces, or of a wealthy Indian civilization or passage to the South Sea just beyond.

3

The Land Itself Would Wage War

Spanish claims to La Florida did not go unchallenged for long. Its rivals never accepted the 1494 Treaty of Tordesillas, which divided the West and East Indies between Spain and Portugal. French king Francis I (r. 1515–1547) is said to have remarked, "I'd like to see Adam's will to find out how he shared the world."[1] In 1524, he commissioned the Italian captain Giovanni da Verrazzano to explore the North American coast, staking French claims to the continent.

Verrazzano crossed the Atlantic with a single caravel, passing through a storm "as fierce as ever man sailed through," or so he described it. His ship stayed on course to reach America somewhere along the Carolina coast. After a short foray south, Verrazzano reversed course, skirting the Outer Banks so as to avoid Spanish patrols.[2] At that point, the captain made one of the most fateful errors in Europe's early encounters with North America. He mistook the narrow islands of Carolina's Outer Banks for an isthmus between the Atlantic Ocean and a vast sea that might stretch westward to the Pacific. Finding no way through the banks, he continued north along the mid-Atlantic and New England coasts. On reaching present-day Rhode Island in mid-May, Verrazzano noted that the land, although on a parallel with Rome, "is somewhat colder, by accident and not by nature." Otherwise, the captain wrote little about the climate, recording mostly positive impressions of the land, its tall forests, and its Native peoples, who had evidently never seen Europeans before. Running short of provisions, Verrazzano sailed past Newfoundland and returned without incident to France.[3]

Over the next century, Verrazzano's discoveries and descriptions— above all his mistaken "Verrazzano Sea"—would entice French, English, and Dutch explorers to eastern North America. Yet for decades Spain's rivals could do little to confront the Spanish Empire directly, either in Europe or in the New World. The Spanish Empire's wealth and its fearsome military reputation, cemented by victories over France in the Italian Wars, had intimidated its adversaries. France and England concentrated their early efforts on finding a route around the continent to the East Indies, avoiding Spanish claims altogether. If there was any consolation for Spain's first half century of failures in La Florida, it was that Spain's rivals were unable or unwilling to take advantage of them.

Starting in the 1560s, the situation shifted in both Europe and North America. The Spanish Empire became overstretched, and even the mines of Peru could not fill the Spanish treasury as fast as it was drained by Philip II's wars in defense of his Catholic faith and imperial borders. Pirates and smugglers from England, France, and the breakaway provinces of the Netherlands undermined Spanish control of Caribbean trade. Spain's failure to colonize La Florida invited rival nations seeking American settlements or a base to plunder the apparently vulnerable Spanish Empire. So even as Spanish colonial officials began to sour on La Florida and its disappointing land and climate, the danger of competition drew them back in. For the rest of the century, the Spanish Empire was forced to react to threats and rumors of foreign incursions.

Climate would play a central role in this phase of European rivalry in North America. Severe winters and tropical storms continued to derail expeditions already ill-equipped for the environments of La Florida. America's stronger seasons, variable weather, and irregular rainfall continued to disappoint explorers and colonists in search of Mediterranean climates and commodities. Extremes of the Little Ice Age—including some of the worst droughts of the last millennium—continued to put pressure on Native American communities confronted with European invaders and their demands for land, food, and other resources. Throughout the sixteenth century, English and French expeditions proved no better prepared for these challenges than had Spanish ones. In time, officials of the Spanish Empire would learn of their rivals' failures and conclude that maybe there was little to fear from foreign competition after all. La Florida's difficult climate and environment might defend the region for them.

France's first direct challenge to Spanish claims in North America was more a matter of opportunism, or even desperation, than a real show of strength. France and Spain had spent most of the century at war in

Europe. A major Spanish military victory at St. Quentin in 1557 forced France into a disadvantageous peace at Cateau-Cambrésis in 1559. Later that year, the death of French king Henry II opened factional struggles for control over his young successors. Meanwhile, sectarian conflict between France's Catholics and its growing Protestant population, known as Huguenots, threatened civil war. France's military weakness in Europe may have encouraged statesmen to undertake a less direct and less expensive strategy of confronting the Spanish Empire in the Americas, where it might prove vulnerable. At the same time, many Huguenots saw the New World as a potential refuge from religious strife and growing poverty at home.[4]

And so in February 1562, sponsored by the Huguenot leader Gaspard de Coligny and led by his officer Jean Ribault, two ships of French Protestants departed for North America. After a long crossing, they reached Florida in early May and first disembarked at the mouth of the St. Johns River. Exploring farther north, they found Santa Elena and set up a colony there named Charlesfort.[5] Ribault took the ships back in order to gather fresh settlers and supplies.

Upon returning to France, however, the Protestant Ribault was caught up in a new outbreak of sectarian violence. Fleeing to England (then allied to the Huguenot cause), Ribault took a year to organize a relief expedition for the colony. In the meantime, he penned a fantastic description of its territory as "the fairest, frutefullest and pleasantest of all the worlde."[6] For the abandoned settlers at Charlesfort the situation went downhill. Expecting timely deliveries from France, the colonists had made little provision for food and were forced to barter away their goods in return for Indian corn. Conditions turned desperate when an untimely fire destroyed what few supplies remained. After an attempted mutiny and the murder of one of the captains, the colonists gave up on La Florida and built a makeshift boat to return to France. Becalmed halfway across the Atlantic, they suffered severe privation, even resorting to cannibalism, before they were rescued off the coast of Europe in October 1563.[7]

Driven out by France's escalating wars of religion, a second expedition of 300 Huguenots sailed the following spring, reaching Florida in early July.[8] The colonists established a new settlement, Fort Caroline, at the site of the previous French landing on the St. Johns River. Again the expedition failed to bring adequate provisions and soon became dependent on Indians. "As regards foodstuffs," one settler recalled, "of which we had high hopes in this new world, there was not a morsel; and if the natives had not shared their food with us every day there is no doubt that some of us would have died of hunger."[9] A late summer heat wave precipitated an epidemic among the settlers. The captain of the expedition, René Goulaine de Laudonnière, fell gravely ill. In the vacuum of leadership, discontent boiled over into mutiny. First a small group of conspirators stole a boat to attack ships in the Caribbean, and then in December a larger conspiracy took Laudonnière prisoner and forced him to authorize more privateering against the Spanish.[10]

By then hunger had become a pressing issue. The indigenous people of that region customarily dispersed in the winter, leaving behind their fields and villages to hunt, fish, and gather in the woods.[11] During the next five months the French traded away everything they could for food, even the shirts off their backs, to the mockery of Indians. The colonists finally resorted to hostage-taking and extortion from nearby villages, but without much success. By then the region was entering a serious drought, and even Indians may have been short of corn.[12] Later testimony described settlers dying of hunger "and the remainder so emaciated that the skin was hanging on their bones."[13] By July 1565, the corn began to ripen and some food became available again, but the colonists had so badly antagonized their neighbors that they no longer felt secure.

The next month, they thought they found an unexpected deliverance. The English privateer John Hawkins, fresh from a raid on the Spanish Caribbean, arrived at Fort Caroline and agreed to sell the French colonists a ship and some supplies so that they could return home. But just as the colonists loaded up their vessels and waited for favorable winds, Jean Ribault's long-awaited relief expedition finally sailed into view, laden with new settlers and supplies.

Since the spring of 1565, Ribault had been in a race to reach the vulnerable colony. The Spanish ambassador in Paris had been transmitting intelligence on French plans since early 1563. In response, Philip II had commissioned yet another effort to colonize Santa Elena, this one led by the son of Lucas Vázquez de Ayllón. That expedition fell apart even before it reached North America. Meanwhile, a separate Spanish fleet found the abandoned remains of Charlesfort at Santa Elena.[14] Further reports of Ribault's preparations in late 1564 and the interrogation of some captured French mutineers soon raised Spanish fears of a Huguenot conspiracy to seize La Florida. In March 1565, the king agreed to a bold plan by Captain General Pedro Menéndez de Avilés to launch an armada from Spain, drive out the French, and establish a base at Santa Elena.[15] Hearing of Spanish intentions, Ribault outfitted his fleet with extra men and guns, and set sail for Florida in May. However, the French ships faced contrary winds and made unaccountably slow progress to Fort Caroline. Meanwhile, Menéndez de Avilés had departed for Florida about a month after Ribault, only to have his ships scattered in a violent storm during their Atlantic crossing. It took another month to reassemble some of his ships and men at Puerto Rico and finally proceed to the French colony.[16]

Thus by a remarkable coincidence both the Spanish and French forces approached Fort Caroline within days of each other in early September 1565. At that point, Ribault made the mistake of keeping most of his men aboard the ships rather than unloading them into the fort. Probably blocked by the low sandbars around the coast, he hesitated to land and attack the Spanish forces, whom he discovered marching along the shore. In the meantime, the weather turned ominous, and Ribault ignored the warnings of Laudonnière and other colonists who recognized the signs of a coming hurricane.[17]

The storm struck on September 20 with furious winds and rain. "Such a great tempest and accompanied by such gales, that the Indians themselves assured me that it was worst weather ever seen on this coast," Laudonnière recalled.[18] Three of the French ships were sunk and the rest swept to shore near Cape Canaveral, leaving a small company of mostly weak and sick men to guard Fort Caroline. The Spanish pressed on through the hurricane, capturing the fort in a bold assault. All the French who could escaped to the remaining ships. The Spanish rounded up hundreds more on the shore, and Menéndez de Avilés ordered them executed as trespassers and heretics.[19] Two years later, a French force raided the Florida coast and massacred more than 200 Spanish soldiers in reprisal. But the French would not settle in the region again for more than a century.

As his correspondence reveals, Pedro Menéndez de Avilés arrived in La Florida with grandiose ambitions, based on a hopelessly inaccurate sense of the region's environment and geography. The Florida of his imagination had a fertile Mediterranean climate, ideal for raising cattle and silkworms, and for growing wheat, vines, olives, and sugar. Criss-crossed with navigable rivers, it lay near an inlet stretching deep into the continent (a misconstrued Chesapeake Bay) and a vast inland waterway that might open a passage to the Orient (a version of the Ver-razzano Sea). Overland, it was only a short distance to the rich mines of Zacatecas. Menéndez de Avilés planned a thriving settler colony that would give Spain control of the region with its routes to Mexico and the Pacific.[20] To that end, the conquistador established two principal settlements, at Santa Elena and St. Augustine, as well as several out-lying garrisons. He opened alliances with neighboring chiefdoms, called for missionaries from Spain, and sent an expedition under Cap-tain Juan Pardo to explore inland.[21]

The realities of the situation soon undermined those plans. The Spanish had arrived short of supplies and had suffered in the fighting and the hurricane. Mutiny and desertion were rife. Hunger and sick-ness carried off as many as half the soldiers by 1566.[22] Throughout that year and the next, the Spanish in Florida complained repeatedly about the threat of starvation and the need for food and reinforcements.

Adding to their troubles, the drought around St. Augustine intensified. By some measures, the region suffered worse drought during the 1560s than any it has seen since.[23] Years of failed harvests undermined food supplies and complicated relations with local chiefdoms. Hungry colo-nists in Santa Elena literally prayed for deliverance.[24] Additional settlers arrived in 1568 and 1569, but they only brought more sickness and more mouths to feed. Regular support from New Spain in the form of money and provisions did not materialize until the 1570s. By then, Menéndez de Avilés had been forced to abandon all the Spanish outposts beyond St. Augustine and Santa Elena and to drastically reduce his garrisons.[25]

The first Jesuit missionaries, who arrived in early 1568, fared no better. Initial optimism waned as the friars found the local Guale and other Indians to be at best reluctant converts. Three basic problems undermined their efforts. Florida's Indians were more ready to adopt outward Christian rituals than to accept Catholic dogma and taboos. Moreover, as the French had already learned, their way of life was to disperse from villages to hunt and fish throughout the winter, which left the missions empty and without food for several months each year. Worst of all, the friars soon brought European diseases to mission In-dians. Their practice of baptizing the sick and dying gave the impres-sion that they were killing their converts with sorcery.[26]

In September 1570, a handful of Jesuit priests and followers sailed into the Chesapeake to establish a mission at a place they called Ajacán, only a few miles from the eventual site of Jamestown. Guiding them was a certain Paquiquineo, an Indian taken from his home by a Spanish exploratory mission to the Chesapeake in 1561. He had come first to Spain and then to Mexico, where he had learned the language and been baptized "Don Luis."

"We found the land of Don Luis very unlike what we thought," wrote one priest after they landed,

> not because there had been any mistake by Don Luis in de-
> scribing it, but because Our Lord has punished it with six years
> of sterility and death, which has left it very depopulated com-
> pared to what it used to be, and because there are many dead
> and also those that have gone to other lands to see to their
> hunger. Few have remained other than the principal people,
> who say that they want to die where their parents died, even
> though now they neither have maize nor are there found wild
> fruits of the forest that they usually eat, nor roots, nor any-
> thing else.

Not only was the land in the midst of a serious drought—now clearly attested in the tree-ring record—but the winters had been severe as well: "They are so hungry that all think they will die of hunger and cold this winter as many have in the past winters because besides that they have difficulty finding the roots that they usually live on, the great snows they have in this land do not let them look for them."[27]

The testimony of a single survivor, a young boy, provides our only version of what happened next. Paquiquineo left the priests to rejoin his people, rejecting Jesuit demands for assistance. Finally in February 1571, when some priests went to his village to fetch him back, Paquiquineo killed them and then led a group of men to the mission to kill the others. A relief expedition finally arrived in the spring, which rescued the survivors and massacred Indians in reprisal. The motivations of "Don Luis" have been a matter of speculation ever since.[28]

Even after Florida's drought of the late 1560s, the settlements of Spanish Florida continued to struggle. Official investigations in 1573 and 1576 found widespread problems with supplies, defense, and settler morale.[29] Deliveries of provisions and clothing from New Spain proved unreliable and inadequate. Settlers testified that the land was sandy and barren. They complained of extreme hunger, which reduced them to eating "herbs, fish, and other scum and vermin."[30] Archaeological investigations of the settlements have found no signs of outright starvation but confirm that the Spanish turned to native plants and wild animals out of necessity.[31] The situation was worst of all in Santa Elena. The island proved too small, and the soil was completely unsuited to raising food. Even the pigs and goats fared poorly. In late 1571, a fire destroyed that colony's supplies, and sickness spread among the hungry colonists.[32]

Above all, Florida failed to provide the Mediterranean climate that settlers from Spain expected. Struggling with Florida's unfamiliar seasons, they felt cheated. As one settler in Santa Elena complained:

> Even if the land were broad and fertile, it does not have the climate nor does the earth ever dry out [for the ripening of grain], except with frosts and excessive cold spells brought by the winter, which is December and January, because in April and May when the [grain for] bread ripens in this island, it does nothing but rain all that time, which is when we sow and gather maize; and so we have suffered and do suffer great hardships. . . . So we have wasted all the means with which we came well supplied and other things, since we were farming people in Spain, and we farmed with all kinds of cattle in Spain, so here we are lost, old, weary, and full of sickness.

Despite the long and dangerous voyage across the Atlantic, and the hardships that had driven them from Spain, many pleaded to return home.[33]

Hunger pushed the Spanish at Santa Elena to demand food from nearby Indian villagers, who were already suffering shortages brought by the recurring droughts. In 1576 Guale Indians, followed by neighboring Orista and Escamacu villages, launched an assault that forced the Spanish out of Santa Elena.[34] Three years later, incited by rumors that the French had returned to support Guale resistance, the new governor, Pedro Menéndez Marqués, carried out a brutal campaign of suppression, burning Indian villages and destroying their food supply. His tactics, combined with another serious drought during the early 1580s, finally starved the Guale into surrender.[35] The Spanish rebuilt Santa Elena, and Spanish Florida recovered slowly for a few more years, until its colonies were brought low by another disaster, this time at the hands of the English.

With the destruction of Fort Caroline and France's descent into political instability and religious strife, England emerged as the principal opponent of the Spanish Empire in Europe and the New World. Driven by religious differences, the conflict took on an emotional and personal character for the rival nations and their respective monarchs, Philip II and Elizabeth I. Until the mid-1580s, it played out in proxy conflicts, the English aiding Protestant rebels in the Netherlands and the Spanish supporting Catholic dissidents in Ireland and England. Meanwhile, the legendary wealth of Spanish America proved an irresistible target for smugglers and corsairs, who carried out an unofficial war on behalf of Spain's rivals. John Hawkins's 1564–1565 Caribbean expedition (the one that had nearly rescued the French at Fort Caroline) launched an open season for pirates and privateers in New Spain. It would culminate in Francis Drake's 1573 seizure of Spanish treasure in Panama and then his 1577 voyage through the Strait of Magellan, which raided far up the Pacific coast and went on to circumnavigate the globe. Drake's triumphant return, loaded with plunder, captured the imaginations of a generation of Englishmen.[36] This Anglo-Spanish rivalry and the fever for privateering shaped the planning and fortunes of what would become the Roanoke settlement, in the land that the English called Virginia.

Despite the romantic place it holds in the American imagination, the famous "lost colony" began as little more than a nest for pirates. Roanoke's architect and promoter, the courtier Walter Raleigh, first sent two ships on a reconnaissance voyage in May 1584. They found an entrance into Pamlico Sound and dropped anchor at Hatteras Island, which they would later come to know as Croatoan. They then continued north to Roanoke Island, where the Secotan Indians gave them an elaborate reception, implying they had traded with European merchant ships before.[37] The expedition arranged to take two Indians, named Manteo and Wanchese, back to England. An English party

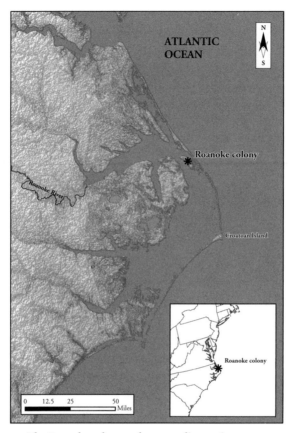

3.1 The Roanoke colony and surrounding region

scouting the mainland fell into an ambush by another group of Indians, and several Englishmen were killed and possibly cannibalized.[38] Despite this episode, the survivors returned with only the most glowing descriptions of the region and its resources: wild grapes, abundant timber, soil "the most plentifull sweete, fruitfull, and wholesome of all the world," and even "the people most gentle, loving, and faithfull, void of all guile and treason, and such as lived after the manner of the golden age."[39]

During that autumn and winter, Raleigh gathered more support for his venture and developed plans for the colony. The next expedition, led by Richard Greenville and Ralph Lane, departed in April 1585, but a storm scattered its ships as they crossed the Atlantic. One pinnace, almost lost at sea, eventually cast up its men on the shores of Cuba, where almost all died of starvation.[40] The rest of the fleet rendezvoused as planned at Puerto Rico, where they bartered for fresh food and live-stock and gathered tropical plants that they imagined would grow in their new colony.[41] Unable to resist some privateering against Spanish ships around Hispaniola, they did not reach the Florida coast until late June.

The following month, the expedition suffered a fatal setback when its flagship, the *Tiger,* ran aground on the Outer Banks. Its passengers escaped, but all its cargo was lost. "And soo with grete spoyle of our provysyones, [we] saved ourselfes," reported Ralph Lane, in his charac-teristically creative Elizabethan orthography.[42] When they reached Roa-noke, they decided Greenville should return to England immediately for fresh supplies while Lane held a colony of about a hundred En-glishmen at a fort constructed near the north end of the island.[43]

The loss of provisions on the *Tiger* left the settlers dangerously de-pendent on Roanoke's Indians. This time, at least, the colony had cer-tain advantages. Manteo and Wanchese served as translators. Thomas Hariot, a scholar serving the expedition, learned enough of the local Algonquian language to study indigenous cultures. Through the summer and fall, the Secotans and their chief, Wingina, were apparently eager to keep trading food for English goods.

By the end of 1585, however, relations began to break down. Ralph Lane remained suspicious of Indians and always ready to avenge the smallest slight, real or perceived. An Old World epidemic introduced among the Secotans—"that they neither knew what it was, nor how to cure it," as Hariot described—may have aroused fears of English sor-cery.[44] Frequent trading may have devalued English merchandise with respect to corn, especially at a time when Secotans were preparing to live off the land for the winter. But the most likely reason of all, as we'll see, is that corn was growing scarce even for Carolina's Indians.

Meanwhile, the expedition's activities in the Caribbean raised alarms among Spanish officials, who had already caught wind of English plans from spies in Europe. As early as April 1584, Spanish intelligence in London reported on supposed designs to colonize Newfoundland. By early 1585, spies had learned some details of the expedition's inten-tions, and they now placed the English colony somewhere in New

England. That summer and autumn officials finally gathered an accurate picture of the venture from the survivors of the lost pinnace washed up at Cuba and from Spanish sailors captured by Greenville's fleet during his trips to and from Roanoke.[45] Suspecting designs on his colony, Florida governor Pedro Menéndez Marqués sent out a ship to spy on the English that December, but it turned back well short of the Roanoke colony. The governor pleaded for help from officials in Spain, who responded with characteristic delay.[46]

In the meantime, events elsewhere seized Spanish attention and raised new fears about English designs. During the mid-1580s Anglo-Spanish conflict broke into open war. Spain colluded in assassination attempts on Elizabeth I. The queen renewed her support for Protestant rebels in the Netherlands and licensed more privateering against Spanish ships and colonies. With royal support, Francis Drake outfitted a large new fleet that raided Spain in autumn 1585 before crossing into the Caribbean. In early 1586 Drake's men looted first Santo Domingo and then Cartagena, laying waste to towns and sowing panic around the Caribbean. Nevertheless, Drake faced growing dissension among his sailors, who succumbed to tropical diseases at such a rate "that every day they threw corpses overboard." In late spring, he turned back toward England.[47]

Before crossing the ocean, Drake's fleet stopped to make one last raid, on St. Augustine. Spanish officials testified to a heroic defense of the fort until, faced with overwhelming numbers, the whole colony was forced to evacuate. English accounts mention only "3 or 4 small shot" before the Spanish took alarm and "with all speede abandoned the place."[48] Indians may have set fire to the town before the English even reached it.[49] Both English and Spanish versions agreed on what followed. "There were about 250 houses in this Towne," wrote one English sailor, "but wee left not one of them standinge." They "burned it and razed it," one Spanish official testified; "spoiled the plantations and carried off everything, even to the trifles," reported another.[50]

This destruction of St. Augustine raised new questions about strategy in La Florida and the threat posed by England's colony at Roanoke. For some Spanish officials, the episode underlined Florida's vital role in the defense of the Caribbean. For others, disillusioned with the land and its ailing colonists, it was time to pull the plug and focus resources on the Antilles and Mexico.

Governor Gabriel de Luján of Cuba was among the latter. He despaired of a colony whose inhabitants had to rely on continual subsidies and still "starve the rest of the year."[51] Another Cuban, Alonso Suárez de Toledo, advised Philip II that after Spain's decades of failure in La Florida, there was little to fear from any English colony there:

> To maintain Florida is merely to incur expense because it is and has been entirely unprofitable nor can it sustain its own population. Everything must be brought from outside. If, although Your Majesty possess Santo Domingo, Puerto Rico, Cuba, Yucatan, and New Spain, the garrison of Florida has nevertheless suffered actual hunger, what would happen to foreigners there who must bring their subsistence from a great distance to an inhospitable coast? The land itself would wage war upon them![52]

Toledo's dramatic prediction was not far off the mark. During the winter and spring of 1586, the environment and climate of Roanoke became the English colonists' worst enemy.

In spite of those first glowing reports about the island's fertility in 1584, Roanoke's soils were in reality thin and easily eroded, probably never capable of producing a great surplus of food. But from the point of view of climate, the English had picked possibly the worst years of the entire past millennium to depend on Indian corn. In a remarkable study of regional tree rings, climatologist David Stahle and colleagues discovered that the English settlement in Roanoke coincided with the onset, during 1587–1589, of the deepest regional drought of the last 800 years. As examined by archaeologist Dennis Blanton, the drought almost certainly hurt crop production and supplies of fresh water, intensifying competition over provisions. Already in 1586, Hariot had observed how "their corne began to wither by reason of a drough which happened extraordinarily."[53]

Ralph Lane, however, construed the Secotans' refusal to deliver food as a mere stratagem. He concluded that their chief, Wingina, was plotting to starve out the English colony. As Lane realized, the challenge for the colonists would be to hold out just a few more months until either relief arrived from England or the new corn ripened in early July. However, as the ships failed to materialize and Indians denied them new food, the colony went desperately hungry. Afraid to spread out his men and open them to ambush, Lane held them in the fort until "the famine grewe so extreme among us" that he had no choice but to send them foraging. Under threat from real or imagined plots by Wingina, Lane eventually allied with the chief's enemies on the mainland. In June 1586, they devised an ambush in which an English soldier shot Wingina and cut off his head.[54]

Barely a week later, the colony was saved by the chance arrival of Francis Drake, fresh from the sack of St. Augustine. His fleet loaded with plunder but his crew decimated by disease, Drake was only too glad to offer Lane the ships and supplies he needed to relocate the colony. Yet the hazards of American weather intervened once more. On June 23, as Drake's ship anchored nearby, "there arose such an un-wonted storme, and continued foure dayes that had like to have driven all on shore, if the Lord had not held his holy hand over them." The winds and tide snapped anchors aboard the ships. One of Drake's sailors described rising seas, thunder, and rain with "hailstones as big as hens' eggs."[55] For the haggard Roanoke settlers, it was the last straw. They gave up the colony and took passage home on Drake's ships. Greenville's relief expedition, which only got under way in late spring 1586, arrived to find the colony already deserted. Greenville left fifteen men to hold the fort at Roanoke and turned back to England with the rest.

English opinion reacted to the failure at Roanoke much the way opinion in Spain and Mexico had responded to the return of the Narváez and Soto expeditions. Back in England, survivors reportedly cursed their leaders and "slandered the countrie itselfe." On the other hand, the venture's supporters worked to build enthusiasm for another expedition, publishing optimistic descriptions of the land and speculation of even greater promise in regions just beyond. Thomas Hariot extolled the "excellent temperature of the aire . . . at all seasons, much warmer than in England" but "never so violently hot" as the tropics. Equating latitudes with climates, he argued that Virginia would soon grow sugarcane, lemons, oranges, and every other Mediterranean crop.[56] Ralph Lane decided that England should pursue even more promising lands in the Chesapeake, a region supposedly unequalled in the world "for the pleasantness of seate, for temperature of

Climate, for fertilitie of soil, and for the commoditie of the Sea." Rumors spread of a fabled land nearby called "Chaunis Temotoan," with rich mines and a passage to the South Sea. Convinced by Lane's descriptions, Raleigh organized a new expedition to the region in early 1587, this one carrying women and families and provisioned to establish a settler colony.[57]

For English sailors, however, the lure of piracy remained strong, and that temptation would derail these plans for a settlement. The 1587 expedition set out in mid-April but made slow progress through the English Channel. Following the southern route via the Canary Islands, the ships took until early July to reach the Caribbean island of St. Croix, where some of the hungry passengers fell sick eating poisonous fruits. As the vessels turned toward the North American coast the colony's intended governor, John White, fell out with the expedition's pilot, Simon Fernandes. At the start of August, they reached the Outer Banks, where they had planned to pick up the small garrison left by Greenville the year before. However, Fernandes and his crew refused to carry White and the settlers any farther north, claiming "the Summer was far spent"—that is, they meant to go privateering in Spanish waters before the sailing season closed for winter.[58]

So John White and his colonists abandoned their plans for the Chesapeake and went to repair their old fort on the north end of Roanoke Island. They found no trace of the fifteen men Greenville had left the year before. Three days later, "divers savages, which were come over to Roanoke," ambushed a settler named George Howe, shot him with sixteen arrows, and then "beat his head in pieces," according to White. A detachment of twenty colonists guided by their translator Manteo went over to Croatan to learn what had happened. The Croatan Indians at first fled, but upon recognizing Manteo they came back, "and some of them came unto us, embracing and entertaining us friendly, desiring us not to gather or spill any of their corne, for that they had but little."[59] The drought was almost certainly taking a toll on food supplies. That fact could help explain the less than friendly reception English settlers received from Secotan Indians and their allies, who were apparently responsible for the massacre of the lost garrison and the murder of George Howe.[60]

By the end of August, Simon Fernandes and his sailors were anxious to leave the colony. However, a sudden nor'easter whipped up squalls and waves and forced the flagship to cut its cables. With most of the sailors still on land, its crew struggled for almost a week before they could safely come ashore. In the meantime, alarmed by their failure to establish peaceful exchange with Roanoke's Indians, the colonists insisted that John White return to England "for the better and more assured helpe, and setting forward of the foresaide supplies." Supposedly with great reluctance—we have only his word to go on—the governor departed with the ships in early September, leaving the colonists behind.[61]

The expedition's failure to reach the Chesapeake had the unintended consequence of throwing off Spanish pursuit. From 1587 to 1589, Spanish interrogation of English sailors gradually built up a picture of English designs.[62] With support from Madrid, Florida governor Pedro Menéndez Marqués set out in search of the English a first time in May 1587, shortly before the new colonists had even arrived. Before his ship could enter Chesapeake Bay, a "tremendous storm" blew it off course, driving it south to the Bahamas.[63] A year later, he sent a second ship in search of the colony. It carried out a reconnaissance of Chesapeake Bay and its rivers, but it, too, turned back without finding evidence of any settlers left over from Roanoke.[64]

Neither the Spanish nor English would ever see those settlers again: they were soon to enter history as the famous "lost colony." Their misfortunes began with White's return voyage in 1586. His ship struggled for weeks to cross the Atlantic under "scarce, and variable windes"; then a northeasterly gale blew it back so far that it took weeks to recover its course. Reduced to "stinking water, dregges of beere, and lees of wine," the passengers "expected nothing but by famine to perish at Sea." They made land on the west coast of Ireland only in late October, and it took White another month to meet with Raleigh and the colony's investors.[65] By the time they could outfit a relief expedition in spring 1588, the Spanish Armada was preparing to invade England, and Queen Elizabeth's Privy Council ordered all able vessels to stand ready for the nation's defense. John White could only manage the dispatch of two small vessels to Roanoke. Predictably, their English captains preferred to go hunting Spanish vessels. In an ironic turn, they fell into the hands of French privateers instead.[66]

Once the Armada invasion was defeated—as much through accidents of weather as English valor—it was too late in the year for another ocean crossing.[67] Raleigh, meanwhile, was distracted from Roanoke by problems in Ireland and quarrels at court. Finally in March 1590, White persuaded another fleet of privateers bound for the West Indies to help him search for the Carolina colony. Once again, chasing after Spanish prizes consumed most of the spring and summer sailing season. The ships did not reach Florida until early August, in peak hurricane season. It soon ran into "very fowle weather with much raine, thundering, and great spouts" that pursued them for almost a fortnight.[68]

They reached Hatarask late on August 25 and spied smoke rising from near Roanoke Island. But, as White recalled, "when we came to the smoke, we found no man nor signe that any had beene there lately, nor yet any fresh water in all this way to drinke": perhaps another indication of the drought. As they took two boats through the treacherous shallows of the Outer Banks, a "great gale" came out of the northeast. One boat nearly wrecked, spoiling its cargo and drowning several men.[69] At Roanoke's abandoned fort, they found only "heavie things, throwen here and there, almost overgrowen with weedes," and some supplies "spoiled and scattered about." In a prearranged sign, the last colonists had carved "Croatoan" into a nearby tree, indicating they meant to move to that island. White and the captain resolved to press onward in search of survivors, but their ship nearly wrecked again when casting off into deeper waters. At that point, "the weather grew to be fouler and fouler; our victuals scarce, and our caske and fresh water lost." A storm brought violent winds out of the west and northwest, driving their ships away from shore, whence they continued back to England.[70] Historians can only speculate on the fate of Roanoke's "lost colony." Incidental testimonies, recorded decades later, indicated the settlers had escaped to the interior, only to be massacred in an Indian conflict around 1607.[71]

Spanish spies and officials took several years to piece together the fate of the English colony. A 1588 report discovered that "the natives proved poor friends to the English and that Francis Drake carried off those who survived."[72] One year later, a Spanish sailor captured by English privateers communicated the full story to officials in Hispaniola: the English had settled a small island a little northeast of Santa Elena and fallen into conflict with its natives. Drake had found them "in poor condition and greatly in need of provisions, for the land produces little to eat, having nothing but maize, and of that little, and the land wretchedly poor." However, the informant believed the English planned to settle further inland, where "there is much gold and so that they may pass from the North to the South Sea."[73] In fact, the English presence at Roanoke lived on much longer in Spanish official correspondence than in reality, periodically inciting new rumors and fears.[74]

Chasing after reports of the English settlement, some officials became interested again in lands to the north and west of Florida.[75] After his 1587 voyage toward Chesapeake Bay, Pedro Menéndez Marqués informed the king it would be "imperative to explore all that coast, for what I saw of it is very different from what the chart shows."[76] Vicente González, who explored the bay in summer 1588, spread tales of abundant gold and copper, a passage to the South Sea, and a short passage to New Mexico just through the mountains to the west. "They have, too, apples, medlars, walnuts, much hunting of all kinds just as in Spain," he claimed, "while the climate of the land is similar."[77]

Conditions in Spanish Florida, however, undercut official enthusiasm for new ventures in the region. Florida's garrisons and settlers struggled to recover from Drake's 1586 raid and from a hurricane that battered Santa Elena the same year.[78] Officials in Cuba responded with unusual alacrity, rushing provisions up to the settlements. But the colony of St. Augustine had lost everything: buildings, supplies, fields, orchards, and animals. "I am reduced to such a situation that I do not know where to begin to relate the hardship and misery which have befallen this land," Menéndez Marqués wrote in the wake of the attack. Given Florida's vulnerabilities, the governor decided to concentrate its meager population and resources in St. Augustine, abandoning Santa Elena in late 1587.[79]

For the next decade, the rebuilt St. Augustine proved a mixed success. Its population stagnated at a few hundred, mostly men in spite of official efforts to recruit Spanish families. It continued to rely on an inflow of new settlers to keep up its numbers. Spanish missionaries made uneven progress in spreading Catholicism. Over time, they set up a number of resident and temporary missions among Indian villages and won converts and allies among Guale and Timucua chiefs, who received gifts and recognition from the Spanish.[80] By the end of the century, missionaries could plausibly claim thousands of converts over an area stretching from present-day central Florida through southeastern Georgia.[81]

The colony depended heavily on its annual subsidy from New Spain, which averaged more than 60,000 pesos per year. Supplies frequently fell victim to delays, English piracy, and official corruption. Scattered historical documents indicate that some of the wealthier inhabitants lived well, but archaeological evidence points to a poor material existence for most, alleviated by a growing acceptance of Native foods, Native culture, and also Native wives. Efforts to acclimatize Old World crops to the seasons and soil of Florida bore fruit, literally and figuratively. Sources testify that melons, figs, onions, and other cultivars grew in St. Augustine's gardens and orchards by the 1590s, even if they did not yield especially well. In good years, the colony was almost self-sufficient in maize.[82]

But not every year was a good year. The sandy soil around St. Augustine failed to hold moisture and so it would not produce crops without timely rains. Droughts proved a recurring problem throughout the region's colonial history, leading to poor harvests in 1588–1589 and again in 1591.[83] In the meantime, Spanish missionaries continued to describe freezing Little Ice Age winters in Florida, perhaps to underscore their suffering in the service of faith. Some emphasized the country's harsh northerly wind, a wind purportedly so cold it left them barely able to walk or to use their hands. One described walking barefoot in up to a foot and a half of snow, which would be an unbelievable amount for the present-day region. If accurate, these descriptions indicate the regional climate was considerably colder than that of modern times. If exaggerated (as seems likely), these descriptions nevertheless gave an unfavorable impression of the country and its climate back in Europe.[84] In short, Spanish Florida adapted and survived, but it hardly thrived in a way that generated new enthusiasm for imperial ventures.

The first French and English attempts to colonize La Florida failed for many of the same reasons that Spanish attempts had failed. Expeditions arrived unprepared for the environments and climates of North America. They faced unexpected storms, shipwrecks, hunger, and droughts that derailed plans and demoralized settlers. As had been the case during the Narváez, Soto, and Luna expeditions, there was nothing inevitable about these disasters, even during the Little Ice Age. Hurricanes and extreme weather struck at critical moments of these ventures in ways that no one could have foreseen. Nevertheless, the underlying climatic changes of the Little Ice Age aggravated the risks of failure. Global cooling and shifts in atmospheric circulation contributed to the severity of winters and frequency of drought in the region.

At the same time, French and English approaches to colonization had been risky all along. Lacking sustained investment, these expeditions depended on inconstant support from aristocratic patrons. Leadership on the ground failed during crucial moments. The lure of piracy clashed with imperatives of security and stability. Suspicion and aggression toward Native Americans provoked unnecessary conflicts. Propaganda designed to raise support for the ventures failed to convey the real challenges of colonization. Overall, they placed too much hope in too few ships in unforeseeable conditions. As Karen Kupperman has concluded of the Roanoke venture, "Everything seems to have hung by a thread, which was easily snapped."[85]

Even more than the failures that preceded it, the mixed success of Spanish Florida underscored the adaptations and compromises that Europeans would have to make to colonize North America. Crops needed to be acclimatized to different seasons. Colonies needed to establish reliable supply lines for years or decades until they could become self-sufficient. Above all, colonial planners and officials had to revise their expectations and take time to find viable locations for colonies. After the high hopes that had launched Pedro Menéndez de Avilés's seizure of Florida, however, the human and natural disasters that struck Spanish Florida repeatedly over the following decades left little enthusiasm for this kind of sustained investment in further colonization. Soon climatic and human disasters on both sides of the Atlantic would call into question Spain's entire commitment to North America and open the way for new French and English challenges.

4

Bitter Remedies

The last years of the sixteenth century brought Spanish Florida back into crisis. Troubles began in 1597, when Guale warriors massacred several Catholic missionaries. Despite investigations, contemporary and modern, the circumstances of the uprising have remained something of a mystery. Spanish interrogations of Indians, some under torture, indicated that Guale men resented missionary interference in their traditional way of life. Just as they did after the 1576 uprising, Spanish soldiers retaliated by attacking Guale food supplies, aiming to starve them into surrender.[1]

Their efforts were aided by another regional drought in 1598, which also caused food shortages in St. Augustine. Then the year 1599 brought the colony one disaster after another. A fire destroyed a number of the town's buildings; summer drought contributed to another poor harvest; and in September a major storm with high winds and flooding destroyed part of the fort and damaged its dikes, raising fears of a deluge.[2]

News of these events brought unwelcome official scrutiny of conditions in Florida. Franciscan friars denounced Florida governor Gonzalo Méndez de Canso, blaming the uprising on his corruption and cruelty. One warned of "the great ruin that threatens this outpost, in spiritual matters as well as temporal." Another called Florida "a wasteland" full of woods and swamps that "renders little harvest after much labor," adding that "the port and sound are worthless." Complaints such as these, coming on top of the conflagration, drought, and flooding, suggested deeper problems in the colony. Observing that Florida offered few resources, no rich mines, and not even an adequate harbor, the new Spanish king, Philip III (r. 1598–1621), sent the governor of Cuba on an investigation to decide the colony's fate.[3]

Following the usual delays in imperial communication, the enquiry got under way in 1602. Testimonies from soldiers, friars, and officials in St. Augustine dwelled on decades-old doubts about the colony. The land was unproductive. Its colonists were poor and hungry. The corn failed to ripen or was eaten up by pests. It lacked precious metals or commodities. The harbor was too shallow. Some argued for staying, others for abandoning Florida altogether. Several supported moving to the north or northwest, bringing up old rumors of fabulous Indian kingdoms in the interior.[4] Governor Méndez de Canso was soon recalled to Spain to answer the charges against him, but the colony was not immediately disbanded. Over the next few years, mission activity continued, and the new governor, Pedro de Ibarra, worked to restore Spanish influence among neighboring Indian nations.[5]

The deliberations on the colony did not end there, however. Philip III asked Viceroy Monterrey of New Spain to make his own investigation. The viceroy noted Florida's value in guarding the Bahama Channel but also emphasized the "great waste" in holding St. Augustine, the "discomfort and suffering of the soldiers," and "the little use that we get from it . . . given that it seems impossible the enemy could colonize it and support themselves."[6] Then the king asked former governor Méndez de Canso for his opinion on abandoning the colony. Philip III's enquiry described Florida as a "sandy waterlogged marsh" without resources, without gold or silver, and with "natives very poor, warlike, and dishonest." The former governor replied, without enthusiasm, that St. Augustine was nevertheless the only acceptable harbor in the region. It might be dangerous to leave the whole coast to Spain's enemies, he warned, and it would be irresponsible to abandon the missionized Indians. As the enquiries made clear, the king tended more and more to view Florida as a worthless outpost he could ill afford to maintain. By 1606, he decided it was finally time to give it up.[7]

Philip III's attitude reflected not only disappointment with La Florida and crisis in St. Augustine but also a deeper crisis in Spain itself. His predecessor's unending wars in Europe and abroad had left his kingdom bankrupt. Silver flowed from American mines, and peasants in Castile supported crippling taxes, yet still the treasury's revenues were mortgaged for years in advance. The new king was ready to abandon his father's inflexible defense of religion and reputation, and to pursue new policies grounded in prudence and *raison d'état*. It was time to cut costs and keep the peace with European rivals, even if that meant ceding ground at the edge of empire.[8]

Recent research demonstrates another, equally powerful reason for Spain's imperial retrenchment: the Little Ice Age. Both physical and written sources give clear evidence of a catastrophic shift in climate during the last decade of the sixteenth century and the first years of the seventeenth. A series of large volcanic eruptions, in addition to a long-term cooling trend, brought some of the coldest global temperatures in thousands of years. A shift in atmospheric circulation contributed year upon year of unseasonable weather, floods, and droughts.

Spain did not suffer alone: historians have long recognized a "crisis of the 1590s" across Europe and beyond. Ongoing research in economic and climate history has only confirmed this picture of famine, mortality, economic depression, and conflict.[9]

Nevertheless, a closer look at the period suggests two important revisions. Countries across Europe experienced disasters, but not all of them underwent a real crisis, in the sense of a turning point. For some countries, with already stressed economies or strained political conditions, these decades of climatic extremes tipped a precarious balance, with lasting impacts. For other countries, these were miserable times, but passing. Different populations also had different perceptions of these disasters—and those perceptions mattered. Similar episodes could produce opposite impressions and opposite reactions, with enduring consequences in policy and diplomacy.

These revisions explain why the turn of the seventeenth century was a turning point for European rivalry in North America, too. The people of Spain, France, and England were more or less equally exposed to these coldest decades of the Little Ice Age. But they were not equally vulnerable to their impacts, both material and psychological. For Spain, these years truly represented a crisis: a crisis of agriculture, of finance, and, in a deeper sense, a crisis of confidence in its empire. For England, by contrast, temporary economic depression and even mass mortality paradoxically provided a further justification to reach for overseas colonies.

To understand this paradox, it helps to step back and take a wider view of death and disasters in sixteenth-century Europe. For people of the time, we might say that death was simply a fact of life. Nearly half of European infants never saw adulthood, and of those who did, most would not live beyond their fifties. Death was a constant presence and preoccupation in religion, art, and literature. In early encounters between Europeans and Native Americans, death, mourning, and burial customs were often among the first items of observation and even the first topics of conversation.[10]

In Europe, however, it seems that fewer people were dying than in the past. For reasons still unclear, the bubonic plague (*Yersinia pestis*), which had decimated Europe's population since the Black Death of the 1340s, went slowly into retreat. Starting at around 1450, populations slowly but surely climbed upward toward their pre-plague numbers.

Unfortunately, without any real improvements in living standards, the retreat of plague only cleared the way for other causes of death. As one of the Four Horsemen of the Apocalypse slowed his work, the other three kept busy. War and religious strife, dearth and famine were constant threats. New diseases came to take the place of *Y. pestis;* they ranged from syphilis, typhus, typhoid, and smallpox to mysterious ailments such as the English sweats. In the mid-1600s, population growth would finally grind to a halt in an era of misery, war, and revolution sometimes called the "general crisis": the age when Thomas Hobbes famously wrote of the "war of every man against every man" and described "the life of man, solitary, poore, nasty, brutish and short." Following those disasters, European trends would diverge. Some countries, notably England, would develop a stronger agricultural and economic base to support a more sustained rise of population into the modern age.[11]

But to return to Europe at the turn of the seventeenth century, we need to turn Hobbes's description around. As the life of man (and woman and child) became *less* solitary—that is, as populations rose and people moved into towns and cities—life became that much poorer, nastier, more brutish, and often shorter. Agricultural productivity and economic opportunity did not rise in step with the number of new people. Prices soared, particularly prices for food, spurred on by growing demand and by the influx of American silver. Real wages declined precipitously. The average height of European men actually fell during the late 1500s and early 1600s, a sign of declining nutrition.[12]

Case studies across Europe tell similar stories. As populations grew, land holdings were parceled out, forests cleared, common lands overused, and more marginal land taken under the plow. Contemporaries observed the beggars on the roads, the hard bread made of chaff and

inferior grains, and the general misery of a growing segment of society. Records of declining tax yields and tithes per household illustrate
the spread of poverty. The deceleration of marriage and fertility rates
reconstructed from parish registers points to shrinking opportunities
and declining health.[13]

Europe as a whole did not simply run out of food as people multiplied beyond the capacity of the continent to feed them. Instead, population growth put pressures—often fatal pressures—on certain people
at certain times. A striking feature of Europe's growth, both to contemporaries and to modern historians, was how the numbers of poor,
landless, and vagrants rose much faster than the population as a whole.[14]
While many Europeans lived on a comfortable margin, many barely
subsisted at all. Conventional obligations, mutual aid, and public and
private charity could soften the blow of hard times for the most vulnerable. However, the close analysis of mortality at local levels indicates
that the poor suffered more and died sooner than the better-off.[15]

Just how much poorer, nastier, more brutish, and shorter life became
also depended on where people lived. Population growth took place
much faster and brought more challenges in some countries than
others. Spain and England, relatively underpopulated at the beginning
of the century, grew much faster than relatively crowded France. Within
countries, regional and local factors made a difference, too. Areas of
more diversified agriculture or better access to markets could prove
more resilient in a crisis, while more isolated regions had no way to
feed more mouths in hard times. Some environments were naturally
healthier, while others encouraged the spread of diseases arising from
poverty and crowding, through contaminated water and air, or shared
fleas and lice. Towns and cities, which grew faster on the whole than
rural areas, often served as a reservoir of infections and a drain on the
population of the surrounding countryside.[16]

Above all, in any given year of the sixteenth and seventeenth centuries,
the poverty, brutishness, and nastiness of life depended on the weather.
And in this sense, the increasingly poor and crowded inhabitants of Europe were profoundly unlucky to be living through the Little Ice Age.

Weather sometimes had a direct effect on mortality. Cold winters
spread respiratory diseases, especially among the elderly. Hot summers
could spread gastrointestinal infections, especially among infants and

young children. Where historians have found the data to test the influence of temperatures on mortality, they have discovered impacts that were modest but significant.[17]

For most Europeans the real significance of the weather was what it meant for the harvest. During the 1500s, the price of food multiplied several times over, driven by the general trend of inflation and rising demand. But what most made life difficult was how the cost of subsistence could fluctuate wildly from year to year depending on the availability of basic grains. In the worst cases, when especially bad seasons ruined the harvest and the price of food moved beyond reach, the poor of sixteenth- and seventeenth-century Europe could still face true "subsistence crises."[18]

Historians have long debated whether or not the poor literally starved when the harvests failed. In most cases this is a moot point: there were so many other ways to die when food ran short. Populations weakened by hunger, traveling in search of sustenance, lacking basic hygiene, exposed to the elements and to the rich microbial environment of early modern Europe were perfect targets for disease. Typhoid, typhus, dysentery, and any other number of microbes could finish the job before starvation. While the poorest might suffer and die first, the worst crises could soon spread epidemic disease to the population as a whole, multiplying already high levels of background mortality. And then, of course, there were the millions who were never born at all, from the marriages that never took place and the conceptions that never occurred during famine conditions.[19]

We cannot blame all of this misery directly on the Little Ice Age. Blights, crop pests, ravaging armies, rack rents, and rapacious taxation could leave the poor without food just as easily as bad weather could. Moreover, good harvests were a matter not just of average climatic conditions but of very particular weather. Depending on the region and crop, too much rain, drought, heat, cold, hail, or frost in almost any season could be fatal. So much could go wrong in an age before modern agricultural technologies that it is almost surprising the crops usually came in at all.[20]

A couple of contemporary manuals on weather and farming illustrate the point. In England in 1591, John Florio warned especially of moist summers ("it is a signe that all fruites, corne, and all kind of pulse shall be blasted and corrupted"), excessive summer rains ("it is a manifeste token of a great dearth"), and spring mists and frosts ("it is a manifest token that all fruits of the earth shall be blasted, rotten, or destroyed").[21] Gerónimo Cortés of Spain in 1598 found just as many dangers in a Mediterranean climate: in humid summers "the fruits rot on the trees"; autumn frosts would destroy the grapes; excessive summer drought meant a poor harvest of grain; in spring, too much cold impeded the ripening of crops, but too much heat threatened worms and diseases. "Finally," he concluded, "the disruption of the natural qualities of the four seasons of the year is a sure sign of sterility, lack of provisions, and a multitude of sicknesses."[22]

Real catastrophe struck when years of unseasonable weather followed closely one after another. Such catastrophic episodes—including the 1310s, 1430s, 1690s, 1740s, and 1810s—were a defining feature of the Little Ice Age in Europe.[23] Yet the longest and coldest stretch of recurring bad seasons and failed harvests across Europe took place from the mid-1580s through the first years of the seventeenth century.

In 1586, a large eruption took place at Colima, Mexico, launching dust and sulfates high into the atmosphere. Ice cores reveal traces of a smaller, unidentified eruption six years later.[24] Then on March 12, 1595, Nevado del Ruiz in present-day Colombia "gave off three deafening rumbles, like bomb-blasts, so great that they could be heard for 30 leagues all around." The mountain split open and erupted stones and hot gases. The great heat melted its snow cover into a flow "that seemed less like water than a dough of ashes and dirt with such a pestilential smell of sulfur that it was unendurable even at a great distance."[25] On the fifteenth of February, 1600, the volcano Huaynaputina in the Peruvian Andes began to shake with increasingly violent tremors. Four days later, the mountain erupted in a series of great blasts, shooting out hot gases, fire, and ash that turned the sky "darker than the blackest night." Thick rivers of lava and meltwater mixed with rock and ash streamed from the slopes, crushing trees and houses. The eruption was felt hundreds of miles away and ashfall reached as far as Panama.[26]

None of these eruptions was on the scale of Tambora, the cause of the famous "year without a summer" in 1816. Nor can any of them compare to the less famous but even larger eruptions of the 530s, 1250s, and 1450s. Yet the cumulative impact of these eruptions between 1586 and 1600, when considered against the backdrop of global cooling during the late 1500s, was probably even greater. Dust and aerosols shot into the stratosphere and spread around the globe, dimming incoming sunlight. The Northern Hemisphere cooled faster than in any other period of the last five centuries, and it was the only time during the past millennium that proxies in both the Northern and Southern Hemispheres demonstrate significant cooling at the same time. The latest tree-ring measurements indicate that summer temperatures in Eurasia and North America during the late 1580s and the 1590s fell almost 1°C (1.8°F) below the last millennium average, and by well over 1°C during 1600–1609. In 1601 temperatures around the Northern Hemisphere were an estimated 1.8°C (3.5°F) below their long-term average, making that possibly the coldest summer, in the coldest decade, of the last two millennia.[27]

We should take the numbers with a grain of salt. Tree rings are not thermometers, and there are uncertainties in these data. As proxy climate reconstructions continue to improve, the results will change. It may turn out that the turn of the seventeenth century was only the second- or third-coldest period of the last two millennia. The tree samples may have been unrepresentative for that decade. Factors other than temperature could have affected tree-ring growth. The summers might have cooled more than winters.

Yet we should be aware that the opposite could just as well be true, and the period could turn out to be even colder than currently estimated. The winters could have cooled even more than the summers, and the tree-ring measurements, mostly from Scandinavia and Canada, could actually understate the climatic anomalies in other parts of the world. The historical observations presented in this book—though only anecdotal—suggest as much. Contemporaries were just as amazed by the freezing winters of those decades as by their miserable summers, and contemporaries wrote of extreme cold far beyond the boreal forests where climatologists usually drill for temperature-sensitive tree rings.

In Europe, the climate of those decades was more than just cold. From a human perspective it was particularly awful, bringing the worst kinds of weather for crops and livestock. Hundreds of contemporary diaries, letters, chronicles, and official and private records describe the climate and its effects: winter freezes, spring floods, cold rainy summers in some regions, exceptionally cold and dry ones in others. Decades of work by dozens of researchers have gone into evaluating, indexing, and quantifying such accounts, then feeding the results into climate models. Thanks to that work, we now have a much clearer and more detailed picture of European climate during the late sixteenth and early seventeenth centuries.[28]

Almost every summer during the 1590s saw low atmospheric pressure across central Europe, bringing cold fronts and wet weather from the northwest. During many winters, high pressure over Eastern Europe drove freezing easterly winds. Putting all this information together indicates that the 1590s brought by far the coldest, snowiest winters and among the coldest, rainiest summers to central Europe since the fifteenth century. Painstaking work on the climate history of the Netherlands, Germany, Switzerland, and the Czech Republic in particular has found that nearly every winter from the late 1580s through 1601 was unusually cold and nearly every summer was cold and wet.[29] Proxy evidence (including tree rings) and observations of natural phenomena (including glaciers, harvest dates, and the freezing of lakes and rivers) confirm that these were among the coldest years in central Europe for centuries if not millennia.[30]

The severe climate reached to almost every part of the continent.[31] Scandinavia experienced its coldest summers and among its coldest winters of the entire Little Ice Age.[32] Intense cold and summer rains ruined crops and fodder, bringing starvation to the poor, and Finnish peasants rose up in rebellion against the rich. Scotland faced cold, dearth, and famine from 1594 to 1598.[33] Northern Italy experienced flooding and a series of bitterly cold winters, when the Arno River in Florence and the lagoons of Venice froze over.[34] Crops died and city relief efforts in that part of Italy were overwhelmed, creating the worst famine in the region's modern history.[35] Even in the Netherlands, with Europe's most advanced agriculture and economy, there were harvest failures and serious hardship—even hunger—for the poor.[36]

Disasters turned into crisis where conflict and political instability undermined the capacity to respond. In those cases, a "fatal synergy" took hold, in the words of historian Geoffrey Parker.[37] Famine and disease, migration, and violence each fed off one another, driving a spiral of destruction and mortality. In the eastern Mediterranean, the climatic shift brought a series of freezing winters and the worst drought of the Little Ice Age. Crops failed year after year; cold and contagious disease wiped out entire flocks of sheep and herds of cattle. The Ottoman Empire, already facing problems of population pressure, landlessness, and banditry, was plunged into famine. Locked in a long war with the Habsburg Empire, the Ottoman regime continued to requisition food and livestock from Anatolian peasants, setting off a widespread rural rebellion. Millions perished or fled from the hunger and violence, and the empire faced generations of economic and political upheaval.[38] Russia had been caught in a succession dispute since the death of its czar in 1598. Then extraordinary cold following the Huaynaputina eruption ruined years of harvests. During the famine that followed in 1601–1603, up to a third of the population died, and the country collapsed into civil war.[39] In Ireland, freezing winters and cold wet summers combined with the violence and displacement of the Nine Years' War (1594–1603), bringing widespread starvation. Edmund Spenser in 1596 famously described the condition of Ireland's poor:

> Out of every corner of the woods and glens they came creeping forth upon their hands, for their legs could not bear them; they looked like anatomies of death, they spake like ghosts crying out of their graves, they did eat the dead carrion, happy where they could find them, yea and one another soon after. . . . [I]n all that war there perished not many by the sword, but all by the extremity of famine, which they themselves had wrought.[40]

Many historians have discussed Ireland as a model for English colonialism in the New World, and that connection has probably been overdrawn. But Ireland undoubtedly offered some preview of the conflict, hunger, and death that Englishmen would encounter in North America.[41]

At first glance, Spain and England appear to have had similar experiences during these decades, and neither suffered catastrophe on the scale of Turkey, Russia, or Ireland. Nevertheless, the disasters were a turning point in their imperial rivalry. The reasons lay in both the nature and timing of climatic disasters and the impressions they left in each country.

Both countries experienced climatic anomalies. In Spain the period brought exceptional cold and extremes of precipitation. In the southern part of the country, the 1590s were probably the second-rainiest decade since the fifteenth century.[42] The Guadalquivir and other Mediterranean rivers underwent more and greater flooding than in any other period since the Middle Ages.[43] At the same time, tree rings in central Spain indicate the late 1590s–1610s brought the most summer drought of any period from the early 1500s until the impacts of global warming in the late twentieth century.[44] The tree rings of 1600–1602 in the Guadarrama Mountains are the thinnest on record, pointing to exceptionally cold dry weather following the Huaynaputina eruption, and the cold in central Spain persisted throughout the decade.[45] In England, the 1590s brought an infamous succession of freezing winters and exceptionally cold, wet summers that rotted crops in the fields. These were probably the same seasons that inspired these lines of Shakespeare's *A Midsummer Night's Dream:*

> The seasons alter: hoary-headed frosts
> Fall in the fresh lap of the crimson rose,
> And on old Hiems' thin and icy crown
> An odorous chaplet of sweet summer buds
> Is, as in mock'ry, set. The spring, the summer,
> The childing autumn, angry winter change
> Their wonted liveries, and the mazèd world
> By their increase now knows not which is which.[46]

Both countries suffered a series of failed harvests. And in both countries, these harvest failures came at a time of population pressure, inflation, and wartime taxes, which meant real famine for the poor.[47] In England, one contemporary described "the heavens['] . . . continual weeping, the earth glutted with waters, into barrenness . . . the sword of the enemy abroad to threaten us, famine and fear of mortality at home."[48] Food prices soared and real wages fell to perhaps their lowest point since the Black Death.[49] In 1596 Hugh Plat wrote a pamphlet, *Sundrie New and Artificiall Remedies against Famine Written upon the Occasion of This Present Dearth,* in which he recommended everything from peapods to acorns to swallowing handfuls of dirt to relieve hunger. "I have heard many travailers deliver of their own knowledge and experience, that a man may live 10, or 12 daies by sucking of his owne bloud," he added helpfully.[50] In Spain, Andalusia was probably the first region to experience real famine, but disasters came in quick succession across the country. The town of Palencia in northern Spain, for instance, was already facing shortages of grain in 1595. Then 1597 brought another bad harvest. In 1598, frosts destroyed the grapevines, and summer hailstorms beat down the fields. The town council warned of "the multitude of poor and migrant people who converge on us."[51]

In both countries, famine was followed by mortality. In England during 1597–1598, the desperate, the hungry, and famine refugees spread diseases such as typhus and typhoid that raised the national death rate almost a third above normal. Marriage and birth rates plummeted. Just as the population began to recover in 1603, one of the worst plague outbreaks of the century reached London and spread across parts of the countryside, bringing mortality just as high as in the worst years of the 1590s.[52] In Spain, the poor harvests and impoverishment of the countryside sparked an epidemic of typhus, which spread from south to north during 1591–1595. As food prices and famine in Castile reached a peak during 1596–1597, bubonic plague erupted from central Spain, especially among the hungry and destitute. As a saying of the time went, "God save us from the plague descending from Castile and hunger rising from Andalusia." The contagion spread around the Mediterranean and through western Europe before crossing the Channel in 1603. England and Spain shared not only similar disasters during the crisis but sometimes the same microbes.[53]

Taking a closer look, however, we can see subtle but important differences in the ways that Spain and England experienced these disasters, and the ways they shaped perceptions and reactions. First, the impacts were distributed differently in each country. In Spain, the disasters overwhelmingly afflicted rural Castile. This concentration made the intense mortality of the 1590s even more visible to the statesmen and intellectuals of Madrid. It added to an already pervasive sense that Spanish agriculture was in crisis, and that the countryside was becoming depopulated.[54] In England, the pattern of mortality was more complicated. The famine and related diseases of 1595–1597 overwhelmingly afflicted poor and isolated regions in the north—marginal populations in every sense of the word. Their deaths were almost invisible to the literate and powerful in London, who focused instead on the influx of dangerous, disorderly vagrants into the city seeking food and employment. They blamed these vagrants for the plague of 1603, which struck predominantly in London and the more urbanized and well-connected southeastern parts of the country.[55]

Second, the scale and swiftness of death were much greater in Spain than in England. With famine and plague striking at once, the mortality of the late 1590s was far more intense than in any subsequent crisis of Spanish history. About 600,000 people, or more than a tenth of Castile's people, perished in just a few years. In some regions, the population would take generations—even a century—to recover.[56] The 1590s marked the start of a deep agricultural depression and a decline in urban industry and finance.[57] In England, too, the mortality was high. However, it came in two separate episodes—the famines of 1595–1597 and then the plague of 1603. Given the country's ongoing economic and demographic expansion, new births soon offset the losses.[58]

The result was that these two nations perceived this era of disasters in very different ways. Historians have spent literally centuries debating whether or not Spain "declined" during the 1600s.[59] Yet commentators at the turn of that century had no doubt: the crisis served as proof of deep-rooted national problems, requiring radical solutions.[60] These so-called *arbitristas* focused above all on the depopulation of the countryside and depression in agriculture they witnessed in Castile. While some blamed the sterility of the soil, the horrendous weather, and the plague, most perceived these problems as only symptoms of Spain's social and political failures. They offered a plethora of solutions to a laundry list of problems: excessive taxation, debt, the growth of the royal court, vagrancy, the decline of marriage and birth rates, a surfeit of priests and nuns, even the excessive breeding of mules.[61] Many used the analogy of Spain as a sick or dying patient in need of drastic surgery. "The illness is critical," wrote Pedro Fernández Navarrete in 1621, "incurable by ordinary means. Bitter remedies work best for the sick, and to save the body sometimes you have to cut off an arm, or to save a cancer patient [you have to] heal with fire."[62]

To many observers, the gangrenous arm that needed to be amputated, or the cancer that had to be burned out, was Spain's expensive and unmanageable empire. As so often happens during crises, it was tempting to blame foreigners, foreign trade, and foreign commitments for problems at home, and many did. Imperial wars proved enormously expensive, and mortality among Spanish soldiers appallingly high.

Sancho de Moncada declared the empire "the first cause of the general damage to Spain." The empire was "extended to such and to so many distant provinces, for whose defense and preservation it is necessary to bleed Spain of its people and silver, such that the conquest of remote nations in the Indies and the preservation of imperial territory . . . has been a worm infecting Spain."[63] For Castile, emigration to the empire, and particularly the New World, represented a real drain on an already plague-stricken population at home. Demographers have estimated that around 3,000 Spaniards emigrated to the Americas each year during the 1500s—not a large number by today's standards, but certainly enough to be noticed.[64] Some in Spain itself thought the number was far higher: one *arbitrista* put the loss of Spaniards to the empire at 40,000 a year.[65] The result was an intellectual backlash against foreign commitments and new colonies. We may doubt whether this disapproval deterred any poor Spaniards desperate to make a better life the New World.[66] But it likely discouraged official support for new colonial ventures in North America, which the Spanish crown was already inclined to see as a waste of precious resources in an age of crisis and retrenchment.

For England, on the other hand, historians have spent almost as long tearing down the myth of an Elizabethan golden age. The rampant poverty, conflict, and instability of the 1580s and 1590s all stand out in the historical record.[67] Yet English writers of the time had no doubt their country was a rising power with a burgeoning, even overflowing population. In the run-up to the Roanoke venture, one promoter had warned of the "multitude of loyterers and idle vagabondes . . . so many that they are readie to eat upp one another." They might all be put to work, he claimed, in the gold mines, olive and orange groves, and silk plantations he envisioned in North America.[68]

The crisis of the 1590s and first decade of the 1600s only confirmed this view, especially in London. The influx of famine refugees, even as soldiers returned to England from wars in Ireland and the Netherlands, heightened a sense of crowding and overpopulation that had been mounting for decades. In September 1603, King James I declared "the great confluence and accese of excessive numbers of idle, indigent, dissolute and dangerous persons" into the capital to "have bene one of the chiefest occasions of the great Plague and mortality." Yet it was not the number of the dead and dying that concerned him so much as the number still living: the "Rogues, Vagabonds, idle and dissolute persons" who "swarmed and abounded every where more frequently then in times past, which will grow to the great and imminent danger of the whole realme."[69]

Popular sentiment apparently agreed. Playwrights were thrown out of work when the theaters closed during the plague, and some turned to writing dark, biting pamphlets about the times. Thomas Middleton imagined the personifications of War, Pestilence, and Famine at a London pub, arguing over who could be the cruelest: "What, is not flinty Famine, gasping death, worthy to be in rank with dusty War and little Pestilence? Are not my acts more stony-pitiless than thine, or thine?"[70] Fellow playwright Thomas Dekker put the thought more pithily in verse:

> For if our thoughts sit truly trying
> The just necessitie of dying
> How needfull (tho how dreadfull) are
> Purple Plagues, or Crimson warre.
> We would conclude (still urging pittie)
> A Plague's the Purge to clense a Cittie.[71]

To take the metaphors of the time, if Spain was an ailing old man who needed drastic surgery, then England was a feverish youth who needed to bleed or excrete his excess. As a sermon proclaimed, "Our multitudes like too much blood in the body, do infect our country with plague and poverty."[72] Or else England was a swarming overcrowded beehive, and it was time for its surplus workers and lazy drones to fly off and found another colony.[73]

Spain and England shared these metaphors and perceptions of each other. Some arbitristas complained that Spain's depopulation was "infamous" across Europe.[74] English and French writers disparaged the infertility of Spain's land and people, drawing on a trope contrasting vigorous northern Europeans to exhausted southerners. They cast themselves in the role of the hearty, prolific barbarians of classical times, overrunning the contemporary Spanish equivalent of a decadent Roman Empire.[75] At the same time, Spanish observers noted the "many idle and wretched people as they have in England," whose only outlet was to go overseas.[76]

As these descriptions make clear, England's excess people could be seen as an asset or liability, a resource or a peril to the country. Beggars, vagrants, migrants, and other "masterless men" were a constant preoccupation of the better-off. Crime, riot, and rebellion were real dangers during times of poverty and dearth. One consequence was the so-called poor law of England: the various acts and decrees charging each parish with control and care of the impoverished. These acts really came into force in response to the disasters of the 1590s. Although as much or more about public order as public welfare, they had the long-term effect of cutting crisis mortality in England by averting starvation and the spread of disease from famine refugees. They ensured that the poor of England would be less desperate, but more numerous, in the coming century.[77]

There is little evidence that it was always the poorest or most desperate who emigrated to America during these years. Some might have been lured by the rosy propaganda about the New World, while others just followed their sense of adventure.[78] Nevertheless, the planners and promoters of American colonies could capitalize on the image of overpopulation to sell their projects, to gather the private investment and public support that made those ventures possible. Disasters of the Little Ice Age came at a particular moment and in a particular way that helped to undermine Spain's commitment to North American colonization but to reinforce England's. That commitment would be critical to the endurance of England's next colony, which was about to make its way to Virginia.

In late 1606, rumors about Philip III's plans to abandon Florida began to reach St. Augustine. In February the following year, the order went out to dismantle the fort, reduce the garrison by half, and remove the missionized Indians to New Spain. It cited "the very little use" in keeping the colony, whose Indians "every day became more obstinate" and whose land "neither bore fruit nor had gold or silver or other metal in it."[79]

The order's arrival at St. Augustine provoked an impassioned response. Franciscan missionaries pleaded for "help to defend these souls from the power of Satan" and warned of dire consequences for the Christian Indians if the fort was removed. Governor Ibarra disputed the negative reports about Florida and claimed recent success converting another 2,000 Indians.[80]

In August 1608, in consultation with his committee for war in the Indies, Philip III finally relented and ordered the fort to stay in place. Yet his decision had nothing to do with enthusiasm for Florida. The king emphasized Spain's "obligation . . . to keep [the Christian Indians] in the faith they profess and ensure the spread of the holy gospel." And, just as important, he cited new reports "of the settlement of Virginia by the English" not far away.[81]

For more than a year, Spanish spies in London had been reporting on new plans for a colony in America.[82] As early as spring 1606, acting on recent intelligence, the Council of War advised Philip III to attempt a diplomatic solution but to be ready to respond with force "because doing otherwise would be to lose reputation." By late that year, after some initial confusion, Spanish agents managed to locate the intended settlement in the Chesapeake, which alarmed Ambassador Pedro de Zúñiga in London. On the one hand, this meant the English "have in mind to obtain from the country . . . the same commodities as from Spain, in the same latitudes, so as to have no need [of Spain]." On the other hand, based on Spain's experience in La Florida, he cautioned that "the land is very sterile, and consequently there can be no other object in that place than that it seems good for piracy."[83]

In October 1607, Zúñiga finally confronted the British monarch directly to demand an explanation. Whether out of cunning or mere embarrassment, James I prevaricated, claiming not to know anything about Virginia, and blaming the whole business on England's unruly Parliament. Zúñiga insisted to Philip III that Spain should "root out this noxious plant while it is so easy." The king's councilors in Madrid pressed for action as well, warning of the spread of Protestant heresy in the New World, "after which comes reputation, which is so important."[84]

Time and again, Philip III agreed to these steps but then failed to mobilize the Spanish fleet in the Indies. Behind the delay was his concern that a Spanish attack would break the peace with England, a price that some felt was just too high to defend Spanish claims in North America. Finally, in November 1608, the king ordered the governor of Florida to send a single ship to spy on the English colony, which they feared would soon reach 1,500 or even 3,000 men. The captain, Francisco Fernández de Ecija, was to report on their fortifications "and of what substance is the land."[85]

Given delays in communication, Ecija did not receive the order until June the following year, and news of his expedition did not reach Madrid for several more months. The voyage managed to gather some information on the location of the fort, which the English called Jamestown. However, Ecija approached during contrary winds and when an English supply ship just happened to be there. Mistaking the ship for a sentinel, and convinced the English would rush to defend the fort, the captain ordered a retreat. It was as close as the Spanish ever came to challenging Jamestown.[86]

In the meantime, some promising intelligence began to reach the king and his councilors. In November 1608, Zúñiga related that settlers from another English colony had just returned from New England "in a sad plight."[87] By April 1609, he heard that the settlement in Jamestown might be abandoned altogether "because it is unhealthy and many have died there."[88] Over the following year, reports came back of the settlers' discontent, conflict with the Indians, starvation, and even cannibalism.[89] By early 1611, Spanish intelligence reported that the pattern of Atlantic winds and currents did not even favor the Chesapeake as a base for privateering in the Caribbean. Virginia was nothing more than a dumping ground for England's unwanted vagrants, and so "it does not appear to be as inconvenient as we thought for the English to have a settlement," the Council of State concluded in April.[90]

During the following years, as more news trickled back through Spanish spies in London, the king and his councilors must have congratulated themselves on their prudence. The English colonists were reduced to extremities, caught in their forts by hostile Indians.[91] It had become a deathtrap, a waste of investment, "an object of ridicule . . . that will never bear fruit."[92] By late 1612, it was common opinion in Madrid that "the business will fall of itself." By letting it linger, they would only let the English waste more lives and money.[93]

Philip III's inaction during this moment of crisis would prove to be a disastrous miscalculation in the end. By letting England's colony survive, Spain opened the way for others to follow, and ultimately for a shift of power in the North Atlantic world. However, given the circumstances and the past century of colonial failures in North America, the prediction that Jamestown would "fall of itself" seemed entirely justified. It was *almost* right.

5

⟡

We Had Changed Summer with Winter

In December 1606, three ships—the *Godspeed,* the small pinnace *Discovery,* and the quaintly named flagship *Susan Constant*—set out from London "ready victualed rigged and furnished" according to the orders of the newly chartered Virginia Company. The ships, commanded by Captain Christopher Newport, carried a full crew of sailors and an estimated 144 colonists to settle on the east coast of North America. As one historian has remarked, "Few epoch-making adventures have begun with less fanfare." What details we know of the voyage's beginnings come largely from a lawsuit occasioned earlier that month when the *Susan Constant* struck another ship moored too close in port.[1]

It had been an inauspicious start, and the voyage only ran into more troubles that winter. "Unprosperous winds" and "great stormes" kept the ships bottled up off the coast of Kent for more than a month. One passenger fell ill, and several others apparently quarreled over religion. A certain John Smith was arrested on charges of mutiny, only to be released a few weeks later. The ships eventually caught favorable winds, passed through the English Channel, and steered south toward the Canary Islands, which they reached in February. From there, they caught the easterly trade winds across the Atlantic to the Caribbean. Meanwhile, the bad omens continued. On the night of February 12, the passengers spied a "blazing Starre, and presently a storme."[2]

Nevertheless, the ships arrived safely in the New World a few weeks later, an accomplishment not to be taken for granted in an age of long and dangerous Atlantic crossings. They sighted Martinique on April 2 and reached the island of Dominica the following day. "Three weekes we spent in refreshing our selvs amongst those west-India Isles," John Smith wrote. They traded with Caribbean Indians and feasted on exotic flora and fauna such as pineapples and plantains, tortoises, iguanas, and parrots.[3]

The three ships sailed in a northwesterly arc along the Antilles, passing by Guadeloupe, Nevis, St. Eustatius, and Puerto Rico. When the fleet reached the island of Mona, in the passage between Puerto Rico and Hispaniola, they were forced to resupply, "seeing that our water did smell so vildly that none of our men was able to indure it," as colonist George Percy later recalled. They ventured onto the island in search of food and clean water; however, "the wayes that wee went, being so troublesome and wilde going upon the sharpe Rockes, that many of our men fainted in the march." These included the first of many eventual fatalities in the expedition, a "gentleman" named Edward Brookes, "whose fat melted within him by the great heate and drought of the Countrey."[4]

In light of the freezing weather and famine the Jamestown colonists would suffer in the months and years ahead, it is ironic that the expedition's first death apparently came from heatstroke. The excessive heat was, however, exactly what the expedition's planners feared all along, not only in the Caribbean but in Virginia and even New England.

English colonization faced the same puzzle of American climates as its Spanish counterpart, but the English came at the problem from the opposite direction. England lay far to the north of the territory it sought to colonize. Moreover, English ideas about the New World were shaped by decades of searching for a Northwest Passage around the continent. In the course of that searching, promoters of English overseas ventures came up with their own interpretations of North American geography and explanations for the unexpected extremes of its climates. Although sometimes perceptive and often inventive, these theories created fatal misconceptions about the continent that England would eventually try to colonize.

Of all the voyages in Europe's Age of Discovery, expeditions in search of a Northwest Passage encountered the clearest obstacles from climate. If current trends continue, we can expect a regular ice-free shipping route over Russia or Canada before the year 2050.[5] During the Little Ice Age, that passage was not even remotely possible. Falling Arctic temperatures and expanding sea ice condemned already impossible voyages to even more swift and certain failure.

As historian Dagomar Degroot has shown, variations from year to year and region to region meant that climate affected each voyage differently. Historical records demonstrate that some explorers were far more fortunate than others when it came to encounters with Arctic storms and ice.[6] Ultimately none was successful, and more than a few never came home at all. Nevertheless, the variation in their experiences—particularly the variation in climate and weather—seemed to leave open a perpetual window of hope that the next voyage might go farther.

To understand how climate change influenced the quest for Northwest Passage, it helps to consider the circulation of the ocean and atmosphere in the North Atlantic. The details can be complicated, but the essential picture is simple. The North Atlantic Current brings warm waters from the Gulf Stream past western Europe. North of Scotland, some of the flow branches west in what we call the Irminger Current and reaches the southern coast of Iceland, keeping that island much greener than nearby Greenland. The rest of the flow continues north as the Norwegian Current. Once it has passed that country's shores, it splits again. Some of the flow veers east into the Barents Sea and toward the islands of Novaya Zemlya in the Russian Arctic, but most continues north through the Fram Strait up to the Svalbard Archipelago high above Scandinavia. At the same time, cold currents descend from the Arctic, the largest being the East Greenland Current and the Labrador Current, the latter of which runs along the eastern shores of Canada and out past Newfoundland.

The interplay of these currents controls when and where sailors might encounter sea ice in any given year. On the one hand, a strong North Atlantic Current can thin out the sea ice at very high latitudes and bring mild weather as far north as Svalbard. On the other hand, a strong Labrador Current can carry the icebergs of Baffin Bay far out into the North Atlantic, and the meeting of its cold waters with warmer seas off Newfoundland creates dangerous fogs.

Above the oceans, variations in air pressure are dominated by a large-scale pattern that goes under various names, but most often the North Atlantic Oscillation (NAO). In simple terms, the NAO describes the difference between sea-level pressure at two points in the North Atlantic: the Azores High and the Icelandic Low. When the NAO is positive, this means the usual high-pressure anticyclone around the Azores is strong and persistent, while the sea-level pressure off Iceland remains low. This mode of the NAO typically means that westerly winds and maritime air prevail over western and northern Europe, bringing wetter and milder winters to those regions. The pressure contrast and stronger westerlies may also contribute to stronger flows of the North Atlantic Current from year to year or decade to decade; that connection remains a topic of ongoing research. When the NAO is negative, this means that the Azores High is weak or even that an anticyclone forms around Iceland, which can bring frigid northerly winds and dry, frozen winters to parts of Europe.[7]

Modern instrumental records of Arctic temperatures, currents, sea ice, and the NAO cover only the last century. The work of reconstructing these conditions during earlier centuries and millennia has inspired some ingenious methods. Researchers have drilled tree rings at the Scandinavian tree line, sampled stalagmites in Scottish caves, and taken cores from glaciers and lakes in Greenland, Svalbard, and Novaya Zemlya. They have tested shells of centuries-old clams. They have identified the varieties of microscopic sea creatures sensitive to water temperatures or ice cover, and have looked for their traces in sediment cores from the ocean bottom. They have measured the growth of coralline algae sensitive to how much sunlight filters through the sea ice. They have compiled and homogenized early instrumental series and sorted through thousands of old ship logbooks reporting wind direction and strength.

Taken together, this research yields a coherent, if still imperfect, picture of past conditions. In general, the Arctic was a colder place during the Little Ice Age, with more sea ice than today. Yet most of the sixteenth century was a relatively mild interval, with some of the least sea ice of the past millennium. Looking more closely, we find variations over time and space. It was probably a strong North Atlantic Current that brought the unusual Arctic warmth of the sixteenth century, as attested in proxies of temperatures and currents around Svalbard. Sea ice cover in Baffin Bay still increased throughout the Little Ice Age. A strong, cold West Greenland Current brought more icebergs into the Davis Strait, and a robust Labrador Current carried them onward past Newfoundland. North of Canada, sea ice grew in the mid-1500s and then fluctuated for the rest of the century. Records from Iceland vary from decade to decade, with especially icy conditions during the 1570s and 1590s.[8]

At the turn of the seventeenth century, conditions changed abruptly, and in unexpected ways. After large tropical volcanic eruptions, the earth cools differently at different latitudes, and that uneven cooling causes shifts in global atmospheric circulation. Often these shifts include a stronger NAO, as well as stronger westerly and northwesterly winds in Europe that often come south of their usual range. These conditions can strengthen the North Atlantic Current and temporarily thin out Arctic sea ice in otherwise cold years. That was the pattern following the Tambora eruption and the "year without a summer" in 1816. Northwesterly winds from the Atlantic brought months of incessant cool rainy weather to western and central Europe, further chilled by volcanic aerosols that dimmed the sun. The poor in Switzerland and southern Germany froze and starved during the summer, but the following winter was actually milder than usual in most of Europe. At the same time, ships in the Arctic found the sea free of ice much farther north than anyone could remember.[9]

Yet all the evidence suggests something different was going on at the turn of the seventeenth century. Both summers and winters from the mid-1580s through the turn of the seventeenth century were exceptionally cold throughout Europe. That period also brought some of the coldest Arctic temperatures of the last millennium, probably second only to those of the early 1800s. Temperatures on Baffin Island, already cold, briefly plunged to some of the lowest levels of the past 1,200 years. Several proxies point to a weakening of the North Atlantic Current as well: winter temperatures fell and ice cover increased in Svalbard and Novaya Zemlya. The first years of the seventeenth century included the very coldest summers in northern Scandinavia for more than a millennium. The frigid Labrador and East Greenland Currents strengthened. Sea ice cover expanded across the Arctic, and particularly around Greenland and Iceland.[10]

The most extreme anomalies were short-lived. However, the shift to colder Arctic temperatures and more sea ice persisted for decades. The latest work in climate modeling offers a plausible explanation. The strengthening of the Labrador Current carried colder, fresher water and more Arctic ice out to the North Atlantic. This weakened the circulation of the subpolar gyre, meaning that the North Atlantic Current no longer transported as much heat to the high latitudes of Europe. Feedbacks from sea ice and sea level pressure sustained this shift in oceanic and atmospheric circulation. At the same time, the buildup of colder waters created high-pressure blocking east of Labrador, which in turn displaced low pressure onto northern and central Europe—a situation that could explain some of the extreme seasons there during the 1590s and first decade of the 1600s. Although still hypothetical, this pattern derived from models appears to match the proxy record and historical observations very well.[11]

For explorers in search of a Northwest Passage, this climatic shift meant that false hopes raised during some relatively mild and ice-free summers of the 1500s would be dashed by the colder, icier conditions that set in at the turn of the seventeenth century. By exploring their stories, we can see the human experience of those larger climate patterns glimpsed through the proxy record.

Arctic exploration during the sixteenth century involved two kinds of journeys. The first were the physical voyages of fragile wooden ships and men into northern waters choked with ice, along uncharted rocky shores shrouded in mist. The second were the conceptual voyages, the leaps of imagination to justify the enterprises with new thinking about weather, climate, and geography. These journeys ran in parallel, and they would continue almost into the twentieth century. Both ran into repeated setbacks and dead ends. But every voyage, physical and conceptual, glimpsed enough new promise to inspire the next adventurer.

It is easy in hindsight to criticize the recklessness that characterized these expeditions. However, the search for a Northwest Passage also represented much of what was most modern and admirable about European discovery during the 1500s and 1600s. The quest was a curious outgrowth of the increasing boldness, optimism, and cosmopolitanism of early modern merchants and explorers. While sophistry and wishful thinking helped promote the ventures, proponents of Arctic navigation also built on the latest geographical theories and expert information, hired the best navigators, and pursued the richest trades.

They also turned to many of the same creative applications of meteorology and cosmography used to help explain weather and seasons in Spanish America. For example, the insistence that the far north could be navigated and even colonized mirrored Columbus's surprising discovery of the habitable tropics. If the dry, burning "torrid zone" of the equator was actually an earthly paradise, then why not the "frozen zone" of the poles? The Spanish chronicler Oviedo argued that the Americas must be habitable all the way to northernmost Canada. If the Arctic couldn't be used, wondered Spanish natural historian Tomás López Medel, then "why would God create so much extra space?"[12]

Of course, arguments like these flew in the face of real experiences by hundreds of European sailors who began to ply the northern waters for fishing and whaling, as well as the first explorers and navigators who began to probe the Arctic coastline. As early as the 1490s, the English king Henry VII commissioned Venetian navigator Zuan Cabotto (also known as John Cabot) to explore the New World across the North Atlantic. Cabot's voyages to Canada brought back rumors of frozen seas, icebergs, and lands covered with snow "where the cold almost killed the entire company, although it was the month of July."[13] In 1527, Englishman John Rut ran into icebergs and freezing temperatures while trying to sail up the Labrador coast. Less than a decade later, an expedition led by Richard Hore spotted "mightie Islands of ice in the sommer season" as it sailed north from Newfoundland.[14] Hugh Willoughby's 1553 expedition in search of a Northeast Passage over Asia was forced by ice and storms to winter on Russia's Kola Peninsula. Crowded aboard their ship, sealed tight against the bitter cold, the entire crew perished, most likely poisoned by carbon monoxide from their stove.[15] Two years later, Englishman Steven Borough discovered the Kara Strait, leading around Novaya Zemlya, but storms, darkness, and "great and terrible abundance of ice" blocked his way forward.[16]

Despite these setbacks and falling Arctic temperatures, English interest in a Northwest Passage to the East Indies grew stronger over the sixteenth century. During the mid-1560s, the adventurer Humphrey Gilbert promoted a Northwest Passage expedition, arguing from sketchy historical evidence that American Indians must have traveled that way to Europe during classical times. However, the Muscovy Company of English merchants in Russia quickly asserted its monopoly on Arctic voyages, blocking Gilbert's venture.[17]

The Muscovy Company waited another decade to organize its own Northwest Passage expedition, led by the veteran privateer Martin Frobisher.[18] Frobisher's first voyage departed England with three ships in June 1576, reaching Greenland the following month. Even in midsummer, it was surrounded by "islands of ice" so tall "clouds hanged about the top of them," as Michael Lok, the Muscovy Company director, recalled. There the expedition ran into a storm that sank its pinnace and forced a second vessel to turn back to England. Frobisher's remaining ship reached Baffin Island in early August, and he mistook the deep bay that now bears his name for a passage to Asia. When Inuit captured five of his sailors, the expedition kidnapped a man from Baffin Island in reprisal and then headed back to London.[19]

The Inuit captive and his kayak proved "such as wonder onto the whole city and to the rest of the realm that heard of it as seemed never to have happened the like great matter to any man's knowledge," Lok recalled.[20] Upon investigation, the Muscovy Company optimistically concluded that Frobisher's supposed "straight" was indeed "a trueth, and therefore a thinge worthie in our opinions to be followed."[21] But most important was that an Italian assayer claimed to have found gold in a piece of "ore" that Frobisher had picked up from Baffin Island.[22]

Queen Elizabeth soon granted funds and a royal charter for a "Company of Cathay" to trade with Asia through the presumed Northwest Passage.[23] Yet it was the rumors of precious metals that inspired most of the new investment for a second expedition. Frobisher set out again in late spring 1577. He tried and failed to make a landing on the Greenland coast, and then tried and failed to find the men taken captive by Inuit the previous summer, kidnapping an Inuit woman and child in reprisal. The weather remained bitterly cold throughout. "We tasted the most boisterous Boreall blasts, mixt with snow and haile, in the moneth of June and Julie," as one passenger recalled. Returning to Frobisher's "strait," they found "cold stormes, insomuch that it seemed, we had changed summer with winter."[24] In August they sailed to the islands off Baffin, gathering tons of new "ore." But upon the expedition's return to England the following month, it all turned out to be nothing but worthless rock.

In spite of this setback, the Cathay Company decided on a third expedition the following spring of 1578—the largest yet to sail to North America, at an incredible cost of almost £25,000.[25] Frobisher's fleet departed late in the season and faced gales and fog in late June while rounding Greenland's southern cape. Crossing over to Baffin Island, they then ran into heavy sea ice driven south by the Labrador Current. As officer George Best recounted, "They were brought many times to extreamest pointe of perill, Mountaines of Ice tenne thousande times scaping them scarce one inch."[26] In mid-July, a sudden storm drove one ship into an iceberg and nearly crushed the rest in floating ice. A storm near Baffin Island brought "so much snow, with such bitter cold aire, that wee coulde not scarce see one another for the same, nor open our eyes to handle our ropes and sailes," recalled Best."[27]

In mid-August, the crew began their real mission: to load their several ships with promising-looking "ore" from around the Baffin coast. In the rush to get out before winter, it is doubtful the men put much care into selecting stones to fill the hulls. Dozens died from cold, sickness, poor provisions, and the backbreaking work of mining hundreds of tons of hard rock.[28] On September 10, a meeting of officers resolved to head home, now that the ships were almost completely loaded and the "verie tempestious weather" made further reconnaissance impossible.[29] The storms, cold, and shortage of supplies dissuaded Frobisher from what could have been fatal plans to set up a colony and overwinter in Canada.

More ice and storms scattered the fleet on the homeward journey. Returning singly or in pairs, the ships unloaded their cargo at the port of Dartmouth, England, where expensive new furnaces had been erected just to handle the expected ore.[30] What the assayers found instead was 1,300 tons of all but worthless rock. Investors abandoned the Cathay Company, some of them ruined by the loss.

In answer to Frobisher's failures, proponents of a Northwest Passage only worked harder to explain away the problem of Arctic cold and ice. One line of argument was to analogize from the tropics. Experience had shown that the equator itself was not as hot as the regions around it, such as the Sahara. So perhaps the polar regions would prove milder than the icy lands of northern Russia or Canada. First ventured in the 1520s, this idea would be repeated throughout the sixteenth century and beyond.[31] Other writers argued that the continual sunshine of Arctic summers would make up for the oblique angle of light reaching the poles: "that therby all the force of freezing is wholy redressed and utterly taken away," as the navigator John Davis put it.[32] Others still denied that the ocean could freeze at all, claiming the water was too salty and the currents too strong. If explorers could only avoid the ice that formed around land, surely they would find their way into open Arctic waters.[33]

Yet the most common way to explain away the Arctic cold was to blame it all on local "accidental" factors, while insisting that the "true"

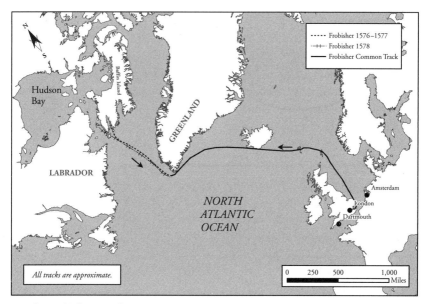

5.1 The Frobisher expeditions

climate of the far north remained mild or at least tolerable. This line of reasoning basically reversed cause and effect. What should have been obvious signs of a cold climate—glaciers, icebergs, frozen seas, and boreal forests with freshwater lakes—were all presented instead as local sources of cold winds and chilling vapors. Though sophistical and self-serving, this kind of justification offered a flexible and apparently convincing way to explain the otherwise inconvenient or confusing realities of Arctic weather. It also held out hope that deforestation and cultivation might one day "improve" the land and its climate, a notion that would live on in North America throughout the colonial era and beyond.[34]

These ideas crystallized in Humphrey Gilbert's next scheme, an English colony in Newfoundland. By the 1570s ships from all over Europe converged on the island's teeming fisheries, typically arriving in late spring or summer to stock up on cod and then departing before the weather turned cold in autumn. Gilbert meant to convince potential investors and settlers that Newfoundland was a place worth staying year-round, with valuable commodities on land as well as at sea.

Unfortunately for Gilbert and his fellow promoters, they faced two contradictory problems. On the one hand, Newfoundland lies south of England, which raised fears that its climate would be unhealthful. Englishmen, like other Europeans at the time, imagined that their bodies were uniquely adapted to their native environment. Warmer latitudes might enervate their constitutions, upset the balance of their "humors" (bodily fluids), and breed disease.[35] On the other hand, actual experience from fishermen and sailors indicated that Newfoundland's winters were extremely cold, and its waters could be frigid and choked with ice.

In response, proponents insisted that Newfoundland's climate must really be like that of Europe on the same parallel—that is to say, France. Some argued that only incidental factors such as local terrain or flows of Arctic ice made it *seem* cold at times. As one wrote in 1578, "You shall understand, that Newfoundland is in a temperate Climate, and not so colde as foolish Mariners doe say. . . . This colde commeth by an accidentall meanes, as by the ice that commeth fleeting from the North partes of the worlde, and not by the situation of the countrey, or nature of the Climate."[36] Humphrey Gilbert's own explanation of Newfoundland's climate was even more confused. Relying on classical cosmology and meteorology, he argued that the heavens' east-to-west rotation around the earth must bring continental heat to northern Europe and ocean vapors to eastern America, giving the more southerly Newfoundland a climate not unlike England's.[37] In other words, he exactly reversed maritime and continental climates.

In the end, Gilbert did manage to raise enough support for a Newfoundland expedition, which foundered in a single disastrous voyage of 1583. His fleet of five ships departed England in late June, making a difficult crossing in the face of persistent westerly winds, mists, and sea ice. Arriving in St. John's Bay, Gilbert summoned visiting fishermen and sailors and read out Queen Elizabeth's orders granting him possession of the land. Making little effort to establish a permanent colony, Gilbert's small expedition set about exploring Newfoundland's shores. Along the way, one of the larger ships was run aground and lost at Sable Island, and with it went dozens of men and most of the expedition's supplies. Short of equipment and food, the whole venture had little choice but to sail back to England. Meeting a storm along the way, the reckless Gilbert (his personal motto was *quid non—*

"why not?") refused to abandon his favorite little vessel, the *Squirrel*. It was lost at sea with all men on board.[38] The English would not try to colonize Newfoundland again for almost three decades.

In the meantime, the accomplished navigator John Davis managed to sell investors on a new expedition in search of the elusive Northwest Passage. In summer 1585, his two ships managed to sail to Greenland through thick fog and ice, "not experienced of the nature of those climates, and having no direction by Chart, Globe or other certaine relation," as Davis later recalled. Rounding Greenland's southern cape, they continued northward along its western coast. The expedition was fortunate enough to enjoy peaceful encounters with Inuit and mild weather as far as about 66°N, "altogether void from the pester of ice." At that point, Davis took the expedition across the strait that would later bear his name and into Baffin Island's Cumberland Sound. In August, Davis and his officers decided to turn back before their luck ran out.[39]

News of his voyage earned Davis investments for a second venture the following spring. He departed in mid-May, his two barques accompanied by a pinnace, the *Northstar,* and a larger flagship, the *Sunshine*. Four or five weeks later they approached the southeast coast of Greenland, "mightily pestered with ice and snow."[40] At this point the fleet split up under disputed circumstances: Davis with his two barques continued west around Greenland, while the *Northstar* and *Sunshine* turned north.[41] The latter ships made a brief stopover in Iceland and then headed northwest in search of a passage over Greenland, a route that soon took them into impassable sea ice. They then turned south down the east coast of Greenland, stopping onshore to trade (and apparently play some football) with Inuit men they encountered. From there they turned back to England, losing the *Northstar* "in a very great storme" three days into the homeward voyage.[42]

Frobisher's two barques rounded Greenland's southern cape, regaining the western side of the island "after many tempestuous storms." They explored the fjords and suffered (or provoked) some violent clashes with Inuit as they made their way up the coast. In late July, just past 63°N, the ships "fel upon a most mighty and strange quantity of ice," as Davis recalled. "The aire at this time was so contagious, and the sea so pestered with ice, as that all hope was banished of proceeding."[43] Davis sent one ship back to England, continuing his explorations with only a single barque. He sailed past 66°N, the weather turning warmer and the way now "void of trouble, without snow or ice." Crossing to Baffin in late August, the expedition found excellent hunting and "the fish swimming so abundantly thicke about our barke as is incredible to bee reported." However, in mid-September their luck ran out for good. Two men died in an Inuit ambush, then "it pleased God further to increase our sorrowes with a mighty tempestuous storme." Suffering damage to their ship, they were forced to head back, reaching England in mid-October with more than 500 sealskins in their hull.[44]

With the failure to find an ice-free passage to Asia, most investment for the venture dried up. Davis managed to secure financing to take one small pinnace exploring in the Arctic, while two accompanying barques would return to the fishing grounds off Baffin. Setting out in late May 1587, Davis again sailed northwest and then north to follow the western coast of Greenland. This time, Davis enjoyed extraordinarily mild weather, reaching almost 73°N by July 10, when he steered west again, hoping he could continue over the top of North America. Instead, as he later recalled, "I fel upon a great banke of ice: the winde being North and blew much, I was constrained to coast the same toward the South." Reaching Baffin again, Davis took his pinnace south along its coast, passing the Cumberland Sound and Frobisher's so-called strait. On August 10, as another member of the expedition recalled, they passed an "overfall" and a "great gulf" with "the water whirling and roring, as it were the meetings of tides." Yet at that point, without exploring any farther into what was probably Hudson's Strait, the expedition turned back to England.[45]

Davis remained convinced for the rest of his life that the strait bearing his name would lead to Asia, but he never found backing for another Arctic expedition. He had come about as close to finding a Northwest Passage as any explorer would again for decades. The cooling Arctic climate at the turn of the seventeenth century ensured that few subsequent voyages would make it as far or come back with as many men alive.

In 1601 the newly formed East India Company commissioned Captain George Waymouth to find a shortcut to the Spice Islands. He was instructed to sail to Greenland, head up the Davis Strait, and then "passe on forwarde in those seas by the Norwest, or as he shall finde the passage best to lie towards the parts or kingdom of Cataya or China

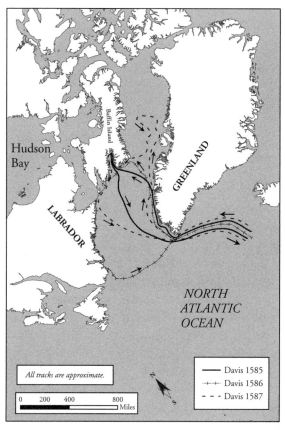

5.2 The Davis expeditions

or the backe side of America." Under no circumstances was he to turn
back before at least a year of trying. The record of Waymouth's expedition of 1602 makes plain how much Arctic waters cooled and sea ice
expanded at the turn of the seventeenth century. In midsummer they
began to pass great islands of ice on the way between Greenland and
Labrador. As they approached the southeast tip of Baffin in early July
they faced fog, snow, and gathering sea ice. After failed approaches to
the island, the ships were nearly crushed amid icebergs, "whereupon
we thought good to take in some of our sayles," Waymouth recalled,
"and when our men came to hand them, they found our sailes, ropes,
and tacklings, so hard frozen, that it did seeme very strange unto
us, being in the chiefest time of Summer." Soon "extreame cold" and

"exceeding great frost" rendered the rigging almost unusable. Demoralized by the weather, the sailors verged on mutiny and refused to sail
onward, according to Waymouth. They turned back at the end of July,
the captain's ship narrowly escaping a capsizing iceberg. After passing
through storms along the Labrador coast, Waymouth finally caught a
favorable westerly late that month to "cleare our selves of the Land and
Ice" and return to England.[46]

While acquitting Waymouth of any wrongdoing, the East India Company discarded plans to send him on a second expedition. The company
waited three more years to gather the necessary funds and then chose
as captain John Knight, recently returned from the first of the Danish
Greenland voyages described in the introduction. In summing up the
Knight expedition, it is hard to do much better than the Victorian historian Thomas Rundall: "The results were the loss of the master and of
some of the crew: peril, excessive toil, with severe hardship to the rest
of the people; and unmitigated disappointment to the projectors."[47]

Knight departed with a single barque from Gravesend in late
April 1606. Even keeping well south of Greenland, the expedition was
"pestered with ice" most of the way across the Atlantic.[48] They sighted
land at around 57°N latitude on June 23, but as they approached the
Labrador coast, the ship was "compassed about with many great Ilands
of Ice" and nearly smashed by powerful gales.[49] Turning south, they
came in sight of the mainland again six days later, but northerly gales
continued to drive ice into their ship. On July 4, a collision with an
iceberg smashed their rudder into the stern, and the crew was forced
to shelter their broken vessel in a cove. The captain took four men
aboard their shallop to a nearby island in search of a harbor and materials to mend the barque.

Knight and his companions were never heard from again. Trum-
peting and shooting their muskets to get the captain's attention, the
men aboard the broken ship instead attracted the fear or interest of a
group of Inuit, who reportedly tried to ambush their lookout that night.
Fearing another attack, the remaining crew performed makeshift re-
pairs on their vessel, and on July 10 hacked their way out of the ice
using picks and axes, "with little hope of recovering our Countrey,"
as one recalled.[50] Barely keeping the leaky vessel afloat, they improvised a new rudder and reached safety among English fishermen at
Newfoundland.

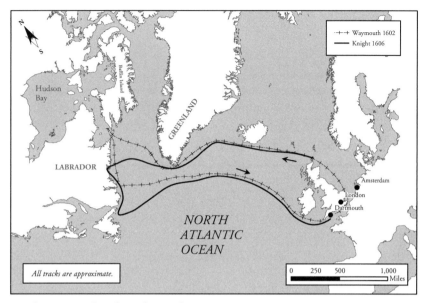

5.3 The Waymouth and Knight expeditions

The failure of Northwest Passage expeditions did little to dampen enthusiasm for other ventures in the New World. By the 1580s, the Americas had become an object of public fascination in England—although that enthusiasm did not always translate into accurate or useful knowledge. Publishers eagerly printed the latest stories of American exploration and discovery, real and fictional. Spanish, Italian, and French works on American geography found a ready market in translation. Collectors assembled "cabinets of curiosities" stocked with such strange New World artifacts as Amerindian spears, tortoise shells, and the tusks of narwhals. Londoners paid good money, as Shakespeare noted, to see a real American Indian, alive or dead.[51]

Following Humphrey Gilbert's Newfoundland fiasco and the lost Roanoke colony, English colonial promoters returned to planning and gathering information with a renewed energy and determination, keen to learn from past mistakes. To a remarkable degree, this monumental task fell to just one person, Richard Hakluyt. A minister by profession, Hakluyt became an enthusiastic amateur geographer, editor, and propagandist for English colonization. He stood at the forefront of a growing community of outward-looking, cosmopolitan Englishmen fascinated by the possibilities of overseas commerce and exploration. Hakluyt's major work, *The Principall Navigations, Voiages and Discoveries of the English Nation,* first published in 1589 and vastly expanded in 1598–1600, captured much of what was then known about the world overseas. It remains an invaluable source for historians even today.[52]

Hakluyt and his fellow colonial promoters propounded an influential vision of English overseas commerce and empire to rival those of Spain. The English (and French and Dutch) of Hakluyt's time preferred to cast themselves as the antithesis of the supposedly cruel and decadent Spanish. There is little doubt, however, that they were inspired by the example and envious of the achievements of the Spanish Empire in the New World.[53] Not least, the English were influenced by what historians now recognize as Spain's commanding lead in many fields of science, including navigation, geography, and natural history.[54] As they worked to make sense of North America's environment and geography, they began largely from Spanish ideas and experience.

Unfortunately, that knowledge was not always easy to come by, even for Spaniards. Much of the vast archives and manuscripts now used to reconstruct the history of Spanish America was then held as state secrets or little disseminated beyond small official and intellectual circles. Even the monumental *Relaciones geográficas de Indias* (Geographical reports of the Indies), painstakingly compiled in the 1570s and 1580s, which included detailed discussions of weather patterns that could have solved so many puzzles of the American climate, was probably never disseminated at all.[55] It certainly never reached Hakluyt, who had to gather information about the Americas from whatever sources he could get his hands on, including Italian, Spanish, and Portuguese travel narratives, as well as interviews with sailors and explorers coming from the New World.

Hakluyt's enthusiasm could lead him to embrace the most optimistic, even improbable, readings of his sources and then exaggerate them even more for promotional effect. In his version of Spanish Florida, the natives supposedly showered Europeans with pearls and food. They had silver in such abundance that they used it to tip their arrows.[56] Nevertheless, for Hakluyt, as for his contemporaries, the greatest obstacle remained a misconception of geography and climates. Hakluyt emphasized repeatedly that North America was not a large continent at all, but narrow and split by one or more passages and inland seas.[57]

More important, Hakluyt imagined "Virginia"—by which he meant the whole coast from the Carolinas to Maine—as an essentially Mediterranean region. His vision, outlined in letters and pamphlets of the 1580s–1590s, was of a warm but temperate country where the English could grow olives, grapes, citrus, and almonds and raise silkworms. The colonization of Virginia would substitute for England's expensive imports of Mediterranean products. The land would have the "climate and soile of Italie, Spaine, or the Islands from whence we receive our Wines and Oiles."[58]

Hakluyt's years of research, both his discoveries and his misconceptions, went directly into the Virginia Company's directives for the Jamestown expedition of 1607. These "Instructions by Way of Advice" guided the venture's fateful first steps in finding a strategic location for a settlement, opening relations with Native Americans, and protecting the colonists' health and morale. On reaching the coast, the leaders of the expedition were advised to "do your best Endeavour to find out a Safe port in the Entrance of Some navigable River making Choise of Such a one as runneth furthest into the Land." They were to keep an eye out for a body of water "which bendeth most towards the Northwest for that way shall You soonest find the Other Sea" (that is, an imaginary passage to the Pacific). The instructions then directed them to search out an appropriate location to settle, preferably on an island or other restricted locale safe from Indian attack. Once landed, they were to divide into three groups in order to accomplish each of the most urgent tasks: erecting fortifications, constructing a storehouse for provisions, and clearing and planting the ground. Yet the colony was not expected to be self-sufficient. The settlers were to seek out trade for Indian corn as soon as possible, before the natives realized the English meant to stay and occupy their land.[59]

Other items in the "Instructions" read like a grim portent of the colony's challenges in the coming years. Conflict with Indians was to be expected or at least prepared for. The settlers were warned not to trade away weapons that might be used against them, and not to water down the value of their merchandise through unregulated exchange. They were advised to "choose a Seat for habitation that Shall not be over burthened with Woods near your town . . . that it may Serve for a Covert for Your Enimies round about You." The instructions also cautioned them not to "plant in a low and moist place because it will prove unhealthful." Regarding climate, however, Hakluyt advised them only to "judge of the Good Air by the People," meaning that the health and appearance of the local population should let the English know whether the climate would unbalance their humors and create disease. They were to report back "what Comodities you find what Soil Woods and their Several Kinds and so of all Other things Else to advertise," but at the same time to censor any bad news so that it would not reach potential investors in England.[60]

To judge from records of their first landing, the leaders of the James-town expedition followed these instructions as best they could. After suffering their first fatality at Mona, the expedition resupplied at the smaller island of Monito, then "disimboged out of the West Indies, and bare our course Northerly," as George Percy recalled. By late April 1607 they had left the tropics and were sailing up the east coast of America. They ran into a "vehement tempest" in early May, but by good luck the storm blew them nearer to the shore. On May 6, they sighted land at Cape Henry and entered the wide mouth of the James River in Virginia. The very first night onshore, they were ambushed by Indians, and one of the captains was wounded. Undeterred, they explored briefly over-land and prepared a shallop in order to explore upriver.[61]

As their instructions specified, they spent several days in search of the best site for their colony. Passing several good candidates, they eventually settled on the tip of a marshy peninsula about 50 miles (80 kilometers) upriver, where they landed on May 24. For generations, the site of the original Jamestown fort was believed to have sunk into the river. It was only rediscovered by archaeologists in the 1990s and is still under excavation, yielding surprising new finds for archaeologists and historians. Even today, more tourists visit a mock-up of the colony a few miles away than see the real site. Standing on its ground and looking out from the now reconstructed walls of the rough triangular fort, one can grasp the appeal of the place. On a fair day, it still feels pristine and healthful, with sandy soil and attractive woods and vegetation. The Jamestown peninsula itself, while not too difficult of access, is more defensible from land than most other sites along the river-bank. It lies up the broad and deep James River around a bend past the low-lying Hog Island (actually another peninsula). This location hides the Jamestown settlement while still offering a lookout for ships entering the river. Strategically, the site seemed ideal.[62]

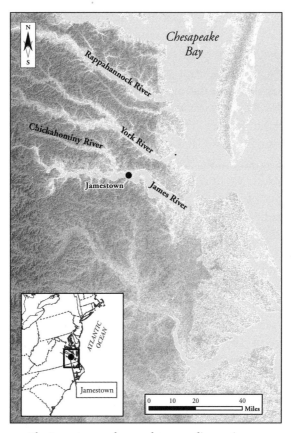

5.4 The Jamestown colony and surrounding region

The few surviving reports from the first couple of months at Jamestown offered little but praise, and little doubt that the country would provide the Mediterranean bounty that Hakluyt had imagined. One extolled its "sandy ground, all over besett with faire pine trees," and the soil "more fertill then can be well exprest." Through May and June, orange trees, cotton, melons, potatoes, and even a pineapple brought from their stopover in the Caribbean all thrived.[63]

Evidently those advantages weighed heavily enough with the leaders of the expedition that they could overlook some unwelcome facts. The site was low-lying and offered no immediate source of water besides the river itself. There was not much land to plant for a colony of hundreds of men. The forest grew close to the settlement, providing enemies with possible cover for ambush, just as the colony's instructions had warned against.

In light of disasters that would fall on the colony in the coming years—its near annihilation from cold, hunger, disease, and violence—it would be tempting to dismiss Hakluyt's instructions or the leaders of the Jamestown expedition as ignorant and shortsighted. However, the truth was just the opposite. They relied on the most highly regarded ideas and best information available to them at the time. In addition to the "Instructions by Way of Advice," the Virginia Company had assembled a whole library of relevant books to prepare the expedition, including volumes on tobacco and silk, works of agriculture, medicine, and even maritime law: a reflection of all their wide-ranging hopes, schemes, and misconceptions.[64]

If anything, their plans looked too far ahead, concentrating on opportunities and eventualities that never had time to materialize—or at least not until long after most of the colonists had died terrible deaths. The site chosen for Jamestown actually put the long term ahead of the short term, privileging its strategic location and potential for trade over the more pressing problems of adequate provisioning and water supply. As one historian has remarked, "There was no conception in Hakluyt's mind of the long struggle for bare subsistence, under primitive conditions," that would follow.[65] The Jamestown colonists were about to find that struggle harder than they, or any of the expedition's planners, could have imagined.

6

Destroyed with Cruel Disease

The Jamestown settlers had two missions once they located their colony: to explore the surrounding country for precious metals and a passage to the South Sea, and to settle peacefully among the Indians.[1] Both went badly from the start.

At the end of May, as the settlers began work on their settlement, Captain Christopher Newport led a party of Englishmen in a shallop up the James River, determined to find its source. For two days they passed Indian villages—"the people in all places kindely intreating us, daunsing and feasting us," as one recalled—until they reached the falls at present-day Richmond. There they met a man whom they mistook for the "chiefe King" of the surrounding country. They exchanged gifts and made declarations of friendship, but then the chief "sought by all meanes to diswade our Captayne from going any further." Meeting impassable rapids and finding no guide to lead them on, the English were forced to reverse course and return downriver.[2]

A day later, something in the behavior of their Indian companions raised suspicions, and Newport's party hastened back to Jamestown. The friendly reception had been a ruse all along. They returned to find that hundreds of Indians had attacked the English camp the day before.[3]

Relations between Virginia's Natives and newcomers had been uncertain from the start, marred by misunderstanding and mutual suspicion. The very first night the English reached the shore, there had been a skirmish, leaving several men injured. As soon as they landed at Jamestown, a local *weroance* (chief) sent them gifts and messages of peace, but his visit to the colony the next day ended in fighting and recriminations.[4] Remarkably, the English had still not properly fortified themselves by the time the assault came in early June. With the aid of their ship's artillery they barely held off the attack. Only later that month did they construct the rough triangular fort whose foundations archaeologists have now rediscovered. In the meantime, Indian archers picked off any men who strayed too far.[5]

The situation soon went from bad to worse. At the beginning of July, Captain Newport had to take the ships back to England, in order to bring news of the colony to the Virginia Company and to fetch more settlers and supplies. Around the time of his departure, delegations of local Indians came offering peace, and the attacks mostly stopped.[6] However, the colonists still feared fresh hostilities, and Indian offers of peace did not come with the generous deliveries of food that the colony's leaders had counted on. In the words of one colonist, Newport left them "verie bare and scantie of victualls, furthermore in warres and in danger of the Savages."[7]

Shortly after Newport's departure, "divers of our men fell sick," as the colony's first president, Edward Maria Wingfield, recalled. Within ten days, "scarse ten amongst us coulde either goe, or well stand, such extreame weaknes and sicknes oppressed us."[8] By mid-August men were dying daily, most of them individuals we know only from their names and dates of death. John Asbie went first "of the bloudie Flixe" and George Flowre on the following day "of the swelling." A week later Francis Midwinter and Edward Moris "died suddenly." The list goes on and on. By late September about fifty men, or nearly half the colony, were dead.[9] And that was only the beginning.

Our most detailed account, from colonist George Percy, gives several reasons for the colonists' deaths that summer of 1607. "Our men were destroyed with cruell diseases as Swellings, Flixes, Burning Fevers, and by warres, and some departed suddenly," he wrote a few years later, "but for the most part they died of meere famine."[10] Unfortunately, this explanation obscures as much as it enlightens. Typical of seventeenth-century writers, he names diseases by their symptoms, leaving us in the dark about their causes. Our accounts of the period mention only a handful of deaths by violence, so "warres" may be an exaggeration. And the statement about "meere famine" seems to contradict the rest.

So what exactly killed them? The horrendous mortality during the first years at Jamestown has long been a favorite mystery among medical and colonial historians. This is not because we are missing an explanation, but rather because there are so many possible explanations.[11] In other words, it is not the sort of mystery where the detective tracks down an elusive unknown killer but the sort where he discovers the body in the middle of the drawing room, surrounded by suspects with means and motives. Jamestown faced so many problems during its first years—poor food, poor water, inadequate shelter, disease, and violence—that it is hard to convict a single culprit.

Yet the case can be made that the underlying source of the colony's troubles was the region's Little Ice Age climate. As tree-ring studies have demonstrated, English settlers arrived at the start of Virginia's longest drought in centuries, from 1606 to 1612.[12] At the same time, evidence from sediment cores in the Chesapeake indicates that these were some of the coldest years of the last millennium—perhaps 2°C (3.8°F) cooler than in the twentieth century.[13] Research on tree rings and cave deposits in West Virginia demonstrates a shift in seasonal patterns at the turn of the seventeenth century that brought colder winters and drier summers.[14]

These climatic extremes overwhelmed the colony during its first and most vulnerable years. Summer drought withered corn in the fields and turned the James River saline and unhealthful. The winter cold proved demoralizing and at times deadly. Most important, year after year of unseasonable weather and poor harvests intensified competition over scarce resources, aggravating conflicts within the colony and between colonists and their Indian neighbors. The first settlers in Jamestown would have encountered challenges in the best of circumstances. But in these, the worst years of the Little Ice Age, they faced continual disasters and near annihilation.

Before assigning an immediate cause for the high mortality that summer of 1607, it helps to exonerate some of the usual suspects. As we have seen, there is no evidence that Old World epidemics had already reached Virginia's Indians. Nor are there signs of malaria. Just as European viruses and bacteria had not spread widely among North American populations by 1607, neither, it seems, had the Old World mosquitoes that carried the malaria plasmodium. The disease would take decades more to become endemic to the Chesapeake area, probably only after the rise of the slave trade during the mid- or late 1600s.[15]

We can also rule out, or at least qualify, Percy's verdict of "meere famine." Rations were scanty—a pint or less of wheat and barley meal per day—and it is unclear what else the colonists found to eat.[16] John

Smith blamed the long and late sea voyage for wasting time and provisions needed during the first summer. The colony planted a field of grain that "sprang a man's height" by late July, but the late sowing left it no time to ripen.[17] The expedition carried guns for hunting and hooks and nets for fishing, but the colonists may have been too afraid or inept to take advantage of game or wild plants during that first summer and autumn.[18]

When witnesses recounted "famine" or "starving" at Jamestown, the image is one of fatigue, cold, and exposure to the elements, not mere want of food. All this time, the poorer colonists lived "on the bare grownd" and even the wealthy inhabited "miserable cottages," eating a monotonous diet and standing constant guard for Indian attacks.[19] Gabriel Archer complained, "Wee watched every three nights lying on the bare cold ground what weather soever came warded all the next day, which brought our men to bee most feeble wretches." He described how one man "starved to death with cold" that September.[20] To modern historians, these descriptions have also suggested malnutrition, particularly beriberi, with its symptoms of inflammation and lethargy. Karen Kupperman has suggested the colonists lived in a state of hunger and shock not unlike mistreated prisoners of war. These conditions might not have killed the colonists outright but probably left them vulnerable to other causes of death.[21]

The most reasonable conclusion is that the colonists brought some infection from England, which flourished in the unsanitary conditions of the fort. The most likely culprit remains typhoid, a disease that preyed on victims of famine. Its incubation period roughly fits the time between Captain Newport's departure and the first deaths at Jamestown. Its symptoms include fevers, fatigue, and diarrhea (Percy's "flixes"). Symptoms can endure several weeks, and an outbreak can take two months or more to spread through a settlement. Past victims can become carriers for years after (such as the infamous Typhoid Mary). It may have been the same illness that had struck a passenger on the voyage to Virginia, but given the outbreaks during famines of the 1590s, any number of the settlers might have been infectious. Transatlantic voyages of the time were grueling ordeals and often unsanitary death traps, but at least human feces could be thrown overboard. At Jamestown, particularly while the colonists feared another Indian attack, their waste probably ended up not far from their water supply.[22]

The water itself was the root of the problem. That summer, the settlers drank straight from the James River. Its contents were, as Gabriel Archer recalled, "at flood verie salt, at a low tide full of slime and filth, which was the destruction of many of our men."[23] Remarkably, it would take the colonists another two years to dig a well in the fort that drew "sweete water."[24] In the meantime, the men at Jamestown drank from a river so brackish it might have sickened and even killed them from salt poisoning. The problem was not that Jamestown was inherently unhealthful. There is archaeological evidence that for centuries Virginia's Natives had lived around the site from time to time.[25] Rather, the exceptionally dry summer had so reduced the flow of the James that saltwater encroached from the sea. In this way, the summer's casualties were the first victims of Virginia's extraordinary drought.[26]

An equally pressing problem was that the colonists—all men—were drinking water at all. For most of them, that was something inherently dangerous and demeaning in the first place. They understood little about water quality and sanitation. Another way to explain the mortality at Jamestown might be to say that they died for want of their usual drink: beer. By recent estimates, a working Englishman of the sixteenth century drank about six pints of the stuff each day—a significant portion of all his daily calories. Put another way, barley was almost as common as wheat in early modern England, and most of it went straight into making beer. Not all of this was weak "small" beer, either: some of it was strong stuff. Malt was basic sustenance and alcohol "an essential narcotic which anaesthetized men against the strains of contemporary life," to quote historian Keith Thomas.[27]

The lack of beer and other alcoholic beverages not only exposed the colonists to pathogens in the water but also left them hungry and demoralized. It became a recurring problem in the early history of the colony. In 1613, reporting on the desperate situation in Jamestown, the Spanish ambassador in London remarked to Philip III that the colonists were "sick and badly treated . . . nor do they drink anything but water—all of which is contrary to the nature of the English."[28] As late as 1620, a Virginia planter observed, "More doe die here of the disease of theire minde then of theire body" because they arrived unprepared for the primitive conditions and poor food, and "not knowinge they shall drinke water here."[29]

Alcohol became a flash point in the rancor and dissension that divided Jamestown's leaders while Captain Newport was away. The Virginia Company had sent the expedition with a sealed box containing the names of the first governing council, not to be opened until the ships reached America. The seven men they had chosen proved entirely unprepared and unsuited for working together. John Smith, the rough former soldier who had already been arrested for mutiny on the outbound voyage, clashed with the colony's first president, the vain and inept Edward Maria Wingfield. Another councilor, George Kendall, was soon arrested as a Spanish spy. Bartholomew Gosnold—one of the most respected men in the colony—just managed to hold the council together until his death on September 1, 1607.[30]

As rations fell and the death rate rose, the remaining councilors turned on Wingfield, accusing him of hoarding food and drink. "The Sack, Aquavitie, and other preservatives for our health, being kept onely in the Presidents hands, for his owne diet, and his few associates," claimed John Smith.[31] Others alleged that he denied even a "spoone-full of beere" to the hungry and sick. In late September, the council removed Wingfield from office and elected a new president, John Ratcliffe, in his place. Wingfield later accused the councilors of dangerously depleting the common stores to win favor with the settlers, and even bribing witnesses with a little corn to testify against him. He claimed that had he given out food and drink as the council demanded, he would have "starved the whole company."[32]

Just when the colony's provisions were almost spent and most of the settlers too weak or sick to work, Indians suddenly began to bring them food. Throughout the autumn, Archer recalled, "those people which were our mortall enemies" came "to releeve us with victuals, as Bread, Corne, Fish, Flesh in great plentie, which was the setting up of our feeble men, otherwise wee had all perished. Also we were frequented by diverse Kings in the Countrie, bringing us store or provision to our great comfort." Some of the colonists interpreted this as a miraculous deliverance. "It pleased God (in our extremity) to move the Indians to bring us Corne . . . when we rather expected they would destroy us," wrote Smith.[33]

These deliveries of food were no miracle, however. Nor did they spell an end to Jamestown's troubles. The decision to first pressure and then relieve the struggling colony was more likely a calculation by Native Virginians to maximize the usefulness of the English presence while minimizing its risks.

As the colonists soon discovered, the real power in Virginia lay with the Powhatan confederacy, a paramount chiefdom exercising hegemony over most of the villages upriver from Jamestown. The Powhatan and their allies, who may have numbered around 15,000 in 1607, inhabited villages spread mostly around the rivers and streams of what is today the eastern part of Virginia.[34] For the paramount chief Wahunsenacawh (also known as Powhatan), the arrival of English settlers, ships, and goods represented both an opportunity and a threat. There was much to be gained if he could force the English to trade their metals, manufactures, and especially weapons for cheap; but much to lose if the invaders grew too powerful or traded those wares to rival Indian nations. As colonist William Strachey later observed:

> It hath been Powhatan's great care to keepe us, by all meanes, from the acquaintance of those nations that border and confront him, for besides his knowledge how easely and willingly his enemies will be drawne upon him by the least countenance and encouragement from us, he doth, by keeping us from trading with them, monopolize all the copper brought into Virginia by the English.[35]

In late 1607, Jamestown's trade fell under the charge of John Smith, who viewed the issue as one of supply and demand. So long as he could control the availability of European metals and manufactures, those should fetch a fair price in Indian provisions. Failing that, as Smith saw it, the English would have to walk a fine line. They would have to act tough enough to make the Indians deliver sufficient food in exchange for their wares, while not acting so aggressive that the Indians withdrew from trade altogether. It was an approach that reflected the increasingly commercial outlook of seventeenth-century English society, as well as Smith's practical background as a soldier and a man of action.[36]

Unfortunately for the colony, this outlook had trouble taking into account cultural differences with Virginia's Indians. At the time, English attitudes toward Native Americans had not yet hardened into

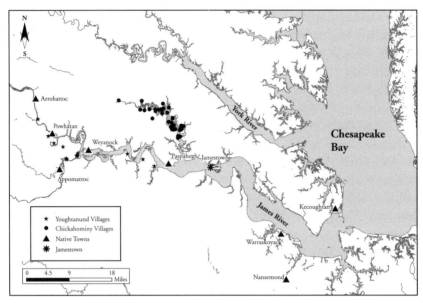

6.1 **The Native world of the lower Chesapeake**

the bigotry and racism characteristic of later centuries. The early 1600s were still an era of relative openness and debate in European understandings of Indians: a moment of lingering hesitation and uncertainty whether to view them as similar or different, rational or irrational, noble savages or just savage. Europeans and Indians alike still endeavored to fit the other into their own cultural categories and preconceptions.[37]

For the Indians of Virginia, as for many Native American communities, trade meant more than just commercial exchange, and trade goods were more than mere commodities. The rules of supply and demand still applied, but they were not the only rules. The practice of trade was wrapped up in issues of status and diplomacy. What Europeans might see as mere trinkets and shiny metals could be prestige goods among the Indians, for whom accumulation and redistribution helped cement political and spiritual authority. The cultural norms and expectations of commerce and gift exchange left considerable room for misunderstandings and recriminations on both sides. A trade deal gone wrong could be mistaken for a theft or an insult. A change in trade relations could be mistaken for a breach of diplomacy or a hostile act. Some historians have put these fatal misunderstandings at the center of Jamestown's early troubles.[38]

Commerce for Indian corn was fast becoming a critical issue, and one that would propel John Smith to leadership of the struggling colony. Cotemporary sources disagree whether the settlers at Jamestown actually risked starvation as the autumn of 1607 turned to winter. More than a decade later, a few survivors would declare they had been "at point of death." Their shelters were still unfinished and the first ground barely cleared for planting, "hunger and sickness not permitting any great matters to bee donne that yeare."[39] Smith, on the other hand, blamed the colonists themselves, "being in such dispaire, as they would rather starve and rot with idlenes, then be perswaded to do anything for their owne reliefe." He marveled at the profusion of plants and wildlife in Virginia during late autumn: "fish, fowle, and diverse sorts of wild beasts as fat as we could eat them." He also boasted of his success in putting the colonists to work and opening exchange with Virginia's Indians, employing a mix of threats and hard bargaining. "At first they scorned him, as a starved man," Smith reported of one trip to the village of Kecoughtan, "yet he so dealt with them, that the next day they loaded his boat with corne."[40]

Smith inevitably looms larger than life in this chapter of Jamestown's history. It is not just that he played an outsized role in events, but that his own writings remain our principal source for those events, and his narration was anything but modest. Much the same goes for our sense of the man himself, most of which comes from his own autobiography, a story that casts him as the protagonist of adventures on the high seas, daring battles with the Ottoman Turks, miraculous rescues by beautiful ladies, and finally the redeemer of a feckless colony. Something of this manly, swashbuckling self-portrait—and something of the legend of Smith as Jamestown's savior—must have been true. Otherwise, it is hard to believe Smith would have succeeded as far as he did, or that the colony would have lasted through its first winters when so many others failed. Yet separating fact from fiction remains a challenge.[41]

Nowhere is this more true than in Smith's most famous story of all, his captivity among the Powhatan Indians and rescue by Pocahontas. After at least three trading voyages up the Chickahominy River, Smith undertook a longer expedition in December 1607 "to finish this discovery."[42] About two days into the voyage, some 200 Indians ambushed Smith's party, killed two of his English companions, and led Smith to Werowocomoco, the seat of paramount chief Wahunsenacawh. This much is confirmed in other contemporary sources.[43] For the rest of the story, we have only the different versions told by Smith himself.

In all of them, the Powhatan Indians led Smith through strange and elaborate ceremonies. Wahunsenacawh feasted the Englishman and marveled at his European technologies and tales of London. Smith learned of Wahunsenacawh's powerful confederacy of Virginia chiefdoms, and heard (or misheard) descriptions of tall mountains and a sea to the west. What differs in these accounts is what happened next. In one version, Smith simply overawed the Powhatans "and got himselfe and his company such estimation amongst them, that those Salvages admired him as a demi-God."[44] In another, it appears Smith was released with the understanding that he would lead the English as another chiefdom within the Powhatan confederacy.[45]

Only years later did Smith tell the story that Wahunsenacawh ordered his execution and that the chief's young daughter Pocahontas stepped in at the last moment to save him.[46] In his defense, historians have argued that Smith did refer to the story in other ways in earlier correspondence; that he might have had other reasons for neglecting to publish it sooner; and, moreover, that the whole episode may have been staged as a sort of "adoption" of Smith as a vassal chief of Wahunsenacawh. On the other hand, there are grounds for skepticism. Prisoners like Smith were usually tortured to death rather than executed; there is no reason a girl like Pocahontas would have been present at such an event; there is no similar "adoption" ceremony in the ethnography of the Powhatans; and the number of last-minute rescues by women in Smith's autobiography already beggars belief.[47]

What we do know is that as soon as he returned to Jamestown, Smith enjoyed an escape almost as timely as any that might have come from the hands of Pocahontas. Gabriel Archer, one of Smith's enemies on the council, had him tried and sentenced to execution for the death of his companions during his last journey upriver. By a remarkable coincidence, Captain Christopher Newport returned to Jamestown that very evening. He immediately pardoned Smith, saving his life.[48]

Newport's first supply arrived on January 12 with two ships and nearly a hundred men, "well furnished with all things could be imagined necessarie, both for them and us."[49] But the relief was short-lived. The newcomers had to crowd into the settlement's few make-shift hovels—still little more than pits in the ground thatched over with reeds. Five days after their arrival a fierce fire broke out among the dwellings and spread to the storehouse and palisades, so "by a mischaunce our Fort was burned, and the most of our apparell, lodging and private provision."[50]

In the meantime, Virginia plunged into an extraordinary winter, even by the standards of the Little Ice Age. Smith described how during his captivity he endured "as bitter weather as could be of frost and snow."[51] One of the new arrivals at Jamestown wrote in a letter that "it got so very cold and the frost so sharp that I and many others suffered frozen feet." The winter grew "so intense that one night the river at our fort froze almost all the way across"—something almost unimaginable in today's climate.[52]

Smith later recognized that Virginia's continental seasons brought more variable weather than in England, and that their first winter at Jamestown was exceptional: "The colde is extreame sharpe, but here the proverbe is true that no extreame long continueth."[53] At the time, however, the situation looked bleak. The fire left most of the English without shelter, supplies, or adequate clothes. Some provisions remained, but Smith blamed the sailors for eating the colony's best food, leaving the settlers with "meale and water" that "little relieved our wants." Men still recovering from the summer's epidemic succumbed to illness, and many of the newcomers to cold and exposure: "Whereby with the extreamity of the bitter cold aire more then halfe of us died, and tooke our deathes, in that piercing winter I cannot deny."[54]

Newport's return raised further complications for commerce between the colonists and neighboring nations. The captain had just come back from an embarrassing trip to London. The supposed ore brought from Virginia in 1607 had turned out to be worthless rocks, and for all the promise reported during the first few weeks of the Jamestown settlement, the captain had almost nothing of value with which to reimburse the investors of the Virginia Company.[55] Now he found more than half the settlers dead or dying, the fort ruined, its provisions wasted, and the upstart John Smith trying to take charge. Newport needed food for the colony, and he needed good news to bring back to London. He seized on Smith's reports of a powerful Indian chiefdom upriver and rumors of gold in the mountains to the west.[56]

In March, Newport led Smith and about thirty other Englishmen on a trip up the York River to Werowocomoco. Smith arrived first with about half the men; he met again with Wahunsenacawh and apparently renewed his earlier agreement to resettle the English as a vassal chiefdom in Powhatan territory. The captain arrived with the rest of the party on the following day. He left an English boy, Thomas Savage, to live among the Powhatans and learn their language. Then he got down to the business of trading English goods for Indian food.

From Newport's perspective, the trip probably looked like a success. After a few days of pleasantries and negotiations, in spite of the "miserable cold," he traded away some copper and trinkets, and then returned to Jamestown. His small barge was loaded with enough corn to last the colony for a few months. In the meantime, his men had the opportunity to collect some more promising "ores." These they loaded onto their ship, and Newport sailed for England on April 20.[57]

From Smith's perspective, the captain's intervention was a disaster. Newport was sailing away with "gilded durt," and now the fever for gold distracted the colonists from doing real work. Moreover, Newport had disrupted the trade between the English and Indians—a relationship that Smith believed he had under control. While waiting at Jamestown, Newport's sailors had traded away English goods and metals for trifles, so now the colonists could scarcely get supplies "for a pound of copper, which before was sold for an ounce." Worse still, in Smith's opinion, Wahunsenacawh had tricked Newport into handling their trade as an exchange of gifts rather than real commerce. "Therefore lay me down all your commodities togither," Smith paraphrased the chief; "what I like I will take, and in recompence give you that I thinke fitting their value." Smith claimed to have salvaged the situation only by driving a hard bargain for some blue beads that had caught Wahunsenacawh's eye.[58]

By the spring of 1608, however, Jamestown faced a still more pressing problem: getting any food from the land at all. The winter had been brutally cold, and Virginia was now in the third year of its epic drought. Even if the colonists could preserve the value of their trade goods, and even if they managed to avoid misunderstandings over trade and gift exchange, they still had to worry whether Virginia's Indians would even have provisions to offer.[59]

Some historians have doubted whether the climate seriously affected Indian food supplies. The subsistence strategies of the Powhatans and other Virginia Indians were diversified and seasonal, combining horti-culture with hunting, fishing, and foraging. As Smith described:

> In March and Aprill they live much upon their fishing weares, and feed on fish, Turkies and squirrels. In May and June they plant their fields and live most of Acornes, walnuts, and fish. But to mend their diet, some disperse themselves in small companies and live upon fish, beasts, crabs, oysters, land Torteyses, strawberries, mulberries, and such like. In June, Julie, and August they feed upon the rootes of *Tocknough* berries, fish and greene wheat.[60]

Late summer and early autumn was usually a time of abundance, as they harvested their corn, beans, and squash; in winter they pursued deer and other game.[61]

Nevertheless, the Indians of Virginia, like others met throughout this book, remained vulnerable to the worst conditions of the Little Ice Age. Their diversified subsistence strategies still exposed them to sea-sonal shortages. "It is strange to see how their bodies alter with their diet," Smith observed, echoing other European descriptions; "even as the deare and wilde beastes they seeme fat and leane, strong and weak." Chief Wahunsenacawh and other elites may have been able to store corn year-round, but most Virginia Indians planted only enough to last a few months, and lived the rest of the year "of what the Country natu-rally affordeth from hand to mouth."[62]

The combination of summer droughts and freezing winters may have been especially destructive. Not only would the maize harvest have failed, but as the first Jesuit missionaries to Virginia in 1570 had observed, heavy snows and frozen ground would have made it difficult or impossible to forage and to dig up the tuckahoe (arum root) that formed such a crucial part of their subsistence. These shortages would have culminated in hunger especially during the spring planting time.[63]

Archaeological evidence confirms this picture, both among the Indians of Virginia and others at the time of contact. By examining stable
carbon and nitrogen isotope ratios in skeletons, archaeologists have
been able to show that these populations had become dependent on
maize for a significant part of their diet. (Isotopically, at least, you are
what you eat.) The same skeletal evidence also reveals pathologies—
enamel hypoplasia and porotic hyperostosis—consistent with episodes of acute and chronic malnutrition, as well as repetitive-stress
injuries and the physical trauma of violence. Although the Indians of
Virginia were taller and healthier than contemporary Englishmen, episodes of hunger and frequent warfare meant they faced a similar or
even shorter life expectancy.[64]

In theory, Native Americans had options to adapt to the changing
climates of the Little Ice Age. They could move their fields to warmer
south-facing slopes and well-drained soils. They could specialize in
crops that were better able to tolerate cold or drought, and they could
store more food to see them through bad years. They could migrate to
warmer lands, or even abandon agriculture and revert to hunting and
foraging.[65]

Some populations did adapt in these ways. There are clear examples
of Indian nations in New England who expanded their range of hunting
and even horticulture during the Little Ice Age. Yet many other
populations—including the Indians of Virginia—evidently did not.
The reasons may have been both cultural and practical. Choices about
diets, crops, hunting, foraging, and settlement could be deeply em-
bedded in religious practices, gender relations, and identity. They were
not easy to change. Among the Powhatans, for instance, horticulture
and foraging were strictly women's work. As anthropologist Helen
Rountree has argued, women probably saw little reason to grow or
store more maize than necessary, when it might be taken as tribute by
male elites. It made more sense to focus their efforts on subsistence
foods such as tuckahoe.[66]

Moreover, conflict within and among groups could limit options.
Elites within communities had to think about maintaining their status
against rivals. In times of environmental stress and shortages, their po-
sitions might be more threatened than ever. Similarly, whole communities had to consider defense against enemies, especially when a
changing climate might aggravate fighting over territory and resources.

In short, ecological pressures and conflict could force societies to become more rigid and centralized, rather than dispersed and flexible, as adaptation to climate change might have called for. Even before European contact, Native American villages of Virginia had been growing larger, more concentrated, and more fortified against enemy attacks. Chiefdoms, it appears, were consolidating their power, and communities were becoming more unequal and hierarchical. The rise of Wahunsena-cawh and his paramount chiefdom formed the leading edge of a larger trend driven in part by ecological pressures and the Little Ice Age.[67]

Once Newport left again for England in April 1608, Smith put the Jamestown colonists back to work rebuilding the fort and preparing the fields for planting. At the end of the month, Jamestown enjoyed a rare stroke of good luck when the final ship in Newport's first supply, which had separated from the others during a storm, arrived in Jamestown at last. It carried enough provisions to last the colony half a year. Its captain, however, declined to undertake any new voyages of discovery in Virginia, and departed for England just days later.[68]

Smith seized the opportunity to lead his own exploration of the Chesapeake. During the summer of 1608, he took a small group of men in an open barge on two trips up the bay, venturing along the Delmarva Peninsula and into the Rappahannock and Potomac Rivers as far as today's Washington, D.C. The voyages revealed a much wider world of Indian nations, and a rich coastal and marine environment. "In diverse places that abundance of fish lying so thicke with their heads above the water . . . we attempted to catch them with a frying pan," one colonist recalled. They also faced Indian attacks, storms, and a stingray that stuck and nearly killed John Smith. More important, despite some of Smith's boasting to the contrary, their search turned up no evidence of precious metals or a passage to the South Sea.[69]

During his absence, the colonists at Jamestown suffered again from misgovernment and summer illnesses, heat, and fatigue. Smith and his companions returned from their second Chesapeake voyage in September to discover some of the men "well recovered" but "many dead [and] some sick," while the previous council president had been arrested for mutiny. A new harvest was gathered, "but the provision in the store much spoyled with rayne."[70] Smith personally took over the presidency and once more put the settlers to work expanding the fort.

In late September, Captain Newport returned to Jamestown with the colony's second supply. It brought seventy new settlers, including Jamestown's first women, but little with which to feed them.[71] Newport also carried a new commission "not to returne without a lumpe of gold, certainty of the south sea or one of the lost company of Sir Rawley," that is, the lost Roanoke colony. All of this infuriated Smith, who was jealous of his leadership, anxious about the colony's day-to-day survival, and desirous of taking credit for any discovery or success in Virginia.

Worse still, from Smith's point of view, the Virginia Company had sent Newport with instructions to return to Wahunsenacawh for the chief's "coronation" as a vassal of the English empire.[72] The Virginia Company conceived this "coronation" as way to legitimate its claims on the territory. From Wahunsenacawh's perspective, it only reinforced his sense of superiority over the struggling Jamestown colony. The paramount chief insisted that Smith and Newport come in person to Werowocomoco for the ceremony, but "a fowle trouble there was to make him kneele to receave his crown." Smith and Newport had to force down the chief's shoulders in order to put the crown on his head. Unimpressed, Wahunsenacawh gifted them some "old shoes and a mantle" as well as a few bushels of corn in return. He refused to help the English make any further exploration upriver. Smith grumbled that the colony "had his favor much better, onlie for a poore peece of Copper till this kinde of soliciting made him so much overvalue himself, that he respected us as much as nothing at all."[73]

Newport returned briefly to the fort before another trip upriver to the unexplored country of the Nansemond Indians. After a "poore tryall" of some ores, his detachment continued overland to Indian settlements past the falls. "Trade they would not, and find their corn we could not," one colonist recalled. They returned to the colony "halfe sicke, all complaining, and tired with toile, famine, and discontent."[74] In the meantime, according to Smith, Newport's sailors back at the fort traded away hundreds of axes, chisels, and other tools and weapons to Indians, diluting their value even further.

In December 1608, Newport and the second supply sailed back to England bearing none of the commodities the Virginia Company demanded. The captain did bring an angry letter from John Smith complaining about the state of the colony, which he described as "but a many of ignorant miserable soules." Despite the two reinforcements from England, little more than 200 settlers remained alive, eking out their existence on "a little meale and water," with "the one halfe sicke, and the other little better." "Though there be fish in the Sea, foules in the ayre, and Beasts in the woods," Smith wrote, "their bounds are so large, they so wilde, and we so weake and ignorant, we cannot much trouble them." The colony was saddled with dozens of assayers and craftsmen, but now it had no way to feed them. To expect Virginia to turn a quick profit, much less to find mines of gold and passage to the South Sea, was a pipe dream.[75]

It must have been about this time that one of the Jamestown colonists, William White, recorded an unusual and revealing episode. "The Wereoance of Quiyoughcohanock," White wrote, referring to a nearby community of Indians, "was so far persuaded, as that he professed to beleeve that our God exceeded theirs, as much as our Gunnes did their Bowes and Arrowes, and many times [he] did send to the President [Smith] many presents, intreating him to pray to his God for raine, for his God would not send him any."[76]

This whole anecdote would be just a curiosity, except that it was far from the only episode of its kind. Other early explorers and colonists—Spanish, French, and English—reported similar experiences all across North America.[77] Cabeza de Vaca wrote that Indians of the Rio Grande valley "begged us to command the heavens to rain."[78] When in 1541, Hernando de Soto and his expedition reached the Casqui, in northeastern Arkansas, their chief supposedly greeted Soto as a "man from heaven" and asked him to end a drought and famine. The conquistador lectured this chief on the Christian faith and then erected a large cross, leading a great procession of Spanish and Indians around it. The chief later thanked Soto, claiming the Spaniards' god "gave [rain] to us in great abundance and saved our cornfields and seed beds."[79]

A year after Pedro Menéndez de Avilés occupied Florida, a Guale chief supposedly pleaded with the captain general to pray for rain and save the Indians' withering crops. Half an hour later a heavy storm came. Indians all across the region heard of the incident and begged the Spanish for Christian crosses.[80] In late 1583, the governor of Florida claimed that "this year a thing has happened that has been a great impulse for the conversion of Indians," namely, that the fields of Christian converts had received abundant rains while among those of the pagans "there has not rained a drop of water in three months."[81]

A few years later, Thomas Hariot wrote that in the Roanoke colony,

> on a time also when [the Indians'] corne began to wither by
> reason of a drouth which happened extraordinarily, fearing
> that it had come to passe by reason that in some thing they
> had displeased us, many would come to us and desire us to
> praie to our God of England, that he would preserve their
> corne, promising when it was ripe we also should be partakers
> of the fruite.[82]

Reports like these continued into the following century. The first French Jesuit priests in Canada related several stories of Huron Indians testing their faith and their spiritual powers by asking for better weather—usually with miraculous results.[83] When the country was "subject to drought even to the withering of their summers Fruits," the Puritans of the Massachusetts Bay Colony prayed for rain, and "at that very instant the Lord showred down water on their Gardens and Fields," and "the Indians hearing hereof, and seeing the sweet raine that fell, were much taken with the Englishmens God."[84]

In 1623, the Pilgrims at Plymouth colony faced "a great drought" that threatened "sore famine unto them," as William Bradford remembered:

> Upon which they set apart a solemn day of humiliation, to
> seek the Lord by humble and fervent prayer, in this great dis-
> tress. And He was pleased to give them a gracious and speedy
> answer, both to their own and the Indians' admiration that
> lived amongst them.

"For which mercy, in time convenient," Bradford concluded, "they also set apart a day of thanksgiving"—the first mention of any Pilgrim "thanksgiving" in America.[85]

Accounts like these are apt to invite skepticism from modern historians. The literature of early European exploration and Native encounters contains its share of fictional and fantastical stories. Many follow a cliché of credulous natives mistaking European invaders for miracle-working "white gods."[86]

There are reasons, however, for taking these particular episodes seriously. They come from authors of many backgrounds—Protestant and Catholic, scholars and soldiers, English, French, and Spanish—and from every kind of source material. It is hard to believe they all meant to follow some common convention. The internal evidence in these sources is also strong. Most are detailed eyewitness accounts; many are corroborated by multiple testimonies; and in several cases, there were interpreters present.

Most important, we have corroborating physical evidence. For almost every encounter described here, tree-ring records confirm that there were indeed droughts. Hernando de Soto reached Arkansas at the start of a dry year. In 1566, when Pedro Menéndez de Avilés supposedly brought rain to the Guale, their country was suffering one of the worst local droughts in the millennium-long tree-ring record. This book has already described the evidence for droughts at Roanoke in 1587–1589, and the seven-year drought at Jamestown in 1606–1612. The tree-ring evidence matches the descriptions of later episodes in Massachusetts and Huron country as well.[87]

Assuming these episodes were not real miracles, then what was actually going on? A closer look at these incidents suggests that, far from overawing Native Americans with the power of their Christian god, these early explorers and settlers may have been playing a dangerous game that ultimately backfired.

The colonists at Jamestown were probably not surprised when the chief of the Quiyoughcohannock asked for divine assistance. Peoples all across the early modern world turned to supernatural powers when the weather failed them. We can find evidence of elaborate rainmaking ceremonies everywhere from Ming China to the Ottoman Empire. In Spain, church ceremonies to end droughts were so common that historical climatologists have used them to reconstruct precipitation during the Little Ice Age.[88] Nor was there anything unusual about asking for help from neighboring communities or even members of other religions: there is evidence, for instance, that contemporary Muslims in the Middle East turned to Jewish prayers and Christian saints to help end droughts.[89]

However, beliefs and rituals regarding supernatural assistance varied from people to people and place to place. For European Christians, the proper response to drought or other adversity was to repent and pray for God's help. Prayer meant to ask something from God, who might or might not answer prayers, depending whether the supplicant was sincere and deserving. Prayer differed from magic, or the use of rituals and incantations to compel some supernatural entity to carry out your command.[90] During the sixteenth and seventeenth centuries, as the Protestant Reformation and Catholic Counter-Reformation hardened Christian doctrine, magic came to be identified with sorcery or witchcraft—that is, dealings with the devil. Witchcraft prosecutions peaked during the late 1500s and early 1600s, reaching thousands per years. Not coincidentally, the most widespread witchcraft scares occurred during the most miserable years—particularly the coldest summers—of the Little Ice Age. Accusations of weather magic, such as conjuring hailstorms and crop blights, were among the most common features of European witchcraft trials.[91]

Therefore, when early colonists and missionaries encountered Indian weather ceremonies, they immediately identified those as witchcraft, too. "Their priests . . . are no other but such as our English Witches are," wrote the early Jamestown colonist Alexander Whitaker. "If they would have raine, or have any thing, they have recourse to him, who conjureth for them, and many times prevaileth."[92]

The first European colonists evidently could not recognize that Native Americans viewed their Christian rituals in the same terms. There is no reason Indians would have shared Europeans' conceptual distinctions between prayer and magic, or even between the natural and supernatural. Instead, most would have seen European prayers as just another set of rituals or incantations that might succeed where their own had failed. We have only indirect evidence about Native American religions at the time of contact, but it seems most communities had priests or shamans believed to influence supernatural entities and to control the weather. Virginia's Indians might have identified the Englishmen's god with their own mercurial god Okeus, who "strikes their ripe Corne with blastings, stormes, and thunderclappes," as William Strachey noted.[93]

Settlers at Jamestown and elsewhere could see no harm in praying for better weather. If their prayers succeeded, then that showed the benevolence of their God and the power of their religion. If their prayers failed, then that meant that the colonists or, more likely, the Indians were unworthy, or even that God was still testing their faith. Moreover, rain prayers have a certain logic. If one prays for rain (or does just about anything) for long enough, then eventually the weather changes. Unfortunately, the Jamestown settlers found themselves in the middle of a very long drought and in a very tense situation rife with cultural and religious misunderstandings.

Nor was this the first time. After Pedro Menéndez de Avilés miraculously brought rain to Guale Indians, the drought in Spanish Florida continued for another four years, through 1569. When he subsequently visited the Hotina Indians, their chief supposedly fled from the governor, sending word "that he had hidden in the forest out of fear for a man who had such power with God."[94] In the Roanoke colony, diseases introduced by the settlers had already aroused Native fears of English sorcery. Hariot's description suggests Indians viewed the terrible drought as another sign of the Englishmen's malevolent powers. However, unlike the mysterious illness that destroyed Indians but left Europeans unharmed, the colonists had no immunity when it came to crop failure and hunger. The poor state of the colony by 1586 revealed spiritual as well as physical weakness. Roanoke's Indians "grew not onely into contempt of us," Ralph Lane recalled, but "nowe they began to blaspheme, and flatly to say, that our Lord god was not God, since hee suffered us to sustaine much hunger."[95]

By offering to pray for better weather, the Jamestown colonists fell into a trap of their own making. From the Indian perspective, as anthropologist James Axtell has put it, "spiritual power was double-edged."[96] If Englishmen claimed they had the power to bring rain, that meant they had the power to take it away, too. When the chief of the Quiyoughcohannock "professed to beleeve that our God exceeded theirs, as much as our Gunnes did their Bowes and Arrowes," that was not because the Christian god had yet brought any rain. Just the opposite: he had brought the worst drought in anyone's memory—a drought that was still ongoing, and which Native ceremonies had been powerless to end. When John Smith and other colonists claimed to have influence with that god, Virginia's Indians could only draw one of two conclusions: either the English were dissembling, or they and their vengeful god had brought this disaster in the first place.

In the meantime, Smith and his fellow colonists threatened Virginia's Indians with a more earthly kind of violence. While Newport was away at Nansemond in November 1608, Smith took twenty men to the villages of Chickahominy to demand corn. At first, "that dogged nation was too wel acquainted with our wants, refusing to trade, with as much

scorne and insolencie as they could expresse," Smith recalled. He ordered an assault on Chickahominy, forcing the villagers to deliver "corne,
fish, fowl, or what they had to make their peace." However, "their corne
being that year bad they complained extreamely of their owne wants."
Smith followed this expedition with another to the Nansemond villages.
"These people also long denied him trade, (excusing themselves to bee
so commanded by Powhatan)," until the English threatened force, "and
then they would rather sell us some, then wee should take all."[97]

With arrival of the second supply and the "coronation" ceremony
that autumn, Wahunsenacawh meant to shut down this exchange altogether, pressuring his subject chiefdoms to embargo the struggling
colony. He would use the threat of starvation to bring Smith and other
settlers into line and force them to deliver what the chief really wanted:
English weapons.[98] Following Smith's trip to Nansemond, Wahunsenacawh invited the colony's president to bring him a grindstone, fifty
swords, and some firearms in return for corn.[99]

Unwilling to comply but unable to refuse the invitation, Smith took
a party of Englishmen upriver to meet the chief. A few days into their
journey, "the extreame wind, raine, frost, and snowe" forced them to
wait out the weather among the villages of the Kecoughtan Indians.
On January 22, they finally reached Werowocomoco, "where the river
was frozen neare halfe a mile from the shore." They broke through the
ice as far as they could, and then Smith supposedly forced the men "to
march middle deepe more than a flight shot through this muddie frore
ooze." (Unfortunately, we have only Smith's word to go upon for all his
supposed heroics during the voyage.)[100]

Wahunsenacawh feasted the cold, hungry Englishmen for a day.
Then he dismissed them, "faining hee sent not for us, neither had hee
any corne, and his people much lesse." The chief might trade them
bushels of corn for an equal number of weapons, "but none he liked
without gunnes and swords, valuing a basket of corne more precious
then a basket of copper, saying he could eate his corne, but not his
copper."[101] After an exchange of thinly disguised threats and accusations, Smith concluded that Wahunsenacawh "but trifled the time to cut
his throat," and he ordered some Indians and Englishmen to help break
a way through the ice to fetch their boats.[102] In the meantime, two
German craftsmen with the Jamestown colony had already colluded
with Wahunsenacawh to smuggle English weapons to Werowocomoco.

Returning downriver, Smith and his men stopped among the Pamunkey Indians. There they encountered Wahunsenacawh's kinsman Opechancanough, who supposedly hatched a plot to ambush Smith and the other Englishmen. Smith (once again, in his own telling) boldly rescued himself and his comrades by seizing Opechancanough and taking him hostage until they all escaped. A few days later, Smith claimed, Indians allied with Wahunsenacawh delivered poisoned food to the Englishmen. Whether these attempts were real or imagined, they marked a decisive breakdown in Anglo-Powhatan relations.

Even had the Indians been more willing to trade food, they had less and less to offer. The drought had diminished the harvest once again. Now another freezing winter—almost as cold as the last—was making it difficult to forage. "Men may thinke it strange there should be this stir for a little corne, but had it been gold with more ease we might have got it; and had it wanted, the whole collonie had starved." Heading upriver again that January, Smith recounted, "we searched also the countries of Youghtanund and Mattapamient, where the people imparted that little they had, with such complaints and tears from women and children; as he had bin too cruell to be a Christian that would not have bin satisfied, and moved with compassion." Meanwhile, in what Smith called "that extreame frozen time," ten of the colonists died attempting to cross the James River to Hog Island.[103]

Smith and his company brought back enough food that January to keep Jamestown fed for a few months, so "the feare of starving was abandoned."[104] He also claimed to have put all the settlers back to work and to have restored order to the fort. Mortality that winter was much lower than the year before. Smith, however, had openly spurned the Virginia Company's directives and was alienating prominent members of the colony. He had also alienated the colony's Indian neighbors, on whom they depended for sustenance and survival. Jamestown's best hopes now lay in new funds and supplies from England. Fortunately for the colony, disasters elsewhere in North America would make even Virginia look like a promising investment by comparison.

7

〜

Our Former Hopes Were Frozen to Death

During late 1606 and early 1607, while the first Englishmen sailed to Jamestown, the weather in Europe turned eerily warm and dry. In parts of Germany, the flowers bloomed in February. Coming after decades of cold, wet seasons, it seemed to some that this year there was "no winter" at all.[1]

That suddenly changed in late 1607, when the continent plunged back into some of the worst cold in generations. The winter of 1607–1608 has gone down in history as one of Europe's "great winters," bringing Arctic cold, snow, and ice. In the Netherlands, the freeze began in late December and continued with few interruptions into late March. Horses and sleighs traveled over the Zuiderzee from Haarlingen to Enkhuizen, and the extraordinary sight would inspire some of the most famous winter landscape paintings of the era. Even Spanish diplomats traveled by sleigh over the ice to broker their truce with Dutch rebels in early 1608. By late winter the rivers were solid and the ground lay under sheets of ice. Birds froze to death; livestock and wild animals starved; fruit trees perished of frost. "In short," Dirk Velius observed from Hoorn, "it was a winter whose like was unheard of in human memory."[2]

His sentiment was echoed all across Europe. In Ireland, wrote one chronicler, "in the winter of this year was a great frost, which began a little before Christmas, and continued till about Midlent. This frost increased with such fervency of cold that all things which grew above the ground died and starved with cold; many beasts, both wild and tame, died and starved with hunger, and so did great numbers of wild fowls. . . . The rivers for the most part throughout all Ireland were so covered over that the people might go to and fro upon the ice as upon dry land."[3] In Germany, heavy rain and flooding in early winter soon gave way to ice: the Rhine froze all the way up to Cologne, and the Main iced up all the way past Frankfurt. "Not only the vineyards but even the trees in warm valleys froze," noted one contemporary.[4] From

Prague, the Venetian ambassador reported snow, ice, and "the greatest, most extraordinary cold."[5] In France, the Loire River froze, and ice floes choked the Rhone. The Seine iced over for nearly two months, so carriages could pass across. Communion wine froze in the churches of Paris and had to be thawed out for mass. Hundreds perished from cold and hunger in the streets of French towns and cities. "The cold was so extreme and the freeze so great and bitter, that nothing seemed like it in the memory of man," wrote the diarist Pierre de l'Estoile.[6]

Even the Mediterranean did not escape. Spain faced bitter cold, frozen rivers, and snowfall as late as May 1608.[7] Northern Italy suffered one of its coldest winters of the Little Ice Age. In Florence, continual rains during December turned to snow and ice throughout January and February; in Milan the ice and snow were supposedly so bad that people could barely go outdoors. In Rome, heavy rains brought frequent flooding of the Tiber.[8] Lakes and rivers froze in Greece. In Anatolia, still ravaged by the Celali rebellion, extreme cold and drought induced widespread famine and reportedly cannibalism.[9]

The most famous image of that winter remains the frost fair on the frozen Thames in London. In December 1607, ice began to pile up at the old London Bridge. The floes began to freeze together about a week before Christmas, and within three weeks the river turned solid from bank to bank. A few brave souls ventured out on the ice, and soon shops, food stalls, and impromptu parties appeared in the middle of the frozen river. "Many fantasticall experiments are dayly put in practise, as certain youths burnt a gallon of wine upon the ice and made all passengers partakers," wrote one contemporary, "but the best is of an honest woman (they say) that had a great longing to have her husband get her with child upon the Thames."[10]

Even in England the winter was not all fun and games. The year 1607 had already begun badly. In January, a tremendous flood of the River Severn had drowned thousands of people and cattle, an event that excited wonder and dread across the country and beyond. In April, long-standing grievances against rising food prices and the enclosure of common lands erupted in rural riots known as the Midland Revolt. Soldiers promptly crushed the ragtag army of "levellers" and "diggers" that June, but resentments lingered.[11] Anger and despair over continuing high food prices are thought to have inspired Shakespeare's descriptions of rebellious plebes in the opening act of *Coriolanus*.[12]

The crisis revealed King James I at his most petulant and insensitive. A series of royal proclamations castigated the rioters and denied any connection between the hunger and rebellion. He prohibited farmers from feeding peas to pigs, brewers from using more malt for beer, and even gentlemen from using starch in their lace collars, claiming these measures would spare enough food for the poor. Other proclamations targeted those hoarding and speculating on grain and the new wave of migrants coming to towns and cities in search of employment or relief.[13] But these steps, and even grain imports from the Baltic, failed to hold back rising prices and hunger. The exceptional cold of early 1608 ruined the wheat crop. An Englishman recalled that as far south as Devon, "an extreme dearth of corn happened this year, by reason of extreme frosts (as the like were never seen), the winter going before, which caused much corn to fall away."[14]

Thomas Dekker penned a satirical almanac that mocked the miseries of the year:

> When Charitie blowes her nailes, and is ready to starve, yet not so much as a Watchman will lend her a flap of his freeze Gowne to keepe her warme: when tradesmen shut up shops, by reason their frozen hearted creditors goe about to nip them with beggerie: when the price of Sea-cole riseth, and the price of mens laboures falleth when everie Chimnye castes out smoak, but scarce any dore opens to cast so much as a marlbone to a Dog to gnaw: when beasts die for want of fodder in the field, and men are ready to famish for want of foods in the citie . . . [15]

Bubonic plague, which had flared up from time to time after the great outbreak of 1603, now made new inroads among the poor, hungry, and vagrants. "You have heard before of certaine plagues, and of a Famine that hangs over our heads in the cloudes," Dekker added wryly. "Misfortunes are not borne alone, but like married fooles they come in couples."[16]

Dekker's epigram was just as fitting for England's twin colonies in North America. It is often forgotten that Jamestown was only one half of the original Virginia Company venture. Its charter of April 1606 called for a pair of colonies: one to the south, with claims from present-day North Carolina to New Jersey, and one to the north, with claims ranging from Delaware to Maine. The former—that is, the Jamestown colony—drew London investors hoping for Mediterranean commodities and lured by rumors of gold-bearing mountains and the Verrazzano Sea. The latter attracted West Country investors looking for a land and climate more like England's, offering goods such as timber and fish, besides the perennial promise of precious metals.[17]

In ordinary times, and given Jamestown's desperate plight, the northern colony might have been the more promising of the two. Poor planning, conflict with the indigenous Wabanaki, and the extraordinary winter of 1607–1608 brought it to an untimely end instead. Around the same time, the voyages of Henry Hudson would bring back new descriptions of extreme Arctic cold and diminish hopes of finding a passage to the Pacific through Canada or New England. These failures would have lasting consequences for English exploration and settlement in North America. For all that the Jamestown colonists suffered during their first winters, experiences farther north would make Virginia—once feared as too tropical for the English—look like the most viable option.

The land that would become New England first entered the European consciousness as "Norumbega." The word is thought to be a corruption of "Oranbega," the name given to Penobscot Bay on an early map of the Giovanni da Verrazzano expedition of 1524. The Italian captain's impression of the region was unfavorable: the same map labels it "the land where there are bad people," probably indicating his difficult exchanges with the hunter-gatherers who lived in those northern latitudes, as opposed to the farming communities he had met before. Verrazzano described the coast as "full of very dense forests, composed of pines, cypresses, and similar trees which grow in cold regions." "We saw no sign of cultivation," he added, "nor would the land be suitable for producing any fruit or grain on account of its sterility." Another voyage that year, by Portuguese explorer Estevão Gomes, would chart the Maine coast in more detail, exaggerating its size and giving it the prominent place it would enjoy for another century on maps of North America.[18]

Norumbega came to occupy an even more outsized place in the European imagination. It somehow became "a New World paradise, a rich, cultivated, civilized spot in the northern wilds of America that had somehow been prepared for the coming of the Europeans," as one historian has put it. Various accounts blessed it with "mineral wealth, navigable waters, agricultural fertility, plentiful game, and pliant natives," based on little more than "a few experiential accounts, passages of recorded hearsay, fabrications, political manifestos, and above all maps." In 1545, Pierre Crignon declared, "The inhabitants of this country are docile, friendly and peaceful. The land overflows with every kind of fruit; there grow the wholesome orange and the almond, and many sorts of sweet-smelling trees." His fellow French cosmographer Jean Alfonse wrote of "a city called Norombegue with clever inhabitants and mass of peltries of all kinds of beasts. The citizens dress in furs, wearing sable cloaks."[19]

During the late sixteenth century, those descriptions grew only more vivid and fantastical. In autumn 1568, coming off a major naval defeat by the Spanish, the privateer John Hawkins put a number of his men ashore on the Gulf of Mexico coast north of Veracruz. One of them, David Ingram, was found a year later in present-day Nova Scotia. Questioned about his adventures, Ingram told his examiners everything they wanted to hear and more. He claimed to have walked the whole way along the coast through kingdom after wealthy kingdom, city after fabulous city. The men were carried in silver chairs, clad in bright colors with giant rubies and other precious stones. The women were draped in pearls and wore plates of gold. The land of Norumbega was wonderfully fertile, growing tropical crops such as cassava. He warned of "the heate of the Climate," which brought tempests and hurricanes. Around the same time, optimistic renderings of the North American coastline began promising another passage to the Pacific somewhere around Maine.[20]

Norumbega kept its allure in part because few Europeans actually set foot there, and then only in the summer. The great resource of the North Atlantic remained the Newfoundland cod fisheries. The abundance of sea life on the New England coast was then hardly known to Europeans. Networks of trade between European manufactures and American furs had begun to spread into the Northeast, but these networks remained in the hands of Native Mi'kmaq intermediaries, already experienced at trading with Basque whalers and French fishermen.[21]

The realities of the Maine coast began to take shape only with a series of small expeditions during the first decade of the 1600s.[22] The earliest to have left a historical record set out in April 1602 under the command of Bartholomew Gosnold. A Suffolk gentleman and former privateer, Gosnold would eventually die in Jamestown that fatal summer of 1607. The discovery of his grave in 2002 has been one the great finds of recent Jamestown archaeology. Gosnold's earlier expeditions in New England have been less well remembered.

The captain took a single small ship with only thirty men, and made swift passage directly across the North Atlantic. They sighted land around the southern coast of Maine, continued down to Cape Cod (which Gosnold named), and in late May sailed into Nantucket Sound past Martha's Vineyard (another Gosnold moniker). In early June they settled on the site of Cuttyhunk Island, where twenty men were to set up camp for the winter.

While off Cape Cod, the expedition had first encountered a shallop of Mi'kmaq traders, some dressed in European clothes, who "spake divers Christian words."[23] Farther on, the only Indians they met had had little or no prior contact with Europeans. The two surviving English accounts suggest that relations began amicably, but after a few days on the island, incidents of theft and fighting raised concerns. The expedition had come undersupplied for its mission, or perhaps its leaders had assumed all along they would live off the land and trade with Indians for food. In any case, the intended colonists were told the ship could spare them only six weeks' provisions. Most immediately refused to stay, and the others soon followed suit.[24] After loading their ship with timber and sassafras, prized as a medicinal root, the entire expedition sailed back to England in late June.

Although it failed to find the purported riches of Norumbega or plant a lasting colony, Gosnold's expedition was encouraging enough to attract funds for another New England venture. Led by Richard Hakluyt and some prominent Bristol merchants, backers of a new expedition raised £1,000 and prepared two small ships under the command of Captain Martin Pring.[25] Departing in late April 1603, they reached land around Penobscot Bay, and turned south to explore the contours of the coast, eventually sailing around the tip of Cape Cod. They were impressed by the forests and "excellent fishing for Cods, which are better than those of New-found-land."[26] Our main surviving account of the expedition, thought to have come from Pring's journal, describes friendly encounters with the Massachusett Indians. Yet the account also casually mentions how the English loosed their terrifying mastiffs "when we would be rid of the Savages company."[27] After about three weeks of exploration, they passed the rest of the summer around present-day Provincetown, taking note of the region's commodities and experimentally planting a few crops. However, the expedition's real priority was sassafras: they loaded one boat full and sent it home in early August. About a week later, a party of Indians surrounded the English camp, and the next day they set fire to the adjacent woods, whereupon the remaining ship sailed back to England.

The Pring expedition temporarily glutted the market for sassafras, undercutting the financial justification for another New England expedition. Reports of rich timber and fisheries nevertheless attracted interest among West Country merchants. At the same time, a group led by a Catholic gentleman began to take an interest in Norumbega as a possible refuge for his coreligionists. These interests combined to sponsor a 1605 expedition led by the Devon captain George Waymouth, the first to seek out a site for a permanent English colony in the Wabanaki homeland, in present-day Maine.[28]

The twenty-nine men of the expedition departed England "well-victualled and furnished with munition" at the end of March. They first sighted land off Cape Cod and made their way up to the Maine coast at the end of May. Initially the English-Wabanaki exchanges remained amicable, and included some trade for skins and furs. "Thus because we found the land a place answerable to the intent of our discovery, viz, fit for any nation to inhabit, we used the people with as great kindnes as we could devise, or found them capable of," recalled one member of the expedition.[29]

Relations soured in mid-June. Coming to trade, "as we had accustomed before," a group of Englishmen encountered a large party of Wabanaki with bows at the ready. "These things considered, we began to joyne them in the ranke of other Salvages, who have beene by travellers in most discoveries found very treacherous," he recalled. The English responded to the supposed treachery of the Wabanaki with real treachery of their own, kidnapping five Indians to send back to England as interpreters on future voyages. After that, the expedition quickly explored up the St. George River, reportedly "the most beautifull, large and secure harbouring river that the world affordeth" with "much di-

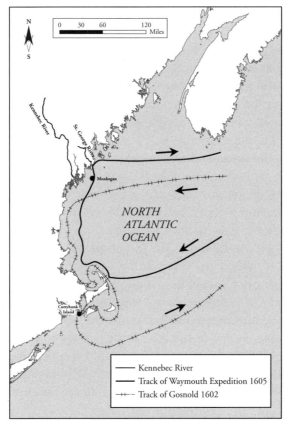

7.1 The Gosnold and Waymouth expeditions

versitie of good commodities."[30] They departed in late June and made a swift crossing back to England.

With these glowing reports, Waymouth sought more funding for a projected colony. Eventually the venture caught the attention of England's lord chief justice, John Popham. Motivated by fears of English overpopulation and poverty, Popham not only subscribed £500 per year but also helped organize the venture and charter the Virginia Company in April 1606.[31]

The recent expeditions had brought back a much better sense of the coastline, flora and fauna, and, to a lesser extent, the Native peoples of

the region that came to be called either "the north part of Virginia" or
(with an eye to colonial promotion) "New England." Published descriptions exaggerated its praise and potentials. The land was "the goodliest
Continent that ever we saw."[32] It overflowed with vines, tobacco, sassafras, and other commodities.[33] Yet exaggerated as they were, these
accounts were not the tall tales of rich kingdoms and tropical fruits
once associated with Norumbega. They gave some feel for the actual
environment of the country.

However, the voyages of 1602–1605 never gave the English a real
sense of the region's climate. Each had stopped over for only a few
weeks during the pleasant months of late spring and summer—probably milder summers during that cold decade of volcanic weather than
the summers we know today. Each left before the first sign of a New
England winter. As historian David Quinn put it, with some under-
statement, "It is clear that the visitors conveyed the impression of a
warmer and more pleasant England to their readers and raised hopes
which were not to be satisfied when year-round settlement took place."[34]

Promoters rationalized that the country must have a wholesome environment and temperate seasons. Some pointed to the fertility of the
soil, the abundance of trees and fruits, and the quickness with which
crops sprouted and grew in late summer.[35] Others reasoned that the
apparent health of New England's Indians meant the climate would be
healthful. "And truely," one observed, "the holsomnesse and temperature of this Climat, doth not onely argue this people to be answerable to this description, but also of a perfect constitution of body, ac-
tive, strong, healthfull, and very wittie."[36] Others still pointed to the
absence of disease among English visitors. James Rosier, on the Waymouth expedition, concluded that "the temperature of the Climate (albeit a very important matter) I had almost passed without mentioning,
because it affoorded to us no great alteration from our disposition in
England."[37] After the failures of the Roanoke and Newfoundland ventures, it was tempting to cast New England as a temperate median between the two—neither too hot nor too cold.[38]

The colony's planners could have learned more about the country
and its seasons from the five Indians kidnapped by Captain Waymouth
in 1605. We know that they ended up in the household of Sir Ferdinando Gorges, one of the principal backers of the colony and one of
our major sources of information about subsequent events. We do not

know how much valuable information the Indians gave him, however, nor how much Gorges and other Englishmen were open to learning. By his own account, Gorges was interested in rivers, navigation, political power, and alliances among the Indians. Evidently incurious about cultures or climates, he lost the chance to gather crucial insights that might have helped the English establish better relationships with the Wabanaki or survive their first winter in New England—assuming his Indian captives had been willing to share it in the first place.[39]

The only indication of New England's stronger seasons appears in a personal letter written by Captain Gosnold to his father. He noted that despite being so far south of England, "yet is it more cold then those parts of Europe, which are situated under the same paralell." They found "the Spring to be later there, then it is with us here, almost a moneth."[40] Otherwise, the published accounts described the region as warm or even hot.[41]

In northeastern North America, it appears that the cold spell that began during the late 1500s never relented. The eruption of Huayna-putina brought even lower temperatures, and these continued more or less the entire first decade of the seventeenth century.

Just how cold remains uncertain. Without the wealth of written descriptions available for early modern Europe, climate reconstruction for early colonial North America has to rely on several types of proxies. This research is not unanimous and it continues to evolve, but the preponderance of evidence at the time of writing is that North American temperatures of the late sixteenth and early seventeenth centuries were truly exceptional. For instance, oxygen isotopes in Greenland ice cores indicate that temperatures throughout the early 1600s remained at their lowest point of the Little Ice Age. Lake sediments from Baffin Island indicate that the coldest and driest conditions of the last millennium began at the end of the sixteenth century.[42]

The strongest evidence for unusual cold comes from the rings of trees in mountains and at the Arctic tree line. Several such studies from northern Canada and Alaska point to falling temperatures during the late sixteenth and early seventeenth centuries. The years following the 1600 eruption of Huaynaputina usually appear among the coldest of the past millennium—in some reconstructions, the very coldest. They look comparable to the infamous "year without a summer" of 1816, when crops failed and it snowed in New England as late as midsummer.

Readers should bear in mind that these kinds of tree-ring data record summer temperatures.[43] And so it is all the more remarkable that most observers commented more on the extreme winters.

The northern venture of the Virginia Company got off to a slower and even less auspicious start than its southern counterpart headed to Jamestown. A first ship under Captain Henry Challons departed in summer 1606. Facing contrary winds and "a greate storme" trying to cross the North Atlantic, Challons had to turn south to catch the trade winds to the Caribbean.[44] While he was sailing off Santo Domingo that November, Spanish authorities seized his ship and imprisoned his crew. Two other English ships, captained by Thomas Hanham and Martin Pring, departed from Plymouth that fall and crossed the Atlantic safely. Failing to rendezvous with Challons as planned, however, they only reconnoitered the Maine coast and returned home.

This delay had important consequences for the whole venture. While reconnoitering, Hanham and Pring discovered that Frenchmen already had designs on the region, and the colony might need to defend itself from French attack. That discovery likely prompted the appointment of George Popham, a veteran soldier and a relative of John Popham, to lead the colony. He was joined by Raleigh Gilbert, the younger son of Humphrey Gilbert, of the ill-fated Newfoundland venture.[45] Furthermore, Hanham and Pring recommended a settlement at the mouth of the Kennebec River rather than the St. George. Finally, the failure of the Challons expedition meant that the whole colony would start later in the year—too late to prepare for the winter.

What would come to be known as the Sagadahoc or Popham colony began with two ships, the *Mary and John* and *Gift of God,* which set out together from Plymouth, England, in June 1607. They carried roughly a hundred settlers and two of the Indians Waymouth had captured in 1605. After some minor mishaps at sea, the ships rendezvoused at Pemaquid, Maine, in mid-August and sailed to the mouth of the Kennebec.[46]

For centuries, the exact location of the settlement was lost. Only in 1888 did historians discover a detailed contemporary map of the fort in Spanish archives, and only in the late 1990s were archaeologists finally able to use that map to locate the site and carry out proper excavations.[47] Much as at Jamestown, standing on the grounds of the former colony gives one a sense of its promises and pitfalls. Sabino Head, as it is called today, is a high promontory on level ground surrounded by pine forests. Its view over the mouth of the Kennebec presents the characteristic charm of the Maine coast. The land had been periodically occupied by Wabanaki Indians, who left behind arrowheads and shell middens later found by archaeologists. A French Jesuit who led a group of missionaries to the site in 1611 remarked, "Now as everything is beautiful at first, this undertaking of the English had to be praised and extolled, and the conveniences of the place enumerated, each one pointing out what he valued the most." "But," he added, "a few days afterward they changed their views." The missionaries saw how the site could easily be cut off by a counter-fort on the next promontory, now the site of the Civil War–era Fort Popham. Moreover, they found scant arable land along the coast, "the soil being nothing but stones and rocks."[48]

Once landed, the English colonists went straight to work building a storehouse for their provisions, a task that occupied most of their time through the end of September. Excavations have confirmed it was a considerable structure, almost 20 by 70 feet (6 by 21 meters). By mid-October they had probably completed Raleigh Gilbert's house and the chapel as well, and trenched and fortified their camp. Then they began constructing a pinnace they would name the *Virginia*, still proudly hailed as "America's first ship" in this traditional shipbuilding region. But the labor proved costly, leaving little time to build adequate shelter for the colonists before the onset of winter.[49] With these steps finished, they sent the *Mary and John* back to Plymouth, bringing word of "their safe arrivall and forwardnes of their plantation" while "importuning a supplie for the most necessariose wantes to the subsisting of a Colony"—a first indication that their expedition lacked provisions.[50] The *Gift of God* stayed behind in case it needed to defend or evacuate the colony during a French attack.

In the meantime, not all went according to plan. Hoped-for precious metals failed to materialize, and exchanges with Indians were difficult, if not (yet) violent. Worse, the weather turned unexpectedly cold. In December, provisions ran low, and travel became impossible as the river filled with ice. The *Gift of God* was forced to leave on the twenty-third, taking more than half the colonists away with it. George Popham sent back an almost farcical letter praising the region's spices and other

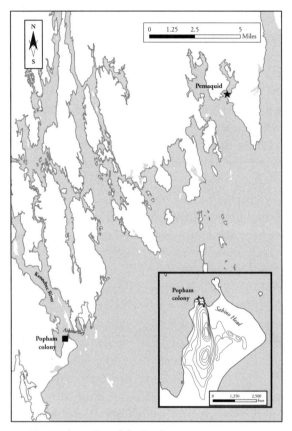

7.2 The Popham (Sagadahoc) colony

tropical commodities.[51] Back in England, Ferdinando Gorges received reports that "the winter, hath ben great, and hath sorely pinched our People."[52]

Following the *Gift of God*'s departure, our detailed written sources from the settlement dry up. Popham, already "ould, and of an unwildy body," apparently passed away in February, leaving the young Raleigh Gilbert in charge.[53] He and the other settlers faced another setback sometime that winter, when a fire burned down Gilbert's house and probably some makeshift lodgings in the fort.[54] English investors, dismayed at the bad news coming out of Maine, put together an expedition "with victualles and other necessaryes" sometime in spring or early summer 1608.[55] That July or August, the *Mary and John* sailed for

Maine again with fresh supplies. It also carried the news that Raleigh Gilbert's older brother had just died, leaving him a fortune.

Based on a later secondhand account, the captain of the *Mary and John* returned to discover George Popham "and some others" in the fort dead of unspecified causes. Otherwise he found "all thinges in good forwardnes." Nevertheless, Raleigh Gilbert had given up on the colony, with "no mines discovered, nor hope thereof," and he was anxious to claim the "faire portion of land" he had inherited back in England. With Popham dead, Gilbert planning to leave, "and the feare that all other winters would prove like the first," the remaining settlers refused to stay in Maine.[56] The survivors reached England by the end of 1608, "returned in a sad plight," as the Spanish ambassador wrote hopefully to Philip III.[57]

Most contemporaries blamed the colony's failure on the extreme cold—what the colonial promoter Samuel Purchas called an "unseasonable winter fit to freeze the heart of a Plantation."[58] Ferdinando Gorges recalled years later how the colonists "were strangely perplexed with the great and unseasonable cold they suffered with that extremity, as the like hath not been heard of since." Between the winter, the shortage of provisions, Raleigh's departure, and the deaths of John and George Popham, "all our former hopes were frozen to death."[59] John Smith later cited "the coast all thereabouts most extreme stony and rocky: that extreme frozen Winter . . . so cold they could not range nor search the Country." Losing their leadership and "finding nothing but extreme extremities, they all returned . . . and thus this Plantation was begunne and ended in one yeere, and the Country esteemed as a cold, barren, mountainous, rocky Desart."[60]

The few historians who have since studied the Popham colony have tried to broaden the range of explanations. Most have acknowledged the "unlucky" winter but pointed to other factors in its failure. The withdrawal of the French from New England in 1607 took away the colony's strategic purpose. The lack of precious metals dried up investment. Deteriorating relations with Native Americans undercut the promise of trade and threatened the lives of settlers. Weak leadership undermined the settlement from within.[61] Yet there are reasons to give contemporaries of the colony the benefit of the doubt. The freezing winter was, if hardly the only reason for the colony's failure, still a crucial factor behind so much that went wrong during the Popham colony's brief existence.

We can infer from proxy climate data and from parallel experiences of that winter in Virginia and in Europe that contemporaries were not exaggerating the cold. Specific details from the historical record back up descriptions of freezing weather. The skies apparently turned "fowle" as soon as early October.[62] In later testimony, witnesses swore that by mid-December ice floes on the lower Kennebec were so thick and struck the *Gift of God* with such force that they could break the planking.[63] Archaeology has also confirmed contemporary accounts of inadequate equipment and provisions—how "their clothes were but thinne and their Diets poore," as Gorges wrote at the time. Excavations of the site have turned up cheaply made supplies and no evidence of substantial shelters for most of the colonists, which supports William Strachey's description of "ill-built and bleak cottages."[64]

The settlement's strategic and economic value also hinged on realities and perceptions of the region's climate. It is not clear whether the French challenge had been an incentive or a liability for the colony in the first place, nor is it clear whether English calculations changed once the French withdrew from the region in 1607. Based on a letter he wrote in early 1608, Gorges thought that the French threat might still be used to drum up royal support for an English presence in Maine.[65] But as the unexpectedly harsh winter chilled English hopes for the region, it altered calculations about whether the French could gain a foothold there, and whether it really mattered if they did. News of recent French failures to colonize New England and Nova Scotia may have allayed English concerns and cast more doubt on the viability of the Maine settlement.[66]

Nor is it clear that the failure to find precious metals would have been decisive had the climate not been so extreme. It seems likely that Raleigh Gilbert, young and impatient, had pinned his hopes on discovering rich mines.[67] Yet when disappointing reports about mineral prospects reached England in late 1607, they did not quell enthusiasm for the venture, nor the willingness of investors to send new supplies.[68] Gorges believed the land still promised furs, fisheries, timber, alum, sarsaparilla, and other worthwhile commodities. After the winter of 1607–1608, however, prospects dimmed that colonists would be willing and able to endure the climate, explore the region, extract the goods, and safely ship them out. Its "multitude of goodlye Rivers and harbours" were less than useful when they froze for months on end.[69]

The question of conflict between colonists and Indians is hard to resolve, given the sparse evidence. Exchanges between colonists and Indians did not start off well. The pair of Wabanaki captives returned from England tried to warn off their fellow Indians from the colonists. One of them attempted to keep the English from even approaching his home village, and when they did, he apparently led the colonists into a tense standoff with armed villagers. In mid-September, when the colonists arranged to meet another group of Wabanaki at Pemaquid in order to conduct some trade, the latter failed to show up. Then in early October, another group of Indians rowed off in a canoe with an Englishman who had been exchanged as a hostage during a trading session. The incident ended without violence, but it contributed to mutual mistrust.[70]

We have little contemporary information about Anglo-Wabanaki relations in the colony in early 1608. Jesuit missionaries visiting the region in 1611 heard from Indians that the English had "maltreated and misused them outrageously" that winter, and how consequently a group of Wabanaki had ambushed the settlers, killing eleven and driving the rest out of their country. However, these Indians would have had good reason to make up such a story. It would have helped them ingratiate themselves with the French, as common enemies of the English, while at the same time warning the French to back off, lest they meet the same fate as the Popham colonists. More important, there is simply no mention of a massacre in any English source, and it would have been difficult to hide a loss of that size, amounting to a quarter of the remaining colonists. Samuel Purchas's account of the colony, written in 1614, mentions only one settler possibly killed by Indians. Otherwise, Purchas suggests that encounters had been nonviolent.[71]

Our best guide to Anglo-Indian relations at the Popham colony comes from a letter written by Ferdinando Gorges in February 1608. In it, he described the settlers "devidinge themselves into factions," each "emulatinge the others reputation amongst those brutish people." This had "much prejudicialled the publique good," because now the Wabanaki were selling their wares very dear, "concealing from us the places, wheare they have the commodities wee seeke." From this description, it appears that the colony's leaders had tried to monopolize the sale of English goods in order to drive up their value, just as John Smith had tried to do in Jamestown. That tactic failed, however, since individuals and groups of colonists competed to cut their own deals with Indians.[72]

There are hints that at least some of the colonists carried on a profitable trade with the Wabanaki. When the *Gift of God* returned in late 1608, the colony supposedly had "many kindes of Furrs obteined from the Indians by way of Trade."[73] Purchas remarked how one Indian traded away 50 shillings' worth of furs for only a straw hat and a knife.[74]

Yet as winter dragged on and the English were "sorely pinched" by hunger and cold, they probably tried to trade away whatever they had in return for food and furs—that is, if they could manage to find Indians to trade with at all. As the colonists discovered, the Wabanaki of the Kennebec dispersed from the coast to "remove their dwellings in Winter neerest the Deere." Purchas described how the Indians of Maine grew "poore and weake" during the winter, but this probably referred to a normal seasonal cycle, not any particular hardship from the cold. As historian Tom Wickman has argued, the Wabanaki thrived in characteristic Little Ice Age winters, when the long-lasting snow cover made it easy to track prey. Centuries of adaptation, including the use of snowshoes, meant that their world opened up and their travel accelerated during the coldest months, while English colonists were trapped in their settlements.[75] Possibly the Popham colonists had more success trading for furs once their new supplies arrived in summer, especially if they planned to finish off their provisions and abandon the colony anyway.

This competition for trade also sheds light on the colony's crisis of leadership. Testimonies by officers on board the *Gift of God* make it clear that by the time they left the colony in December, there was very little food to spare. Even the beer had run out, and those sent back on the ship "were reduced to such penury that they were inforced to drink water," which apparently resulted in three deaths.[76] As at Jamestown, the want of food, alcohol, and housing must have aggravated quarreling and resentment over inequalities and arrogant leadership. From the excavation of the fort, it is apparent that Raleigh Gilbert had imported luxury goods from England and had his house rebuilt immediately after the fire that winter, while most of the colonists continued to suffer the cold in makeshift shelters.[77]

Whatever the reasons for the colony's abandonment, the impressions of that freezing winter had a lasting impact on English North American settlement. For more than a decade, as Ferdinando Gorges wrote, "the Country it selfe was branded by the returne of the Plantation, as being over cold, and in respect of that, not habitable by our Nation."[78] Major investment in New England stalled until the Massachusetts Bay Company of the 1620s. In the meantime, colonial promoters such as Samuel Purchas pushed for more investment in Jamestown, despite the colony's obvious failings. "But your eyes wearied with this Northerne view, which in that winter communicated with us in the extremitie of cold," he concluded his history of the Popham colony, "looke now greater hopes in the Southerne Plantation, as the right arme of this Virginia body, with greater costs and number furnished from hence."[79]

At the same time, Henry Hudson's voyages in search of a new passage to the Indies confirmed the worst fears about Arctic cold and dashed hopes of a sea route through New England to the Pacific. Recent biographies have revived some interest in Hudson's explorations, and research by historian Dagomar Degroot has now placed his voyages in the context of a changing Little Ice Age climate.[80]

In 1607, the Muscovy Company first commissioned Hudson, an experienced navigator, to voyage past the Svalbard Islands above Norway and then straight over the North Pole. Bizarre as it now seems, the notion of a sea route over the top of the earth grew naturally from two widespread notions about climate: that the pole itself was warmer than the high latitudes around it, and that ice came from the land rather than the sea. Putting these ideas together, it seemed logical to look for an ice-free polar ocean, a feature that would periodically appear and reappear on world maps from the late 1500s even into the late nineteenth century.[81]

The unusually warm weather of early 1607 took Hudson farther north than he would ever reach again. Sailing out of London in early May, his small barque *Hopeful* pressed on northward to the high coast of Greenland, where it encountered ice, snow, and dangerous winds. Eventually steering away toward the Northeast, Hudson passed through rains and thick fog to Spitsbergen, the principal island of the Svalbard archipelago. However, he found it "covered with fogge, the ice lying very thick all along the shore." By mid-July, his sails had frozen and the weather turned "searching cold." After Hudson had spent another two weeks exploring the island, reaching past 80°N, the impassable sea ice forced him to return.[82]

On his second expedition with the Muscovy Company the following year, Hudson tried to find a Northeast Passage over Novaya Zemlya. He departed with another small crew aboard the *Hopeful* in early May 1608, still feeling the extraordinary cold of the past winter. While the vessel was sailing over Norway, the air turned dry and frigid, and sailors began to fall sick, "I suppose by meanes of the cold," Hudson recorded in his journal.[83] Pressing on, they reached Novaya Zemlya in early July. They found the weather ashore "calm, hot, and faire" and the island abounding with fowl and marine life; yet their way was blocked by "the great plenty of ice." After a couple weeks of trying to sail through, the expedition turned back, "voide of hope of a northeast passage."[84]

Only a year later, Hudson agreed to undertake the exact same journey again, this time at the invitation of the new Dutch East India Company. For a tiny fraction of what it might cost for a new trading voyage around Africa, the company directors were willing to take a chance that Hudson might yet find a shortcut to the Spice Islands.[85] Hudson's motivations for agreeing were more questionable. Departing Amsterdam in early April 1609, he sailed his ship, the *Half Moon*, up the coast of Norway, where he found seas full of ice and "stormie weather, with much wind and snow, and very cold." Mutiny among the crew reportedly forced Hudson to ignore his instructions and propose another voyage instead: the search for a Northwest Passage at the more moderate latitude of 40°N.[86]

This may have been what Hudson had in mind all along.[87] He had doubtless read the theories of cosmographers and cartographers who insisted there was a more southerly route through the Americas. The legend of the Verrazzano Sea lived on in various guises, as did an equally mythical "Strait of Anian" a little to the north.[88] Early descriptions of the St. Lawrence and other east coast waterways raised hopes of deep channels thrusting into nearby arms of the Pacific. Influential maps of North America presented hopeful pictures of a continent crisscrossed by navigable passages and inland seas.[89] More recently, the "idea had been suggested to him by some letters and maps which his friend Capt. [John] Smith had sent him from Virginia, and by which he informed him that there was a sea leading into the western ocean, by the north of the southern English colony."[90]

And so the *Half Moon* crossed the Atlantic in June, first reaching the Newfoundland banks and then turning south. Our one eyewitness account of the voyage suggests the expedition enjoyed fair or even hot weather as it explored the New England and mid-Atlantic coasts. Finally, it discovered New York harbor in early September: "a very good land to fall with, and a pleasant land to see."[91] Although it seemed a promising site for trade and settlement, Hudson soon found that the river above Manhattan, which would later bear his name, did not offer the hoped-for Northwest Passage. Too short of provisions to over-winter and fearful of another mutiny, Hudson turned back in early October, sailing to England rather than Amsterdam. When officials learned of his return, they immediately ordered Hudson to remain and prepare another voyage for England, this time in search of a mysterious "indraft" or "over-fall" that previous expeditions had encountered south of Baffin Island.[92]

Hudson's final expedition departed in mid-April 1610. It sailed first north to the Faroe Islands and then to Iceland, where they saw an

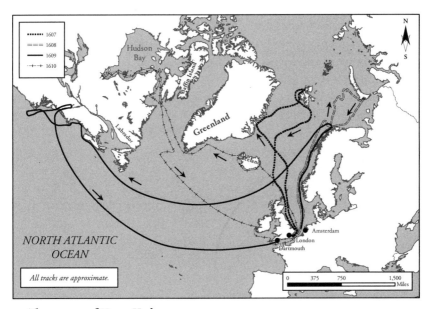

7.3 The voyages of Henry Hudson

eruption of Mt. Hecla: "a signe of foule weather to come in short time."
Turning west, they passed Greenland in early June and continued until
they spied land at Resolution Island, off the southeastern tip of Baffin.
From late June into early July they pressed slowly ahead into the passage
later known as Hudson's Strait, the whole time "troubled with much
ice." Forced onto land by a storm, Hudson apparently showed his maps
to the crew "and left it to their choice whether to proceed any further; yea
or nay." After sharp disagreements, they continued into Hudson's Bay,
working their way south through an icy "labyrinth without end," as one
officer described it. In early November they brought the ship aground in
James Bay, "for the nights were long and cold, and the earth covered
with snow." Within ten days they were frozen in.[93]

"To speake of all our trouble in this time of winter (which was so
cold, as it lamed the most of our company, and my selfe doe yet feele
it) would bee too tedious," wrote the navigator Abacuck Pricket. Hudson
had trusted that by sailing south he would bring them into a warmer
climate, but the truth was just the opposite. The narrow James Bay,
which hangs like an appendix from Hudson Bay, actually freezes longer
and endures more severe continental seasons, with winter lows aver-
aging under −25°C (−13°F). English explorers found its winters so
extreme that their experiences would later feature in scientist Robert
Boyle's New Experiments and Observations Touching Cold (1665). They
would inspire Boyle's observation that the coldest winter weather actually comes from the interior of continents, and not always from the
north, as Europeans assumed.[94]

Hudson's crew forced their way out of the ice that spring. But they
were running short of provisions, and some of the crew still simmered
with resentment at their captain. The story of the mutiny that followed
in June 1611 has been told in detail elsewhere.[95] After breaking into
Hudson's cabin at night and setting him adrift in Arctic waters with
his son and loyal followers, the mutineers struggled through heavy sea
ice to work their way out of Hudson's Bay. Running low on food, they
attempted to trade with some Inuit ashore, but fell into an ambush in
which several of the leading mutineers were killed. (Or at least the
survivors later found it convenient to blame the mutiny on those who
had already died.) Several more members of the expedition perished
of "meere want" on the voyage home. The remaining men resorted to

eating seaweed, animal skins, and "garbage" until their ship was finally rescued off Ireland in September.[96]

Even this miserable outcome would not halt new efforts to find a Northwest Passage. Many more voyages, mostly English, would follow throughout the seventeenth, eighteenth, and even nineteenth centuries. Nevertheless, many promoters and investors must have realized by Hudson's time that there was even less chance of profit in the attempt to sail around North America than in the struggle to colonize it.

8

Winter for Eight Months and Hell for Four

English interest in Virginia also benefited from events halfway across the continent, in the remote and little-known territory that had come to be called New Mexico. In 1587, while promoting the Roanoke colony, Richard Hakluyt had come across an account by one Antonio Espejo, recently returned from a Spanish expedition into that region. Read selectively, it inspired Hakluyt with notions of a favorable climate, friendly Indians, and mineral wealth. Based on his vision of a narrow North American continent, Hakluyt expected the English would find New Mexico just a short distance away through the Chesapeake Bay, "on the backe parte of Virginea," as he put it. Then in 1605, not long before he prepared his instructions for the Jamestown voyage, Hakluyt received news that another Spanish invasion had overawed the Indians of New Mexico and there discovered "certaine very rich Mines of gold and silver." A wealthy, populous civilization was reported to lie not far to the north. Such fabulous rumors could only have added to English enthusiasm for American exploration and inspired confidence that the Virginia colony would soon find gold in the mountains to its west.[1]

Yet even as the first English settlers reached Jamestown, the Spanish conquest of New Mexico was itself coming unraveled. As it turned out, Hakluyt's information had come at just the wrong moment to form a realistic impression of the country. By the 1580s, the bitter disillusion of Spain's first contact with New Mexico had been forgotten. In 1605, reports of new disasters had not yet reached the wider public. Hakluyt may have had no way to know the truth: that a handful of settlers from Mexico and Spain found themselves just then in a land without resources, cursed with a miserable climate, and "remote beyond compare" from any ocean or any outpost of Christian civilization, as many would come to see it.[2] Just as in La Florida, and at just the same time, the whole Spanish enterprise in North America was being thrown into doubt. Once again, confusions over climate and extremes of the Little Ice Age were at the heart of the problem.

No other episode so clearly illustrates the subtle and pervasive role of climate in shaping Europe's encounter with North America as does the Spanish conquest of New Mexico. Geographical mistakes compounded the usual misconceptions about seasons and weather. Those who invaded the country from Mexico took with them vague or misleading notions of where New Mexico was, what to expect there, and just how far it lay from the Atlantic and Pacific Oceans. Lessons learned in one expedition were usually forgotten by the next.

As with La Florida, New Mexico's Mediterranean latitude belied a climate very different from that of Mediterranean Europe. Invaders from Spain often arrived expecting the same cool rainy winters and hot dry summers as in their home country. Instead they found a desert climate with greater variability and seasonal extremes. In New Mexico, air descending from the tropics brings heat and high pressure most of the year, while the Rocky Mountains capture moisture coming from the Pacific. Yearly precipitation varies from only about 10 inches (25 cm) in the desert plains to a few feet at the mountain peaks. About half of that falls in the annual "monsoon" of midsummer thunderstorms, which come to relieve the intense heat of June, when temperatures regularly top 38°C (100°F). Much of the rest of the year's precipitation arrives as snowfall during winter, when temperatures in the uplands can fall below –18°C (0°F), even in our present era of global warming.[3]

During the Little Ice Age, it may have gotten much colder and much drier than that. The American Southwest is the birthplace of dendroclimatology—the study of past climates through tree rings—and it remains one of its leading research centers. In this arid region most tree growth depends on available moisture, and extensive studies of tree-ring width have made it possible to reconstruct past drought in New Mexico at very fine resolution. At high elevations, both the width and cell density of tree rings also vary in part according to spring and summer temperature. Over centuries, the tree line itself has shifted according to long-term warming and cooling. Lake sediments and speleothems (stalagmites) can provide some information about changes in aridity and temperature from decade to decade.

Taken together, this proxy evidence gives some consistent and striking results for the 1500s and early 1600s—results that confirm Spanish descriptions. Tree-ring records indicate that it was unusually cool and wet when Coronado led the Spanish Empire's first invasion of New Mexico in the 1540s. Then in the late 1500s, when adventurers from New Spain rediscovered New Mexico, the region was in one of its coldest and driest periods of the past millennium. This included a "megadrought" across what is now northwestern Mexico and the southwestern United States—the worst since at least the thirteenth century. During the 1570s and 1580s, precipitation failed for years on end in both summers and winters. The turn of the seventeenth century, when Juan de Oñate invaded and conquered New Mexico, saw two simultaneous disasters: the years 1598–1601 brought another major drought to the upper Rio Grande basin, and the first decade of the 1600s was probably the coldest of the past millennium in the western United States. The colonists of Oñate's expedition had the misfortune to feel the full climatic impacts of the 1595 Nevado del Ruiz and 1600 Huaynaputina eruptions.[4]

Even in the best of times, New Mexico's desert climate would have disappointed the intended conquerors and colonists of the Spanish Empire. Like La Florida, it could never provide an ideal home for Mediterranean crops and livestock—the "New Andalusia" that Spaniards hoped to discover in the New World. Nor was it a land with the population and agriculture to offer rich encomiendas, or assignments of Native labor and tribute to the conquerors. It presented no easy mineral wealth and lacked ready resources to support a large mining community. In these respects, the expeditions were bound to fail. Nevertheless, it was accidents of Little Ice Age climate that turned mere failures into disasters, both for the invaders and for New Mexico's Native peoples, particularly the nations known as Pueblos.

The Pueblos of the contact period represented diverse communities united by some common cultural traits. They cultivated maize and beans, raised turkeys, and grew cotton where the climate permitted. Most striking to the Spaniards were their *pueblos* (towns) of flat-roofed houses stacked on top of each other, built for defense against enemies and the elements. The presence of these tight-packed communities surrounded for many miles by mostly nomadic hunter-gatherer peoples gave rise to rumors of exaggerated wealth and numbers. Estimates of their population during the sixteenth century range from only 40,000 up to 100,000, with no sign of impacts from Old World pathogens.[5]

Unlike Spaniards, Pueblo Indians had for centuries adapted to the climate and environment of New Mexico. They stored enough food to last out one or two years of crop failures. Some pueblos dug irrigation channels, while others relied on rain-fed fields or a mix of the two. They supplemented their diets through hunting and trade with bison-hunting nomads. They wove warm cotton *mantas* (blankets or shawls) and robes of turkey feathers, and traded for bison hides.[6]

These adaptations did not render the Pueblos immune to all the vicissitudes of New Mexico's climate, however. Fields planted with maize in the spring could turn to dust before the late summer harvest if the monsoon storms failed. Irrigation channels could run dry without winter snows to refresh the rivers and streams. Corn and cotton in the uplands faced the threat of early or late frosts, particularly during the Little Ice Age, and particularly where the growing season might last little longer than the time it took for maize to reach maturity.[7] The study of human bones from precontact pueblos reveals high mortality, episodes of acute malnutrition, and chronic anemia.[8]

Nor did past centuries of adaptation prepare the Pueblos for the invasion of hundreds or thousands of soldiers and settlers from Spain and Mexico. The Pueblo world had not been a peaceful place even before Spanish soldiers arrived. Nevertheless, Spanish imperial invasions brought a new kind of warfare and made new kinds of demands on the land and its peoples. This combination of violence, climatic extremes, and ecological pressures underlay much that went so badly wrong during the conquest of New Mexico.[9]

In 1536, eight years since the Pánfilo de Narváez expedition had departed for La Florida, soldiers on New Spain's northwestern frontier came upon its last four survivors. "They experienced great shock upon seeing me so strangely dressed and in the company of Indians," Álvar Núñez Cabeza de Vaca later testified. "They remained looking at me a long time, so astonished that they neither spoke to me nor managed to ask me anything." Brought before Viceroy Antonio de Mendoza, Cabeza de Vaca and his three companions gave their testimony.[10]

Their travels had taken them hundreds of miles through lands that probably no European had ever seen before. They had lived for years as captives among Indians of the Texas Gulf Coast. They had watched some of their comrades die of hunger and mistreatment, and others disappear trying to reach New Spain. Those four—including Cabeza de Vaca and a Moroccan slave known as Esteban—had finally escaped together in the summer of 1534. Among the Indians of south Texas they gained a reputation as healers or shamans, and that reputation earned them a privileged place among the peoples of the region, who were then suffering from mysterious ailments. Indians of various nations led the four men south and west across the Rio Grande. Then, for reasons unknown, Cabeza de Vaca and his comrades did not continue into New Spain, but traveled northwest as far as the meeting of the Rio Grande and the Rio Conchos. From there, Native guides took them west into the mountain valleys of Sonora and finally south to their encounter with Spanish soldiers.

The survivors' testimonies helped rekindle Spanish imperial interests in the vast unknown territories of North America, including La Florida. But what most captured the imagination of would-be conquerors and colonists were rumors of a wealthy, populous civilization somewhere north of Sonora: a place called Cíbola, with metalworking, turquoise, and "pueblos with many people and very large houses."[11]

During the years of Cabeza de Vaca's travels, New Spain's northwestern frontier had become a violent place, even by the standards of the Spanish Empire. Raids, pillaging, and slaving had brought the borders of New Spain all the way up to Culiacán in today's Sinaloa. The infamous excesses of Nuño Beltrán de Guzmán, governor of the new state of Nueva Galicia, had recently ended in his arrest for treason. Viceroy Mendoza meant to hold the discovery of this "new land" described by Cabeza de Vaca to higher moral standards. And so in late 1538, he instructed a Franciscan friar, Marcos de Niza, to take the slave Esteban and "reconnoiter the farthest extremities of this continent which stretches to the north."[12] Fray Marcos was to travel as an apostle of peace, Christianity, and the benefits of Spanish rule, and to report back on the land's "quality and fertility, its temperateness, the trees, plants, and domestic and wild animals."[13]

Historians still disagree on how close Fray Marcos ever came to Cíbola. In his official report, he testified their expedition made rapid progress, with Esteban and his party of Indians in the vanguard. Along the way, a succession of Native peoples welcomed them with food and answered all their questions about Cíbola: a land of seven rich cities, tall houses, and civilized peoples who wore cotton clothes. In late May 1539, Esteban approached the first city of Cíbola in his customary guise as healer or shaman. But instead of his usual welcome, he was taken prisoner and then killed as he tried to escape. At that point, Fray Marcos claimed to have traveled ahead to just within sight of Cíbola, and to have a spied a place "grander than Mexico City."[14] Back in New Spain that August, he spread rumors of a fabulous civilization, where the women carried gold necklaces and men wore belts of gold.[15] By holding out the prospect of precious metals and civilized subjects, his reports all but guaranteed the country would be among the next targets for one of the empire's ambitious conquistadors.

In January 1540, that privilege fell to the next governor of Nueva Galicia, Francisco Vázquez de Coronado y Luján—better known in English simply as Coronado. The popular legend of the Coronado expedition depicts a small band of Spaniards in a romantic quest for cities of gold. The reality was a large and expensive enterprise of conquest. Coronado himself, the viceroy, and hundreds of Spanish and creole soldiers of fortune invested their life's savings—hundreds of thousands of pesos altogether—in the hope of carving out encomiendas from the conquered population. They brought along hundreds more women, servants, slaves, and above all their *indios amigos,* or Mexican Native allies, who composed most of the expedition and performed most of the work and fighting.[16]

Signs of trouble came well before Coronado even reached New Mexico. In late 1539 the viceroy had sent a second reconnaissance to check on the reports of Fray Marcos. Once that mission had traveled a hundred leagues (about 300 miles or 480 kilometers) north of Culiacán, they "began to find the land cold and to experience hard frosts." Some of the Indian allies froze to death, and two Spaniards barely survived. "It is impossible to cross the unsettled region there is between here and Cíbola because of the excessive snow and cold," they reported. Forced to wait out the winter near the present Mexico-U.S. border, they learned from local Natives that the pueblos of Cíbola were only "crudely worked buildings made of stone and mud," with some fields of maize and beans, in a land too cold even to grow cotton.

The towns of the Rio Grande valley were just "the same sort [of place]" except a little more populous.[17]

Coronado encountered that reconnaissance mission on its way back to Mexico City in early spring 1540. Consequently, he decided to leave most of the baggage and settlers behind at Culiacán in late April, and to set off north with a swifter detachment of soldiers and Native allies.[18] His lightly provisioned advance party took almost eighty days to reach their destination. "During that [whole] time God knows we lived on reduced rations," one Spaniard later recalled. Black and Indian members of the expedition "died of hunger and thirst" along the way.[19]

What Fray Marcos had called Cíbola turned out to be the Zuni pueblos in the western part of today's state of New Mexico. By the time Coronado's party encountered the first of these, Hawikku, the invaders' provisions were nearly or entirely spent. "We did not have [anything] to eat unless we took it from them," one Spanish soldier recalled. Careful to follow official procedure, Coronado had a chain of translators read out the *requerimiento,* or official order demanding obedience to the Church and Crown: undoubtedly a bizarre and threatening display to the Zuni. According to later accounts, their warriors responded by trying to warn off the invaders, shooting arrows and throwing stones. "But because there is no way to resist the utmost fury of the Spaniards," wrote the soldier Castañeda de Nájera, "in less than an hour the pueblo was entered and taken. Food supplies were found, which is what there was utmost need for." Even Coronado confided in a letter to the viceroy that he had ordered an immediate attack "because the hunger we were suffering did not permit delay."[20]

The rest of the Zuni pueblos soon surrendered and offered up provisions to the invaders. Nevertheless, Coronado remained sorely disappointed with Cíbola, its climate, and its lack of precious metals: all "completely the opposite" of what Fray Marcos had described.[21] While waiting for the main body of the expedition to arrive from Culiacán that autumn, the general sent out scouts to follow rumors of richer lands to the west and east. Finally in December, having assembled his followers and baggage train, Coronado sent the entire expedition to winter in the land they called Tiguex—that is, the Tiwa pueblos along the Rio Grande, not far from present-day Albuquerque. "It did not stop snowing in the late afternoons and almost every night, so that in order to establish quarters wherever they reached, they had to pry up a cubit of snow and more," one soldier later recalled of the journey to Tiguex. "When it fell all night, it covered the baggage and the men-at-arms in their bed[roll]s in such a way that if someone had come upon the camp unexpectedly, he would not have seen anything other than mounds of snow and horses."[22]

By the time they set up winter quarters, freezing weather threatened the expedition's very survival. The snow continued for months on end.[23] The winter of 1540–1541 apparently brought more persistent cold than any recorded in modern instrumental records. One soldier described "such hard frosts that the river froze over one night and remained so more than a month. Loaded horses crossed over on top of the ice."[24]

In a sign of the enormous geographical confusion that beset the expedition, relief supplies had been sent by sea rather than land. The ships tried their best to reach New Mexico by sailing up the Gulf of California and into the Colorado River, only to discover that they were still more than a week's journey overland from their destination. The ships had been sent specifically "in order to carry the clothing the men-at-arms could not take [with them]," as Castañeda de Nájera recalled, "so that all of the clothing went astray." The Native allies, less well equipped and camped in a light makeshift shelter, suffered most of all from the cold: "[They] were in such peril because they were from New Spain and most of them from hot climates."[25]

In the meantime, the brutality of Spanish officers embittered relations with the Pueblos. Some chained and tortured prominent men among the Cicuique (the Towa-speaking pueblos of the upper Pecos River). Another officer raped a Tiguex woman. Indians of the middle Rio Grande retaliated by killing Spanish horses and livestock. They then fortified themselves in two of their pueblos to resist the invaders. Spanish and Mexican soldiers stormed the first and gave its people no quarter, even burning men alive. The second surrendered after a long siege that ended in another massacre. The invaders destroyed another ten or so abandoned Tiguex pueblos in early 1541. Word of these "great cruelties" would soon reach as far as Emperor Charles V, who called for a criminal investigation of Coronado in 1544.[26] The testimonies at his prosecution reveal the underlying role of New Mexico's unexpectedly harsh climate and extreme winter in the outbreak of violence.

Shelter was the first bone of contention. Even before the expedition arrived in the Rio Grande Valley, Coronado had sent ahead the *maestre de campo* (quartermaster) to find winter quarters. Realizing it was too cold to bivouac outdoors, the officer cleared out one pueblo to make room for Spanish soldiers. After the outbreak of hostilities, there are indications the Native allies provoked conflicts or chased away Pueblos in order to seize their houses and burn their timbers for warmth. One testimony stated that they set fire to pueblos "to provide insurance against the cold, which was extreme, especially for people without clothing, [as] everyone in the camp usually was." Coronado claimed in his defense that his expedition was "dying of cold," and they sought abandoned pueblos and "would have burned them and the wood to save themselves."[27]

Shortages of food, fodder, and fuel sparked more hostilities. The Tiguex pueblos simply did not have the ecological capacity to feed the stomachs and campfires of nearly 2,000 invaders for months on end. As archaeologists have observed, Indians already used most of the wood from New Mexico's sparse lowland forests, and they relied on their cornstalks for extra fuel. The long winter also made it difficult to find enough pasture for the invaders' animals. Spanish soldiers aggravated fuel shortages by grazing their horses in harvested maize fields—a common practice in parts of Spain, but an outrage among Pueblos, which some witnesses blamed for the Tiguex uprising.[28]

Most testimonies, however, blamed the struggle over clothing and blankets to keep alive through the freezing winter. One witness at Coronado's prosecution recalled how "many of the people of the company were without clothing and, because it was the month of December, were suffering the cold badly." The *maestre de campo* approached what he took for the leaders of the Tiguex pueblos and demanded that they deliver up *mantas* as tribute; as another witness described, "it was done very clumsily and against the will of the natives." The Pueblos had no time to prepare more clothing, and Spanish soldiers took whatever caught their eye, leaving Indians to suffer in the cold. "It is certain that they rose up because persons from the company went to their houses and, against their will, seized and carried off hides and mantas," one soldier testified at the inquiry. "With regard to [the uprising] in Tiguex province," recounted another, "it was said publicly that it occurred because [Spaniards] took provisions, clothing, and hides from their houses against their will."[29]

In spring 1541, after the suppression of the Tiguex uprising, Coronado set out with a detachment of soldiers to follow rumors of a wealthy, populous civilization far to the east, at a place called Quivira. That journey took them deep into the Great Plains of Kansas past "such a multitude of bison [*vacas*] that to count them is impossible," as Coronado wrote to the king. Quivira turned out to be a modest Indian town "as uncivilized as all those I have seen and passed until now." After seven weeks of travel, with supplies running low, the Spanish officers confronted their Pueblo guide and probably tortured him into confessing that he had misled the Spaniards all along, so they would get lost and die in the plains.[30]

Returning to the Rio Grande in late summer, Coronado found his expedition in even worse shape than before. According to the defense testimony at his trial, the horses were dying, men falling sick, food running out, and "the whole army in danger of being lost because of the unproductiveness of the land and shortage of medicine, foodstuffs, and clothing with which to cover their bodies."[31] Some blamed the illness on the cold, but a more likely culprit is typhus: a deadly fever spread by lice, and the bane of sixteenth-century soldiers, prisoners, and famine refugees. According to Castañeda de Nájera, the soldiers were still trying to take more clothing from the Pueblos during the summer "because the men-at-arms at this point were going about very poorly dressed, in a poor state, and full of lice. Obviously, they were not able to eliminate or get rid of the lice."[32]

By that autumn, Coronado had given up on Tiguex as too remote and too poor to conquer. In a letter to the king, he complained, "The land is so cold that it seems impossible that we can pass the winter in it, because there is no firewood or clothing with which the men could protect themselves from the cold, except hides with which the natives could clothe themselves and some cotton mantas in short supply."[33] He planned to hold out until the spring and then gamble on one more journey of exploration even deeper into the plains, in the hope that those rumors of a rich and populous Indian kingdom turned out to be true after all.

"[But] because nothing in this life is under the control of humans, but rather under the direction of all-mighty God, it was His will that ours did not come to fruition," wrote Castañeda de Nájera. Just before

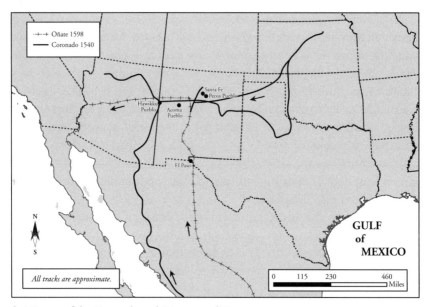

8.1 Routes of the Coronado and Oñate expeditions

setting out on his journey, Coronado suffered a serious injury in a fall from his horse. During his "long and fearful" convalescence, Coronado's officers clamored to return to New Spain, "because nothing of wealth had been found and there was no settlement in what had been reconnoitered where *repartimientos* [i.e., assignments of Indian labor] could be made to the whole expedition."[34] When news arrived of an Indian uprising in Nueva Galicia, it provided the justification Coronado needed to give up and turn back to Mexico. His trial three years later ended in acquittal, and the verdict affirmed Coronado's decision to abandon New Mexico, because "some lands were unsettled by people and others were uninhabitable, being very cold and sterile, lacking foodstuffs, and unsuitable for this purpose."[35]

Just like Hernando de Soto's failed invasion of La Florida, and at just the same time, the return of Coronado's soldiers and settlers left an exaggerated impression of North America's harsh climate and sterility. No Spanish expedition would go back to New Mexico for almost three decades.

Yet also like the Soto expedition, Coronado's invasion left a lingering sense of a task unfulfilled and a vague promise of fabulous riches just waiting for another, more determined conqueror. During the 1550s, some rosier descriptions of New Mexico, including Marcos de Niza's fabrications, found their way into print. The initial shock of disappointment faded, and new rumors and embellishments generated new schemes for conquest. As the expedition's veteran Castañeda de Nájera wrote around 1565:

> Very often the things the common people have heard by chance from someone who does not have even rudimentary knowledge of them are turned into things that are greater or lesser than they are. . . . I mention this because some people turn it into an uninhabitable land, others make it contiguous with La Florida and still others with Greater India, which seems no small nonsense.

Looking past the miseries and failures of the invasion, some members of the expedition even found much to admire about the country. "Although they did not find the wealth of which they had been told, they found the beginning of a good land to settle and the wherewithal to search for wealth and to go onward," Castañeda de Nájera concluded. "Their hearts weep because they have lost such an opportunity of a lifetime."[36]

Meanwhile, the nature of imperial interests in northern New Spain underwent a transformation. Coronado had invaded New Mexico not so much to conquer a land but to conquer its people. He meant to pay for the enterprise by encomienda, or the distribution of tribute-paying subjects. During the mid-1500s, growing criticism of colonists' cruel treatment of Indians began to put the practice of encomienda into disrepute. More important, the destruction of New World populations undermined the system from within. Once populous Amerindian towns and villages became extensive Spanish and creole estates, their environments transformed by the invasion of Eurasian fauna and flora, sheep, and wheat. Hopes of conquering another vast indigenous empire with millions of tribute-paying subjects gradually slipped away.[37]

The new wealth of the New World was not to be found in its Native peoples but in its mines of gold and especially silver. In 1545, according to popular legend, a Native of Peru on the mountain of Potosí uprooted a tree and found silver ore in the ground beneath. Within a decade, his discovery had launched the largest mining operation in the world and made Potosí a byword for fabulous wealth. "Its surroundings are dry, cold, and very bleak and completely barren," wrote the scholar José de Acosta in 1590. "But the power of silver, desire for which draws all other things to itself, has populated that mountain with the largest number of inhabitants in all those realms; and silver has made it so rich in every sort of foodstuff and luxury that nothing can be desired that is not found there in abundance."[38]

As Acosta's description attests, mining and processing New World silver proved a hungry operation: hungry for supplies, provisions, human and animal labor, and ultimately humans themselves. Thousands of tons of food, fodder, timber, fibers, and iron tools had to come by pack mule and wagon to feed ever larger and more sophisticated operations. Tens of thousands of workers—free, slave, and drafted—toiled in difficult and dangerous conditions. The mercury amalgam process used to smelt the ore released so much of that toxic metal into the air that the Potosí operations have left an unmistakable mark in glaciers hundreds and even thousands of miles away.[39]

Both the fame and infamy of the Potosí mines overshadowed the equally productive and equally demanding operations launched in New Spain at the same time. From the 1540s to the 1570s a silver mining frontier extended northwest along Mexico's Sierra Madre, from Guanajuato and San Luis Potosí to Zacatecas and Durango. It drew in tens of thousands more miners, speculators, and settlers, who clashed with the region's Indians. Dozens of smelting operations, also using the mercury amalgam process, cleared the hillsides of wood for fuel. By the end of the century, the silver output of New Spain rivaled that of Peru.[40]

As with any mining frontier, Mexico's silver boom left some men rich and powerful and others bold and desperate. In this situation, the old rumors of mineral wealth in New Mexico found a new and receptive audience among northwestern New Spain's renegades, gamblers, and the next generation of would-be conquistadors.

For decades, circumstances held them back. Starting in the 1550s, the northwest advance of miners and settlers between Guanajuato and Durango met fierce resistance from indigenous nomadic and seminomadic peoples, whom the Spanish called Chichimeca. This conflict gradually wound down in the late 1580s as Spanish officers gave up on outright suppression and offered Indians food for peace. At the same time, the region between Durango and New Mexico was just coming out of the decade-long "megadrought" described earlier in this chapter. Despite being the region's worst drought in perhaps two millennia, it has featured very little in the historiography of colonial Mexico. One study has blamed it for the outbreak of a deadly epidemic in 1576. Its role, if any, in the so-called Chichimeca War remains unclear.[41]

Meanwhile, King Philip II attempted to bring some order to his empire's violent frontiers with a set of new ordinances issued in 1573. No one was to invade territory "on his own account," according to these orders. Viceroys were to license only peaceful "discovery" and settlement in the most favorable regions, based on their geography and climate. Expeditions had to report back faithfully on "the substance and quality of the lands and people" and always prioritize the safety and religious conversion of Indians.[42]

In 1581, the first official expedition since Coronado's departed for the region now called New Mexico. Just nine soldiers and nineteen Indian servants accompanied three Franciscan friars on a supposedly peaceful mission to convert the Pueblo nations. They spent only one winter in the country, and the most detailed testimony of the expedition claimed the pueblos contained abundant food. However, when the expedition ran out of provisions, they had to resort to violent threats before any Indians would offer them corn. Some members described heavy snows in December and January but otherwise left no indication of severe weather. "The climate tends to be cold, although not too much," claimed one returning Spanish soldier; "it is a climate like that of Castile."[43]

Two of the friars stayed behind as missionaries among the Pueblos, but word soon reached New Spain that they had been killed. For Antonio de Espejo—a frontier cattle rancher, fugitive, and adventurer—that news provided the excuse he needed to take an unlicensed expedition into the country. In November 1582, he recruited a few soldiers and another Franciscan friar and traveled down the Conchos River and along the Rio Grande into Pueblo country. A testimony by one of the soldiers recounted how the expedition first encountered freezing weather in January as it traveled north past today's El Paso, Texas. By

March, as they reached Zuni country, Espejo's party met heavy snows. During the summer, as the expedition turned back from a long excursion into present-day Arizona, they began to meet resistance whenever they approached pueblos. Many Indians had never forgotten their experience with Coronado and were determined to starve out the invaders. The great drought, too, may have given them little choice. In July 1583, a Pueblo they called Pocos refused to share provisions with the Spaniards: "They said they did not have any, that there was a lack of rain, and they were not certain they would gather any corn." The expedition returned to New Spain empty-handed that autumn.[44]

Espejo's adventure failed to provoke any official action on New Mexico. Then in 1590 a fugitive former governor of Nuevo León, Gaspar Castaño de Sosa, gathered a band of followers for a second unlicensed foray into the territory. They traveled up the Pecos River that summer and then southwest and south along the Rio Grande. They were finally arrested by Spanish soldiers in March 1591, probably somewhere around present-day Albuquerque.

By the time the party had crossed the present-day border between Texas and New Mexico in late November, "the weather was beginning to get very cold," one witness testified. Two days before Christmas, above the confluence of the Gallinas River with the Pecos, the group was forced to take shelter "because the weather was so bitterly cold and the land was covered in snow." A fortnight later, the travelers cut westward to the Santa Fe River. "It was bitterly cold and snowing," Sosa's people recalled. "When we emerged from the sierra we came to a river, frozen so hard that the horses crossed the ice without breaking through." The cold and heavy snow pursued them through January and February as they struggled along the Galisteo Basin and past the site of modern Santa Fe. Finally "the cold was so intense that our people were freezing and unable to travel this one league with the wagons; and so we left them in a ditch." The cold and scarcity of food may explain why Castaño de Sosa and his followers failed even to attempt a settlement before they were arrested and hauled back to New Spain.[45]

The official testimonies taken from these expeditions were hardly encouraging. Nevertheless, Castaño de Sosa's arrest made it clear that the empire needed to sanction some official "discovery" of New Mexico before any more adventurers tried their luck in unauthorized invasions. The only question was who could lead that expedition. "I do not find a man in this kingdom who may be charged with doing it as is required for the service of God and your Majesty and the good of those natives," complained New Spain's viceroy, Diego Luis de Velasco, to the king in early 1595:

> What is especially required of the person who is charged is that he shall take it up with such earnestness that it drives him to pursue it to the end, so that the undertaking may not be a failure in case he does not encounter riches in mines of gold and silver, which is what they most earnestly seek; [failing to find them] they immediately desist and return to their homes, abandoning the land and the conversion of the Indians and leaving many children baptized, who, as there is afterwards no one to instruct them, remain in their heathenism, all of which is very objectionable. And when they go out and assault, harass, and even kill them to make them give what they do not have, they make hateful the name of Christians, as has been seen in other explorations.[46]

The man who finally answered the viceroy's call was don Juan de Oñate: scion of a mining family from Zacatecas, descendent of conquistadors and of Aztec nobility, and "one of the leading men and the richest in this kingdom," as Velasco wrote to the king that October.[47] Even with that enthusiastic endorsement, the enterprise did not get under way for nearly another three years, as Oñate dealt first with the arrival of a new viceroy, then rival claims to New Mexico, political intrigues back in Spain, and finally two long rounds of official inspections. Meanwhile, his money and supplies were wasted, and followers deserted.[48]

By the time Oñate's expedition finally departed in February 1598, more than half of those who had originally enlisted failed to show up. This still left some 560 individuals: a mixed group of soldiers and settlers, friars and slaves, single men and families, Spaniards and creoles, Africans, and Indians of diverse nations.[49] As the inspections revealed, the delay had left them short of some crucial provisions, including livestock and grain. But as the new viceroy, the Count of Monterrey, later explained to the king, he agreed to send them along anyway because "the frontier provinces are wearied and even exhausted by the long detention of the governor's army there and because most of the people who go on new discoveries are troublesome."[50] Three years later, a defector from New Mexico would blame these shortages for the disasters that followed:

> The first and foremost difficulty, from which have sprung all the evils and ruin of this land, is the fact that this conquest was entrusted to a man of such limited resources as don Juan de Oñate. The result was that soon after he entered the land, his people began to perpetrate many offenses against the natives and to plunder their pueblos of the corn they had gathered for their own sustenance. Here corn is God, for they have nothing else with which to support themselves.[51]

Juan de Oñate has been called America's "last conquistador." We usually think of the years around 1600 as the time when Miguel de Cervantes wrote *Don Quixote:* the age when medieval notions of chivalry and adventure finally gave way to modern disillusion. Yet Oñate—a son of Mexico's violent mining frontier and a veteran of the Chichimeca War—still truly believed in the possibility of courageous conquests and marvelous riches in New Mexico. One of Oñate's captains, Gaspar de Villagrá, even wrote a chronicle of the expedition in epic verse, full of dramatic embellishments that subsequent historians often took seriously. The quixotic visions with which his expedition began, and the cruelty and bitterness that brought it to an end, have rendered Oñate an especially controversial figure, both in his time and in the twenty-first century. Those lofty visions may also explain Oñate's persistent refusal to admit failures and change course in time to avert disaster.[52]

The invasion of New Mexico set out almost due north across the desert to present-day El Paso and then northward along the Rio Grande. At the end of April 1598, the expedition crossed the river and Oñate claimed formal possession of the country for Church and Crown. Trouble began in May during a detour from the Rio Grande, later known as the *jornada del muerto* (dead man's journey). Away from the river, the lumbering expedition suffered a month of hunger and thirst. Alerted to the Spanish presence, Indians began to abandon their pueblos ahead of their arrival.

A Spanish advance guard moved ahead, and in July they set up a base about 30 miles (50 kilometers) north of present-day Santa Fe, at the pueblo of Okhay Owingeh, which they renamed San Juan. There they began to erect houses and a church, and dug channels to irrigate wheat. The main body of wagons and settlers crawled along more than a month behind, arriving hungry and exhausted. By August simmering resentment over their hardships threatened to boil over into a mutiny, but Oñate put a stop to it with a few arrests and pardons, and then by hunting down a couple of deserters in September.[53]

Now acting officially as *adelantado,* or military governor, Oñate made his first priority the pacification of the Indians of New Mexico so that he could explore the region for gold and silver. Late that summer and autumn of 1598, Oñate's soldiers took tribute and "acts of obedience" from the pueblos. One detachment led by Oñate's nephew Vicente de Zaldívar set off eastward into the plains, and in October Oñate himself led a second detachment west, to travel through Cíbola and search for the South Sea. He ordered another of his nephews, Captain Juan de Zaldívar, to join him with relief supplies the following month. In that way, Zaldívar and a small group of Spanish soldiers found themselves at Acoma, about 60 miles (100 kilometers) west of modern Albuquerque, in early December.

It must have been an intimidating sight. Acoma pueblo forms a natural stronghold atop a high, steep mesa overlooking the surrounding plains and dry fields of maize. Probably unknown to Zaldívar and his men was that the Acomas had met and resisted previous Spanish expeditions under Coronado and Espejo. Nevertheless, the pueblo had given its act of obedience to Oñate's soldiers that autumn, and so the captain and quartermaster climbed into the pueblo in order to demand tribute of food and supplies.[54]

What happened next would become the subject of a trial, of official inquiries, and of debates ever since. The basic facts seem clear. When Juan de Zaldívar's party first came to the pueblo, the Acomas sent them away, claiming they needed more time to gather provisions. The soldiers returned a while later and then split up to collect or barter for what they needed. Fighting broke out, leaving several Spanish wounded and killed, including Juan de Zaldívar. The survivors brought word of the incident to San Juan, and Vicente de Zaldívar returned to Acoma a month later with a larger force. They took the pueblo by

storm, massacring scores of inhabitants and capturing the rest. A
Spanish trial convicted the pueblo of treason and rebellion, and sentenced all Acomas to servitude. The Spanish cut off one hand and one
foot from every man. The brutality was typical of early modern warfare, and the punishment had been a usual one for Indians captured
during the Chichimeca War.[55]

To understand why this happened, we need to look deeper into the
documentary record and the role of Little Ice Age weather. At the trial,
Spanish soldiers nearly all gave the same testimony: that Acomas had
attacked them with "premeditated treachery." Some Acomas, by contrast, claimed the Spanish assaulted them first and they fought back in
self-defense. Yet the evidence points to another story altogether: that the
fighting broke out when a soldier tried to take away an Acoma woman's
turkey.[56]

An official, Alonso Sánchez, described the incident in a letter
written just three months after the event:

> When the *maestre de campo* reached the pueblo of Acoma and
> asked for provisions and water so that he could go on, they
> furnished them at first. The next day, before setting out, he
> asked for more provisions and water, as there was no other
> place to get them. At this time there arose a minor incident
> when a soldier named Vivero [i.e., Martín de Biberos] took
> two turkeys from the Indians, and they killed him from one
> of the terraces. The entire pueblo then rose in arms and killed
> the maestre de campo, don Juan de Zaldívar, and ten soldiers
> and captains.[57]

Sánchez had lost his own son in the fighting, so his letter was in no
way making light of the situation. A close reading of trial testimonies
backs his version in several ways and undermines claims of premeditated treachery by either side. Witnesses explained how Zaldívar had
divided up his men to go house to house collecting provisions. Some
attested that Biberos or the men around him were the first casualties
of the fighting. Only a few of the Acoma were armed, suggesting they
had not prepared an attack; many, including women, reportedly threw
sticks and stones. Most telling of all is the testimony of an eighteen-
year-old Mexican servant among the soldiers, who did not give the same

account as the others. Instead "all he saw was that one soldier, whose name was Martín de Viveros, held a turkey in his hand, and that an Indian woman was complaining about him because of the said turkey, which was at the time that the Indians were shouting." An Acoma witness later gave much the same story about a fight over a turkey, and many later Spanish accounts of the disaster emphasized the way Spanish soldiers had taken turkeys and *mantas* from the Indians.[58]

The disaster at Acoma reprised the same conflict that had led to the Tiguex uprising almost six decades before, and for the same reason. The start of another freezing winter meant the Pueblos needed all the feathers and *mantas* they could get in order to keep warm, or even survive the cold. When Spanish troops returned to Acoma in January, one recalled the Indians pelting them with "sheets of ice." Another described it as "so cold that the guns never overheated although firing ceaselessly the whole time."[59]

Worse still, that summer of 1598 had been the first season of a drought that would go on for another three to five years throughout New Mexico. In time, many members of the expedition would come to understand the cycle of seasons in the country, and the way that unirrigated maize fields such as those at Acoma depended on the rainstorms of the summer monsoon. During their first years in New Mexico, however, many Spaniards seem still to have thought that the dry summers and snowy winters were normal. One officer assumed that "the country depends for its moisture on snowfall, which is very frequent from September on until April."[60] Many equated the climate of New Mexico with that of the Mediterranean at the same latitude. Drawing on a little classical learning, Villagrá's epic poem celebrated the expedition's arrival "into the temperate zone and the fourth clime," on a parallel with Jerusalem.[61]

Oñate was quick to dismiss the incident at Acoma, and during the following months he took care that only the most favorable reports made their way back to New Spain. His letters to the viceroy hinted at rich silver ores and promised new expeditions to find Quivira and the South Sea. Rumors of New Mexico's "populous cities with three-storied houses, civilized people, [and] very fertile and fruitful [land]" soon reached all the way to the imperial court in Madrid, from where Hakluyt eventually received those enthusiastic reports of the country described at the start of this chapter. Viceroy Monterrey still harbored doubts about the venture, but in the spring of 1600 he agreed to send additional supplies and several hundred more soldiers and settlers to join Oñate's colony.[62]

By the end of 1601, nearly all of those reinforcements and many of the original colonists would be back in New Spain, fleeing hunger, cold, and a colony that had all but collapsed. Dozens of letters and official testimonies, from both defectors and loyalists, offer a detailed record of the disaster. They bear witness once more to the overwhelming impacts of New Mexico's Little Ice Age climate.[63]

Problems began as soon as the new soldiers and settlers arrived. Intolerant of dissent and angry about the viceroy's decision to replace some of his officers, Oñate supposedly told the new appointees they could go "wipe themselves" with the viceroy's commissions. Even more of a problem was how to feed those hundreds of new mouths when the expedition was already short of food. Oñate loyalists would later point to the success of their irrigated wheat fields as proof their colony was viable. However, even the most optimistic estimates put their harvest in 1601 at only about 5,000–6,000 bushels, not nearly enough for all the new arrivals.[64]

While the heavy snows of recent winters had kept the irrigation channels full, the long run of dry summers had withered Pueblos' rain-fed fields of maize. In 1601, after the reinforcements' arrival, the drought grew even worse. As a Spanish officer wrote that autumn, "This country has now reached the point at which there is no sustenance, because there has been no rain this year." Testimonies from defectors emphasized the cruelty of Spanish soldiers who extorted the last reserves from the pueblos. Others recounted how Pueblos were tortured to make them reveal hidden caches of food. Even loyal officers testified that soldiers took as much as 8,000–10,000 bushels of maize and beans from Indians. According to one witness, "The said Indians give it to them very reluctantly and tearfully, more by force than of their free will . . . and if they drop a few kernels of the said maize, the said Indians go about collecting them one by one." Several claimed that by autumn 1601 the Indians' last stores of food were gone, and both they and the colonists faced imminent starvation.[65]

Even worse than the drought was the extraordinary cold. Eyewitness accounts confirm what the tree-ring record already indicates: that 1601, in the wake of the Huaynaputina eruption, was one of the coldest years in western North America for centuries if not millennia. The Rio Grande froze over again, and communion wine froze in the church. Untimely frosts destroyed fields of cotton and maize that had survived the drought. Livestock perished in "the snow and freezing weather and miserable pasture," as one witness recalled. Archaeological evidence indicates the colonists were already eating the animals faster than they could reproduce.[66]

Once again the soldiers and settlers from Spain and New Spain had come unprepared, physically and psychologically, for the long cold months during their first winter. Just as during the Coronado expedition, soldiers went from pueblo to pueblo demanding hides and *mantas,* and "if they said that they had no blankets to give, the said soldiers took them from the women, who were left naked," as one friar later testified. Several defectors gave tearful descriptions of shivering women and children bereft of their last cloak or blanket. By the time most of the colonists deserted, it was a widely shared opinion that the conflict over food and blankets had been the real cause of hostilities at Acoma and other pueblos.[67]

A testimony by returning official Ginés de Herrera Horta in 1601 summed up the situation:

> They said that they had been sent reports and words and letters about many great things and riches, and they found themselves to have been deceived, with their property spent, deprived of the quiet life they had led in New Spain, and fearing they would lack food and clothing for themselves and their wives and children and relatives. . . . [A]nd the reason for this is the harshness of the said country, with the cold that lasts for eight months of winter, to the point, as he has stated, that the rivers freeze over, and people are always shivering by the fire and there is little firewood . . .
>
> And this witness has heard that after the said winter come four months of summer, when the heat is almost worse than the cold in winter; and so the saying there is, winter for eight months and hell for four [*ocho meses de invierno, cuatro de infierno*].[68]

The colonists' defection and their accusations against Oñate set off an excruciatingly slow process of official investigation and decision-making while documents crossed back and forth from Mexico to New Mexico and Spain to New Spain. It took another four years to revoke Oñate's governorship, seven to recall him from New Mexico, and twelve to put him on trial for "excesses, disturbances, and crimes" during the Acoma massacre and the harsh winter of 1601.[69]

In the meantime, New Mexico's drought came to an end, and there is no sign that any of the following winters were quite as bad as that of 1601. The colony carried on, though much reduced, and missionaries set out to convert the Pueblos. Probably in 1607 and certainly by the middle of 1608, Oñate granted colonists land to set up a new town called Santa Fe, which means that New Mexico's later capital was an almost exact contemporary of Jamestown.[70]

At just that time, however, the future of the colony was thrown into doubt. As had happened with so many previous Spanish expeditions into North America, the first high hopes for New Mexico gave way to exaggerated disappointment once the country failed to live up to expectations. The new viceroy, the Marquis of Montesclaros, was scathing, calling the land "worthless" and all the reports of its wealth "a fairy tale." "I feel quite sure that both the time and expenditure put into it will be wasted, or that at the best it is a gamble," he concluded in a letter to the king.[71] Among the charges leveled at Oñate during his trial, he was accused of hiding how "poor and sterile" New Mexico really was, because he "threatened and ill-treated those who spoke truth in the matter." Captain Villagrá was convicted "on the charge that, from that place he wrote a letter to the viceroy of New Spain, praising highly the quality, richness, and fertility of the provinces of New Mexico, when the opposite is true, as it is a sterile and poor land, sparsely populated."[72] In early 1608, the viceroy announced intentions to abandon New Mexico altogether, "owing to the limitations and poverty of the land."[73]

During the drama of Juan de Oñate's governorship, the fate of a second discovery hung in the balance as well. Back in 1513, Spanish explorers had first sighted the Pacific Ocean from Panama. Following his invasion of the Aztec Empire, Hernán Cortés obtained permission from Charles V to send out new voyages along that coast in search of further conquests and a passage between the Pacific and Atlantic. During one of these expeditions, in late 1533, mutineers seized a ship and landed it by accident at what is today La Paz, capital of Baja California Sur. Those who survived hunger and Indian attacks that winter sailed back to New Spain, bringing reports of pearls and other riches from the newly discovered island (or so it was supposed) that the Spanish began to call California.[74]

In 1535 Cortés took around 300 colonists to the site. However, storms and contrary winds delayed the arrival of supply ships, and the arid land and its poor indigenous people offered nothing to feed the settlement. Starvation, illness, and complaints eventually forced Cortés to evacuate its survivors back to New Spain. One later testified that it was "the most sterile and the most perverse and evil land in the world."[75]

Undaunted by this failure, and excited by news of Cabeza de Vaca's return in 1536, Cortés sent still more expeditions to reconnoiter California. The last of these, led by Francisco de Ulloa in 1539, sailed along the Mexican coast almost to the mouth of the Colorado River and then back down the Baja Peninsula. Rounding the southern tip at Cabo San Lucas, he continued northward. By November, Ulloa's three ships ran into rough weather and contrary winds. Storms and gales plagued the voyage until, in January, they reached the Isla de Cedros, roughly halfway up the Baja coast. There "a north wind blew so cruelly and so hard, bringing on such darkness and clouds and fog," that the fleet was trapped and two of its ships damaged. After two months of fruitless efforts to continue the voyage, Ulloa gave up and sent one of his ships, the *Santa Agueda,* back to New Spain to report their discoveries. Disheartened by the weather and the shortage of provisions, the captain pronounced the Pacific coast "ugly and sterile and of wretched aspect." An officer aboard the *Santa Agueda* gave a more positive testament of the land, comparing its appearance and its winter rains to those of Castile. However, he brought back descriptions of bone-chilling northerly gales that pierced their warmest clothes "like demons let loose in the air," and air so cold "that we froze alive." And so, concluded the contemporary chronicler Francisco López de Gómara, "they were gone on this voyage a whole year, and brought no news of any good country."[76]

The long voyage and shortage of provisions may have contributed to those miserable impressions of the country. However, the next voyage to reach California two years later gives even stronger evidence of unusual Little Ice Age cold. Under commission from Viceroy Antonio

de Mendoza, the three ships of Captain Juan Rodríguez Cabrillo departed the port of Navidad in early July 1542, "to make discoveries in
the South Sea." Crossing the Gulf of California and rounding Cabo
San Lucas, they sailed along the Pacific coast, passing the Isla de
Cedros in late August. By early October they had reached the present
U.S. state of California. They found the land there more inviting and
the coast thickly settled with Indians.

By the time they passed the Channel Islands later that month, however, the ships felt the first storms and gales of winter. As the expedition continued north along the coast in November, its official narrative
described heavy winds and rains out of the southwest. Then came "a
storm as great as any could be in Spain," with "the sea so high that it
was frightful to see." In the rough weather, the fleet missed San Francisco Bay. When they eventually spotted land again at Point Reyes, they
saw "great sierras covered in snow" and felt "air so cold they froze."
Turning south, the expedition found shelter in what would later become
known as Monterey Bay. The mountains on its south cape were "covered with snow to the summits," so they named the cape Cabo de Nieve
and the mountains the Sierra Nevada. Although temperatures in Monterey Bay still occasionally fall below freezing, nothing like that kind of
cold and snow has been seen since daily records began in the twentieth century.[77]

Heading south, the expedition was forced to take shelter in the
Channel Islands from "the rough weather, cold, and snow," as a later
chronicler explained. Injured in an accident, Cabrillo died there in
January 1543. In February and March, his successor led the expedition
north again past Point Reyes, only to be forced back by another storm
and "such cold that they froze." The ships eventually returned to Navidad in April. The official record does not inform us whether the sailors
subsequently fulfilled their vows to Our Lady of Guadalupe, made
during a moment of peril at sea, that they would "go to their church
stark naked."[78]

Thus through accident and misfortune, Spanish explorers failed
to recognize the real "New Andalusia" in North America. Southern
California's climate today is famous for just those qualities that so many
sixteenth-century expeditions sought in vain in Florida, Virginia, and
New Mexico. California's cold coastal currents and high pressure
help generate cool, wet Mediterranean winters to water the fields and

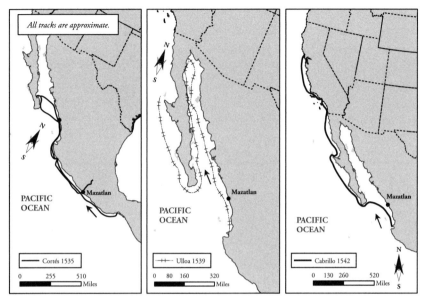

8.2 Early Spanish exploration of the California coast

dry, sunny summers for the ripening of wheat, grapes, and olives. Nevertheless, the distance around the Baja Peninsula meant that two expeditions in a row had now arrived late in the year, only to face winter storms and extraordinary cold—apparently the same extraordinary cold observed in New Mexico and the southeastern United States during the early 1540s. Just as in the Soto and Coronado expeditions and at just the same time, a couple of terrible winters were enough to give an exaggerated picture of harshness and sterility.

The proxy record indicates that California only grew colder during the late sixteenth century. A decline in the mountain tree line and narrower tree rings, particularly in the Sierra Nevada, point to lower summer temperatures. Several studies have found a Little Ice Age cooling trend culminating in the early seventeenth century, and they have identified the years following the 1600 Huaynaputina eruption as California's coldest for centuries or even millennia.[79]

Following the disappointing outcome of the Cabrillo expedition, the initial burst of interest in California faded away. Three factors drew explorers back to the region starting around 1565. In that year, Andrés de Urdaneta charted the return route from the Spanish Empire's new "discovery" in the Philippines. By catching the mid-latitude westerlies across the Pacific to California and then sailing along the coast to the port of Acapulco, he had managed to make the voyage in a respectable four months.[80] Soon the Philippines developed into the center of a large and lucrative transpacific commerce, exchanging Spanish American silver for Chinese luxury goods. The viability of the "Manila galleon" route across the Pacific became a serious imperial concern. Climatic conditions in the Pacific, especially the famous El Niño effect, could alter wind patterns and sailing times. Many voyages would take far longer than Urdaneta's to reach Acapulco, lasting six or even seven months, leaving their ships in peril of winter storms and their crews decimated by scurvy.[81] These dangers raised some new interest in colonizing California, so that passing Spanish ships could find a harbor to rest and resupply.

In the meantime, Spanish shipping in the Pacific faced a more immediate threat from English privateers. Francis Drake's famous assault on the Peruvian treasure fleet in 1579 and his subsequent circumnavigation of the globe raised fears for the safety of Spain's Pacific interests, including the Manila galleons. As Spanish intelligence later learned, Drake's voyage had continued far up the California coast, until contrary weather forced them back. Drake's chaplain recalled, "Wee found such alteration of heate, into extreame and nipping cold, that men in generall, did grievously complaine thereof." Even in midsummer, the ropes stiffened in the freezing rain, "so that wee seemed rather to bee in the frozen Zone than any way unto the Sunne, or these hotter climates."[82] In November 1587, Spain's worst fears were realized when the privateering expedition of Thomas Cavendish actually captured the Manila galleons off Cabo San Lucas.

Finally, during the late 1500s rumors resurfaced of a passage from the Pacific to the Atlantic. Particularly influential was Lorenzo Ferrer Maldonado's apocryphal 1588 account of a voyage through a "Strait of Anian" connecting California to Labrador. Even the notion that California was an island—a notion that should have been killed by the Ulloa expedition back in 1539—began to reappear in some European maps and works of geography.[83]

Nevertheless, few ships making the long journey back from the Philippines to New Spain had the time and inclination to explore the California coast for a promising site to colonize, much less a passage to the Atlantic. A voyage led by Pedro de Unamuno in 1587 reported the discovery of a fine harbor, probably Morro Bay, roughly halfway between present-day Los Angeles and San Francisco, where the Manila galleons could restock and recover. There they found "infinite fish of various kinds, trees for ships, water, timber, and much seafood"; there were even signs of pearls and land "well suited for wheat and maize." However, some of Unamuno's sailors were wounded in a skirmish with Indians, and his ship was endangered by thick fog, heavy seas, and gales. Already suffering from the long trip across the ocean, the expedition continued onward to New Spain without exploring much further.[84]

Seven years later, King Philip II issued a new royal decree calling for exploration of the Pacific, and Viceroy Velasco ordered the next available ship from Manila to make another reconnaissance of California. That voyage, navigated by Sebastián Rodríguez Cermeño, reached the north coast of California in November 1595 and sailed through rough waters into Drake's Bay. Fortunate to find Indians who offered them food, his men recovered from their long ocean crossing. However, the heavy waters wrecked their battered flagship, the *San Augustín,* upon the rocks. Having lost all their provisions, the eighty survivors had to crowd aboard the expedition's small launch and make the long trip down the coast to New Spain, too plagued by cold and hunger to properly reconnoiter the country.[85]

Real plans for a colony in California began only with the appointment of New Spain's next viceroy, the Count of Monterrey. Although lukewarm about Oñate's New Mexico expedition, Viceroy Monterrey saw value in exploring the California coast and establishing a new port to receive the Manila galleons. For that task, Monterrey appointed Sebastián Vizcaíno. Like so many other expedition leaders described in this book, Vizcaíno began his career as a soldier from the hardscrabble Spanish country of Extremadura. He had served as an official in New Spain and in Manila, and in the early 1590s had taken over a venture to search for pearls in Lower California.[86]

Vizcaíno's first voyage to the Baja Peninsula in 1596 ended in disappointment. In Vizcaíno's telling, the natives were "bestial and barbarous" people who ate lizards and snakes, "which for us was a clear sign of the poverty and sterility of the land." The ships made slow progress against storms and contrary winds, and they had to turn back for want of supplies before reaching the upper California coast. The only positive news was of the sheer size of California and "its climate and sort of winter [which] is the same as that of Castile."[87] Viceroy Monterrey was so discouraged by the hostility of the Indians and "sterility of the land" that he wrote to King Philip II to warn of the "small hopes and little profit" in trying the voyage again.[88] What probably saved the venture were new rumors of English privateers along the California coast and another apocryphal story of a voyage through the "Strait of Anian."[89] It would take until 1601 for Vizcaíno to secure an appointment for another expedition to California, with instructions to explore the coast "considering the quality of the land and its climate and the people."[90]

Narratives of Vizcaíno's expedition to upper California provide still more evidence of anomalous Little Ice Age cold. The voyage, which set out in May 1602, spent the summer and early autumn making slow progress up the Baja Peninsula in the face of headwinds and occasional storms.[91] In November the expedition sailed passed San Diego Bay, and in December it reached a fine harbor that the captain shrewdly named Monterey, for his viceroy. Vizcaíno immediately sent back one of his ships bearing a glowing report of their discovery: a port "secure from all the winds," with a "great quantity" of timber, and the land "very fertile of the climate and landscape of Castile."[92] However, the remaining ships soon ran into harsh northerly and northeasterly winds that brought a bitter chill to Monterey Bay. As Vizcaíno recalled:

> The weather was extremely cold and the men worked very hard gathering wood and storing water for our voyage to Cape Mendocino. The cold was so intense that when dawn broke on Wednesday, New Year's Day, 1603, all the mountains were covered with snow. They looked like the volcano in Mexico. The well from which we had been taking water was frozen to the thickness of a palm's width. The earthen jugs filled with water had been left outside during the night, and they were so frozen that even if they rolled them around not a drop leaked out.

The description indicates weather far colder than any observed during the past century in Monterey Bay.[93]

Later that month, the ships continued as far as Drake's Bay in northern California. However, they soon ran into a powerful storm and weather "so cold that there was no one who could withstand it." As Vizcaíno later testified, "We agreed it did not make sense to continue farther, since the people were sick, the weather contrary, and the winter cold." And so the expedition turned back to Mexico, a number of men dying of scurvy along the way.[94]

As Vizcaíno may have calculated, Monterrey was particularly taken with the descriptions of his namesake bay. The viceroy and captain pressed the new king, Philip III, for a permanent settlement to receive the Manila galleons. Their efforts eventually paid off in a royal decree of August 1606 that parroted all the site's supposed virtues: it was "the most suitable for the ships of China and as a stopover for the Philippines trade," "very protected from the winds," and "well populated with a gentle people." Overlooking the severity of the winter, the king even praised it as a "land of a moderate climate and very fertile."[95]

Unfortunately for Vizcaíno, Monterrey's term of office had already expired by then. Subsequent viceroys showed little of the same enthusiasm for the bay named after their predecessor. Viceroys Montesclaros (1603–1607) and Velasco (serving his second term from 1607 to 1611) dragged their feet and threw up a number of objections. The colony would be expensive, they claimed. It would be a target for English pirates. And by the time ships reached California they were almost home anyway. In a sprawling empire where bureaucracy moved slowly and compliance was never easy, their stonewalling threatened to kill the plan. California's very first historian, the Jesuit Miguel Venegas, put it more colorfully: "Governor Vizcaíno, who had more than enough courage to fight the storms and calms of the Sea, lacked it to fight the calms and variable winds that stall and move the currents at Court."[96]

In this way, matters came to a head over the question of colonies in New Mexico and California at just the same time as they had in Florida. In all three cases, King Philip III and his viceroys had to confront the same underlying problems of colonizing North America. The lands in question had proven less than ideal and their climates less than perfect. And now the Spanish Empire was bankrupt.

The immediate problem of cash flow in precisely the years 1607–1609 came from an accident at the mercury mines of Huancavelica, Peru.

Their output was essential to smelt the silver ores at Potosí, and so silver deliveries from the Americas fell off for the first time in decades.[97] Meanwhile, Spain's New World colonies suffered disasters brought on by fluctuations of the El Niño Southern Oscillation—that is to say, years of unusual cooling and warming in the eastern tropical Pacific. Potosí and the central Andes experienced years of drought in 1607–1611. The Valley of Mexico, on the other hand, suffered a flood in 1607 "so violent and forceful that every part of the city was about to drown, and all of it seemed inundated," according to an official report. The disaster marked the beginning of an expensive decades-long effort to drain the capital and keep it dry.[98]

Yet problems in Spanish finances at the turn of the seventeenth century ran much deeper than that. Corruption, sale of offices, and rising local expenditures in New Spain and Peru meant that the colonies sent back less and less of their silver output anyway. Heavy military expenses in Europe far outran imperial incomes. Moreover, the run of extreme seasons, bad harvests, and crisis mortality back in Spain had crippled imperial revenues from already overtaxed Castilian farmers.

Modern science has confirmed what contemporaries already suspected: that silver now flowed out of Spain even faster than it flowed in. Precious metals, like many elements, possess different isotopic signatures (that is, different ratios of atoms with different numbers of neutrons) depending on their provenance. Using these isotopic signatures, a study has found that the silver mined in the Americas scarcely shows up in Spanish coins minted during the early 1600s, illustrating just how fast that silver escaped to the rest of Europe and Asia. Spaniards found themselves awash instead in bonds, promissory notes, and inflationary issues of copper coinage—a disturbing world of fluctuating values in every sense.[99]

Against this backdrop of real and perceived crisis in the empire, the colony of New Mexico managed to achieve the same reprieve as Florida and in almost exactly the same way. Once news reached the colony that Viceroy Velasco meant to abandon New Mexico, its Franciscan missionaries rushed to claim "that more than seven thousand persons had been baptized and so many others were ready to accept baptism that it seemed the Lord was inspiring them to accept Him to gain salvation." Any retreat now would put Indian lives and souls in peril. Although the claim was surely exaggerated, the viceroy relented, mindful of his monarch's religious devotion. However, the country's reputation for cold and sterility, and the lack of funds for further support, ensured that New Mexico would never become a significant imperial priority or a major settler colony.[100]

The projected settlement at California had no such missionaries and converts to make a claim on the king's Catholic conscience. Moreover, the difficult voyages and extraordinary cold experienced by past expeditions had scarcely done justice to a region with some of the most welcoming seasons and fertile soils in the world.[101] California was still regarded as at best a convenient stopover for the Manila galleons, and at this time of fiscal crisis, even the Manila trade had come under attack as a dangerous "drain of silver" from the Americas.[102] Instead, in late 1608, Philip III acceded to his viceroys' request to give up on Monterey Bay altogether and set his sights on wondrous islands in the middle of the Pacific said to be called Rico de Oro and Rico de Plata.[103]

It is almost incredible that Europe's most powerful monarch would give up on a promising colony in order to chase after mythical isles named "Rich in Gold" and "Rich in Silver." The decision offers some of the clearest proof of Philip III's vacillation and naïveté, and of his empire's desperate straits. Rico de Oro and Rico de Plata had entered popular legend from an apocryphal account of a 1585 Portuguese voyage. Blown off course between Malacca and Japan, the ship was supposedly driven to an island with a great walled city, teeming with vessels in its port. Its welcoming inhabitants took the Portuguese captain to the lord of the island, who showered his men with gifts and "let it be understood by signs that the land was very fertile and rich with silver and other things." It is not clear whether Viceroy Velasco even believed in the islands' existence, or whether he was only seeking an excuse to abandon the Monterey Bay venture.[104] In any case, Vizcaíno dutifully went along with Spain's first official embassy to Japan in 1611, and from there he set sail in search of Rico de Oro and Rico de Plata. His fleet zigzagged hopelessly across the western Pacific for two months without sighting land, until finally a typhoon forced them back to Japan. Vizcaíno returned to Spain late in 1613 and failed to raise funds for another venture to Monterey Bay. Spanish colonization of California would have to wait for more than another century.[105]

9

———⌣———

Death Follows Us Everywhere

While English colonies struggled through their first winters and Spanish colonies hung in the balance, the first enduring French settlement in North America was under way in Quebec. The survival of that settlement illustrates how French colonial expeditions managed to revise expectations about the climate and environment of the region and, however imperfectly, to learn from trial and error. In most cases, the worst of those trials had come from the country's unexpectedly harsh seasons.

Two decades ago the Quebec archaeologist Norman Clermont could still question, in the title of an article, "Did New France experience the winters of a Little Ice Age?"[1] The answer is not simple. The first instrumental measurements of weather in Canada date back only to the 1740s, and the first regular observations of agricultural activities, frosts, and lake and river freezing go back only a little further than that. In these records, Quebec experienced its lowest temperatures, and its coldest winters, during the early nineteenth century. The mid-1700s were on average no colder than the twentieth century, with cooler springs and autumns, but warmer summers and milder winters. In these measurements, the Little Ice Age stands out more for its dry seasons and variable temperatures than for extreme cold.[2]

However, several kinds of climate proxies can extend our reconstructions back into even earlier centuries. The width and cell density of growth rings in Quebec's pines reflect the temperature of each year's growing season, especially in the far north near the tree line. In general, this evidence reveals warmer temperatures during the Middle Ages, followed by Little Ice Age cooling starting in the late 1200s. The coldest years—that is, the narrowest rings—tended to follow large volcanic eruptions. Temperatures fell during the late 1500s, with a clear trough during the first decade of the 1600s. Depending on the site and the type of measurement, however, the studies disagree on just how cold this period was. Some show the turn of the seventeenth century as the very coldest or second coldest period of the last thousand years, with summer temperatures falling 2°C (3.6°F) or more below the long-term average. Others find a cooling trend, but only compared to a relatively warm period during the mid-1500s. Others find a serious cold anomaly during the 1590s and the first decade of the 1600s but capture no long-term trends at all.[3]

Different kinds of proxies, less sensitive to year-to-year variations but more revealing of changes over decades or centuries, also find a milder climate during the early to mid-1500s, with falling temperatures late in the century. Some lakes of western Quebec, eastern Ontario, and the upper Midwest can capture and bury pollen in a way that preserves rapid changes in the surrounding vegetation. Research into these sources has often (but not always) found evidence of abrupt cooling in the late 1500s. Hardwoods such as birch and beech declined, replaced by more cold-adapted trees including pines and red spruce. In some of these studies, the turn of the seventeenth century stands out as the coldest period of the past millennium, while in others the early nineteenth century appears colder still. The data suggest the period may have been especially bad for plant growth: perhaps seasons that were dry as well as cold, perhaps unusually short growing seasons, and perhaps overcast weather or sunlight dimmed by volcanic eruptions.[4]

These findings from climatology, imperfect though they are, have some important implications for the early history of New France. As cold as Jacques Cartier's expeditions found the country during the 1530s and 1540s, nothing they experienced would have been unknown to inhabitants of Quebec in the nineteenth or early twentieth centuries. By the time of Samuel de Champlain's voyages six decades later, average temperatures had fallen, marking a second and colder phase of the Little Ice Age. The climate experienced by Champlain and his companions during the early seventeenth century was extraordinarily cold even by Canadian standards, probably second only to that of the early 1800s. It remains unclear whether the winters he faced were truly worse than any in the memory of today's residents of Quebec. Most of our data concern summer temperatures, and although twentieth-century winters were on average milder, Quebec saw little overall warming until the 1990s. Since then, and in spite of the harsh winters of 2014 and 2015, temperatures have begun to rise. Global warming could well

remake the climate and national image of Canada as much as those of
any other country in the world.[5]

The cooling during the late sixteenth century and the unusual cold
of the early 1600s had significant consequences for the French enterprise in Canada. Even as expeditions tried to adapt to Canada's climate, that climate was itself changing, growing more difficult and
unpredictable. The challenge of Canada's climate came less from the
cold itself than from the long winters and short growing seasons, which
posed challenges for diets and health that Europeans of the time scarcely
understood. Moreover, the sudden cooling of the late 1500s transformed
the world of Quebec's First Nations as well, with fatal consequences
for some early French colonists.

Reports of Canada's cold climate must have reached France decades
before King Francis I took an interest in exploring or colonizing the
country. By 1500, news of the Cabot and Corte-Real expeditions were
already circulating throughout Europe, bringing descriptions of cold
weather and frozen seas.[6] In the meantime, fishermen and whalers,
including a few Frenchmen, had discovered the rich catches off
Newfoundland.

It fell to the Portuguese, however, to attempt the first European settlement in Canada. No firsthand account survives from the colonizing
expedition of João Álvares Fagundes, but later descriptions say much
about early impressions of the country and its climate. Apparently in
1520 or 1521 King John III granted Fagundes rights to settle Newfoundland, which at the time was thought to lie within Portuguese claims.
Fagundes secured investments for the voyage and set out with two
ships carrying colonists. According to a Portuguese account written in
1570, "finding the region to which they were bound very cold," they
sailed on instead to Cape Breton. There, according to the same account,
they lost their ships but discovered fertile soil on which to settle, and
Basque fishermen brought news of their colony back to Europe. The
French cosmographer Jean Alphonse, however, wrote in 1542 that Cape
Breton's Indians had massacred the settlers. He described Newfoundland as "a high land without any benefit except its fishery" and "nothing
but a rock without any useful plants or land." Although his source of
information is unclear, Samuel de Champlain later wrote that the Portuguese attempted to settle in Cape Breton and even spent a winter

there, "but the harsh weather and cold made them abandon their colony."[7]

The French king took an interest in the country only after Giovanni da Verrazzano failed to find a more southerly passage to Asia in 1524. During Verrazzano's voyage, Francis I had led a disastrous campaign challenging Habsburg forces in Italy. Captured on the battlefield of Pavia in 1525, he spent a year in captivity until France paid (literally) a king's ransom in treasure and diplomatic concessions. Canada promised another possible route to the Indies, and a French expedition there would be a less risky way to challenge the Spanish Empire than another war in Europe.

In 1532 Francis I was introduced to a mariner named Jacques Cartier, who was supposed to have traveled to Brazil and Newfoundland. He evidently impressed the French king with his proposals for further exploration. Cartier's original commission is now lost, but a royal grant of 6,000 livres (about £500) specified he would "voyage from this kingdom to the New Lands to discover certain islands where it is said a great quantity of gold and other precious things should be found."[8] Francis I even blocked the other ships from leaving the French port of St. Mâlo during early 1534 so that Cartier could gather the supplies and crew he needed for the voyage.[9]

Cartier departed with two ships and sixty-one men at the end of April, crossing the North Atlantic to Newfoundland in only three weeks. He found the island's harbors still blocked "by the great amount of ice that was along the land," that is, the icebergs brought down by the Labrador Current. Sailing north, the expedition found the Strait of Belle Isle between Newfoundland and Labrador but could not press on until mid-June, trapped "by the adverse weather and great number of icebergs." Dismayed by the rugged, barren Labrador coast, Cartier called it "the land that God gave to Cain."[10]

Through late June, stormy winds from the northeast took the ships down the west coast of Newfoundland to St. George's Bay. Early the following month, still facing "bad weather, overcast skies, and wind," they passed through the Gulf of St. Lawrence past the Magdalen Islands and along Prince Edward Island, which impressed Cartier with its fertile soil and rich wildlife. A turn northeast along the New Brunswick coast brought the expedition into further heavy weather and adverse winds. The ships sought shelter in a bay that Cartier named Chaleur

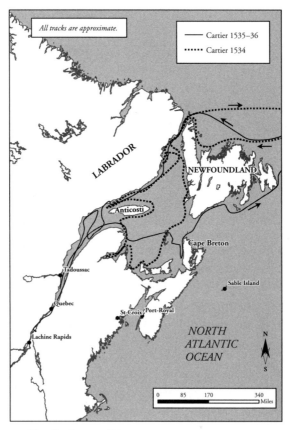

9.1 The Cartier expeditions

Bay for the unusually calm, warm weather and the surrounding countryside, which he judged "more temperate than Spain." The French
traded with Native Americans there, but they had nothing to offer that
Europeans valued: American furs were not yet the prized commodities
they would soon become.[11]

Working their way north from Chaleur Bay, the expedition again
ran into contrary winds, forcing them into Gaspé Bay. There they encountered a different nation, led by a chief, Donnaconna. When Cartier
erected a giant cross with the fleur-de-lys to claim the land for France,
Donnaconna objected, "as if he wished to say that all this region belonged to him." Nevertheless, the chief sent two of his sons to board

the French ship, which caught favorable winds and sailed away at the beginning of August. Crossing to Anticosti, the two ships sailed counterclockwise around that large, forbidding island, mistaking it for a peninsula of Labrador. At that point, fearing contrary storms or winds would trap their ships and force them to overwinter in that unfamiliar land, the expedition's leader chose to turn back home.[12]

In itself, the first Cartier voyage proved a disappointment. It had found no gold or other riches, and no passage to Asia. Yet like so many other early explorations of North America, it hinted at great prizes just over the horizon. It had come tantalizingly close to finding a route into the interior or, just maybe, to Asia. More important, the two sons of Donnaconna came bringing tall tales of a rich land to the west they called Saguenay. Cartier had little trouble securing another royal grant of 3,000 livres and an immediate commission for a second voyage "to complete the navigation already begun." This time he was to bring three ships, more than a hundred men, and provisions for fifteen months— meaning this voyage would pass the first French winter in Canada.[13]

This second expedition departed in late May 1535, thereby avoiding the spring cold and ice met on the previous voyage. However, the ships ran into storms during the Atlantic crossing and lost sight of one another. The flagship reached Newfoundland in mid-July and did not rendezvous with the others until early August. The fleet then made good progress through the Strait of Belle Isle and past Anticosti, which they now understood to be an island. At this point, Cartier sailed down and back across the wide mouth of the St. Lawrence before continuing up the river he now called Hochelaga and into the country he now called Canada. On September 11, the expedition arrived at what would later become the French trading post of Tadoussac.[14]

The expedition continued up the St. Lawrence to Stadacona, the home village of the chief Donnaconna, at the site of present-day Quebec. At that point their amicable relationship with the chief and his two sons began to fray. The latter refused to come back on board the ships or to guide the French any farther upriver. As Cartier made to leave the village, some of its people shouted after him, explaining that their god Cudouagny had warned them "there would be so much snow and ice that all would perish." "At this point," Cartier recalled, "we all began to laugh and to tell them that their god Cudouagny was a mere fool who did not know what he was saying; that they should tell his messengers as much; and that Jesus would keep them safe from the cold if they would believe in him."[15]

Yet as the Canadian historian Marcel Trudel observed, Cudouagny was no fool.[16] The French had only a brief time to prepare for the onset of a Canadian winter. At the end of September, the captain and some of his men continued up the St. Lawrence, taking about two weeks to reach the village of Hochelaga, in present-day Montreal. Cartier's narrative describes an enthusiastic reception by the villagers. They supposedly indicated that past the rapids and up the Ottawa River, the French could find a country rich with gold and silver, which Cartier took for the legendary land of Saguenay. At that point, however, the expedition lacked any guide or translator to make sense of Hochelagans' descriptions and directions. Returning to Stadacona in mid-October, Cartier's party found that the rest of the expedition had settled in a temporary fort up the St. Charles River, a short way from the village. For the next two months, the French exchanged beads and small manufactures for fish and occasional game. Blaming Donnaconna and his sons for the Stadaconans' "stingy" trading, the captain continued to grow more mistrustful.[17]

Then winter came. Starting in late November, Cartier recalled, "we lay frozen up in the ice, which was more than two fathoms in thickness"—that is, more than 12 feet (3.5 meters)—"while on shore there was more than four feet of snow." Not only the narrow St. Charles, but the whole St. Lawrence had frozen. The ship, abovedeck and below, was covered with inches of ice. Their drink, probably hard cider, froze in its casks.[18]

During the past two centuries, the first frosts around Quebec City have usually come around late September and the first snowfall in October. The whole frost-free season has lasted on average only about 150 days. A hard freeze, so that people could pass over the St. Lawrence at Montreal on foot, usually commenced at the start of January, and before the modern engineering of the river, ships could not usually reach Montreal through the ice until mid- to late April.[19] Thus the first French winter in the country was probably colder than most in modern times, but not necessarily exceptional for the Little Ice Age. Cartier and his men, however, were exceptionally vulnerable.

The real threat to their expedition—and nearly every European expedition to Canada for another century to come—was scurvy. Without consuming ascorbic acid (vitamin C), the human body cannot produce collagen, cannot heal wounds, and cannot maintain the growth and repair of tissues. Vitamin C deficiency, or scurvy, would begin with weakness, fever, and irritated skin. The victim would develop blackened, swollen gums with loose teeth, then inflamed and swollen joints and limbs. After more than three months without sufficient vitamin C, the disease brought fevers, hemorrhaging, and finally death.

The smallest quantities of ascorbic acid can cure scurvy within days. However, typical sixteenth-century shipboard fare lacked even that. A sailor's monotonous diet of biscuit, dried and salted meat and fish, some cider or beer, and no fresh fruit or vegetables would leave him at risk of the sickness within a few months. The disease had already become a serious problem on voyages to the East and West Indies, and it had all but destroyed the crew of Ferdinand Magellan's famous 1522 circumnavigation. In France at the time, it became known as the "land sickness" (*mal de terre*), perhaps after victims among the sailors and fishermen who frequented Newfoundland (Terre Neuve).[20] Accounts from the first permanent English colonies on that island, during the 1610s, suggest that after a winter without fresh food settlers began to fall mortally ill by March.[21]

For Cartier, "the disease was a strange one," suggesting he had never encountered acute scurvy on his voyages before.[22] Yet his description leaves no doubt about the diagnosis:

> The sickness broke out among us accompanied by most marvelous and extraordinary symptoms; for some lost all their strength, their legs became swollen and inflamed, while the sinews contracted and turned black as coal. In other cases the legs were found blotched with purple-colored blood. Then the disease would mount to the hips, thighs, shoulders, arms and neck. And all had their mouths so tainted, that the gums rotted away down to the roots of the teeth, which nearly all fell out.[23]

Through ignorance and mistrust, the captain mistook the source of the illness and nearly got his entire expedition killed. During the month of December, he recalled, "we received warning that sickness [*mortalité*] had broken out among the people of Stadacona to such an extent, that already, by their own confession, more than fifty were dead." Fearing infection, the French tried to isolate themselves in their fort; "but notwithstanding we had driven them away," members of the expedition soon fell ill.[24] Some historians have assumed that the Stadaconans' "sickness" was scurvy brought on by a long winter. Yet this interpretation makes little sense for two reasons. First, the timing does not work. There is no reason Indians would have fallen mortally ill of scurvy so early in the season, or that they would have succumbed before Frenchmen eating a poor shipboard diet. Far more likely, contact with the French had spread some Old World disease, deadly to Indians but not the French themselves. Ironically, by quarantining the fort, Cartier almost certainly hastened the onset of scurvy, by cutting off his men from fresh food brought from Stadacona. This would explain why the illness had devastated his expedition by late February.[25]

Second, the people of Stadacona would never have suffered an epidemic of scurvy because they already knew a cure. The French would have learned of it sooner had Cartier not been so fearful of an attack and so desperate to hide their illness. As his men died in the fort, Cartier had them buried out of sight in the snow "for at that season the ground was frozen and we could not dig into it, so feeble and helpless were we." Whenever Stadaconans came near, the captain "whom God kept continually in good health" would come around with two or three of the healthiest men, claiming the rest were at work inside the ships. They learned of the antidote, almost too late and almost by accident, from a villager, Dom Agaya, whom they saw suffering from scurvy but who later returned to them in perfect health.[26]

The identity of Dom Agaya's cure has long been a subject of speculation. Cartier called it *annedda* and described it as a sort of tea made from boiling tree leaves. Its effect on the men seemed almost miraculous, and they drank it so fast "so that in less than eight days a whole tree as large and as tall as any I ever saw was used up."[27] Many large pine needles (and it must have been an evergreen at that time of year) contain at least a little vitamin C. For decades, the best guess for *annedda* was *Thuja occidentalis,* or white cedar, because it was supposedly identified as the "tree of life" brought from America to French botanical gardens during the sixteenth century. More recently, historian Jacques Mathieu has cast doubt on this theory. The documents used to identify white cedar as the "tree of life" are less than conclusive. Moreover, each

100 grams of white cedar needles contain a mere 45 milligrams of vitamin C, a lot less than many other evergreens. A far better candidate would be the balsam fir, *Abies balsamea,* which is much richer in ascorbic acid, and which forms a part of local Native and folk medicine, thanks to its antibacterial properties.[28] Whatever the source of the cure, it appears all but about twenty-five of the French survived the winter.[29]

Cartier's account of what happened next is less clear. Around April, as the ice began to break up but the snow still lay deep enough to track game, Donnaconna went off on a hunting expedition. Other Indians of the village came back to the French ships bringing venison and fish. "They bartered for a good price," Cartier recalled, "or otherwise preferred to carry them away again, because they needed provisions for themselves, on account of the winter, which was long; and they had eaten their stores."[30] Donnaconna returned around the beginning of May but, claiming to be sick, refused to meet with the French. Cartier became convinced in the meantime that Donnaconna was gathering warriors to ambush his expedition. About two weeks later, Cartier's men finally tricked Donnaconna and several others into boarding one of the French ships, seized them, and sailed off down the St. Lawrence, to the evident anger and bewilderment of the Stadaconans. The expedition returned to St. Mâlo in late July 1536.

In material terms, Cartier's second expedition was no more successful than his first. Worse still, it suffered high casualties in a harsh winter. However, Cartier came back this time with even more fantastic rumors about Saguenay. Before he died in France, Donnaconna "assured us that he had been to the land of the Saguenay where there are immense quantities of gold, rubies, and other rich things," Cartier wrote, "and that the men there are white as in France and go clothed in woolens. He told us also that he had visited another region where the people, possessing no anus, never eat nor digest, but simply make water through the penis."[31]

Donnaconna most likely wanted the French as far away as possible from Stadacona, and told them whatever nonsense they wanted to hear.[32] Why Cartier believed him, or pretended to believe him, remains unclear. Based on a conversation reported by the Portuguese ambassador, Francis I fell for the legend as well. The king conceded there was no Northwest Passage along the St. Lawrence and that in Canada "the summer is short, and winter long and exceedingly cold." Yet he confidently expected to find gold and tropical spices, and even bought into Donnaconna's legends of flying men.[33]

Moreover, the timing proved just right for a third and even larger expedition into Canada. France reached a peace with the Habsburg Empire in 1538, and England was then distracted by Henry VIII's disputes with the pope over doctrine and divorce. For Francis I, it was a chance to stake a claim in the New World, and at the same time to rid France of hundreds of newly unemployed soldiers, as well as vagrants and convicts.[34]

The grand expedition that the king had in mind meant far more money, provisions, people, ships, and equipment—and so far more preparation, confusion, and delay. The venture acquired the combined character of an exploratory mission, military outpost, agricultural settlement, and penal colony. It would absorb some ten ships, hundreds of passengers, livestock, seeds, trade goods, arms and armaments, and enough food for two years. Its contingents of soldiers were outfitted with "arquebuses, crossbows, and bucklers," according to Spanish intelligence, "because the savages, who are the people of the country, shoot with bows and can swim a good two leagues underwater." The ships were loaded with "fifteen hundred sides of fat pork, eight hundred oxen and cows, salted and air-cured, a hundred barrels of wheat . . . two hundred pipes of flour, twenty pipes of mustard, twenty of oil, and as many of butter," as well as biscuits, wine, cider, and dozens of pigs, goats, sheep, and horses.[35]

It would take until 1541 to assemble all the passengers and supplies, at an expense nearly ten times that of Cartier's original expedition. Struggling to recruit enough funds and colonists, officials emptied French prison cells throughout the Loire valley, even accepting money to commute death sentences into exile. Convicts would eventually make up about a quarter of the passengers, both men and women, aboard the ships. English intelligence estimated that Francis I had personally sunk some 30,000 livres (more than £2,500) into the project, reckoning that "if the same do take good effect the said money to be well bestowed" but if the colonists "never return again yet that to be no great loss of them especially."[36]

The new character of the expedition also meant dividing its leadership. The king retained Cartier as captain general and master pilot. To govern the settlement, however, Francis turned to veteran officer Jean-François de la Rocque de Roberval for his "good judgment, sufficiency, loyalty, and other good and praiseworthy qualities." The king granted him broad powers to rule over the new colony, subdue the Indians, and "make them live in law and order and fear of God."[37]

Facing continual delays in provisioning, the governor authorized Cartier to depart while Roberval finished the preparations. Cartier sailed off in late May 1541. Habsburg intelligence claimed that his five ships carried at least 1,500 passengers, but the real number was probably much lower.[38]

From that point on, our only direct testimony of Cartier's voyage comes from an anonymous report related to Richard Hakluyt. "And we sailed so long with contrary winds and continuall torments," it records, "which fell out by reason of our late departure." After the difficult crossing and weeks of waiting in vain for Roberval at their rendezvous in Newfoundland, Cartier's fleet did not reach Stadacona until the beginning of September. Passing the site of today's Quebec City, the expedition continued a few more miles upriver to Cap Rouge, where they found a better mooring for their ships and "very good and faire grounds, full of as faire and mightie trees as any be in the world," including the source of *annedda*. Better still, they found what they took for precious ores, diamonds, and "certaine leaves of fine gold as thicke as a mans naile." With little time left before the winter freeze, Cartier sent two barques upriver to try the rapids between Hochelaga and the legendary Saguenay, but they soon found the way impassible and turned back.[39]

Our sense of what happened to Cartier's venture that winter turns on just a few brief testimonies, and on our judgment about the captain's character and intentions. What we know from written sources is that on their way back to France the following June, the expedition encountered Roberval's fleet, which had finally arrived at Newfoundland after a year's delay. According to our only eyewitness account of Roberval's voyage—another anonymous narrative translated by Hakluyt—Cartier "enformed the Generall [Roberval] that hee could not with his small company withstand the Savages, which went about dayly to annoy him: and that was the cause of his returne into France." However, the writer judged that Cartier's men really meant to abscond from Canada with all its supposed gold and diamonds, "mooved as it seemeth with ambition, because they would have all the glory of the discoverie of those partes themselves."[40]

Incidental evidence suggests Cartier had not lied about his conflicts with Indians. Of the Stadaconans and Hochelagans brought back from his previous expedition to Canada, only one—a young girl—had survived the voyage to France and the years of exposure to Old World pathogens. That left Cartier's latest expedition without any translator or cultural intermediary, and it probably raised suspicions at Stadacona that the French had killed the other Indian captives. Our anonymous narrative also suggests Cartier had developed a deep mistrust of Canada's First Nations. According to the contemporary chronicler André Thevet, while Cartier ventured upriver in search of Saguenay, some hotheads among the well-armed militia attacked Indian villagers at Cap Rouge without provocation. Throughout the following winter, warriors from Stadacona picked off French colonists in reprisal. Spanish intelligence later gathered rumors among Newfoundland fishermen that Indians had killed more than thirty-five of the French.[41]

Cartier may or may not have meant to steal the credit for finding Canada's diamonds. Yet he undoubtedly sailed off with the country's most precious resource of all: the secret of the antiscorbutic *annedda*. Roberval had spent the intervening months back in France trying to gather sufficient supplies for the colony, until bad weather forced him to overwinter in port again, nearly provoking a mutiny among the passengers. His expedition of around 200 men and women finally departed in late April 1542, laden with supplies and fortified against assaults and harsh weather. But, as soon became apparent, they were completely unprepared for an attack of scurvy during the long Canadian winter. Bizarre as it seems, neither Cartier nor his officers told Roberval how to cure the disease that had nearly destroyed their own expedition only six years before.[42]

After their meeting with Cartier's ships, Roberval's fleet sailed up the St. Lawrence in early August 1542 to the same site at Cap Rouge. They built a "fayre Fort" with a "great Towre" to safeguard an elaborate colony, including barracks, chambers, dining halls, and kitchens. In late September, they sent two ships back to France to deliver news of their progress and to request further supplies. They remained in the settlement from the onset of cold weather, probably in October, until the "ice began to breake up" in April or early May 1543. The cosmographer Jean Alphonse, who served as the expedition's pilot, later described Cap Rouge as a cold place, "very prone to snow," where the ground lay covered in two to three feet (60–90 centimeters) of ice—a description that suggests another cold winter. Meanwhile, the colony lived on the food aboard their ships: bread, beans, butter, bacon, dried beef, and salt cod—a rich diet, but lacking in vitamin C. "In the ende," according to Hakluyt's anonymous informant, "many of our people fell sicke of a certaine disease in their legges, reynes [kidneys], and stomacke, so that they seemed to bee deprived of all their limbs, and there dyed thereof about fiftie."[43]

A recent excavation of Cap Rouge has shed more light on the colony's conditions. The settlement was evidently constructed with defense in mind, and we can imagine that after hearing of Cartier's experiences Roberval was keen to keep the Stadaconans out. Our narrative source mentions some trade with villagers for fish, and archaeologists have turned up traces of maize on the site. Otherwise, there is no evidence of indigenous plants or of fresh food grown within the colony. The architecture and artifacts left behind also point to a steep hierarchy of status and living standards among the colonists. This suggests that the burden of work, exposure, poor diets, and sickness fell on the poor majority, while the colony's leaders were insulated from the worst effects.[44]

The following spring, once the river thawed, Roberval made his own trip up the St. Lawrence in pursuit of the mythical Saguenay. Of the seventy men he took with him, eight were lost when their boat capsized, and the rest eventually turned back, having failed to pass the rapids. In the meantime, the ships sent to France to fetch provisions had returned to Cap Rouge carrying news that their kingdom was once again at war with the Habsburgs. It remains unknown whether that fact alone obliged Roberval to return or whether he was already inclined to abandon the colony, but by September, he was back in France trying to liquidate his fleet and recoup his considerable expenses.[45]

Portuguese and Habsburg intelligence had followed the Cartier and Roberval voyages from the start. The trend of diplomatic correspondence shows a gradual shift, from outrage over what was assumed to be a piratical expedition to incredulous realization that the French really meant to colonize such a cold and remote part of the New World. In the summer of 1541, a Spanish diplomat finally concluded that the threat of piracy was "nonsense": "Their motive is that they think, from what they learn, that these provinces are rich in gold and silver, and they hope to do as we have done; but, in my judgment, they are making a mistake; for if there are no fisheries, this whole coast as far as Florida is utterly unproductive. In consequence of which they would be lost, or at best would make a short excursion, after losing a few men and the greater part of all they took from France."[46]

By November 1541, the Spanish ambassador in France expected the whole enterprise to "go up in smoke." He noted that the weather had been bad when Cartier departed in the spring, and that he must be icebound in Canada already. He gathered that if Roberval failed in his mission, too, "this daring and enterprise would be abandoned in the future, since few persons would take the risk, seeing the labor and difficulties the said Captain Roberval has encountered in equipping the said vessels with the necessary provisions and a suitable crew." Spanish interrogation of Newfoundland fishermen in late 1542 brought out descriptions of dangerous waters, extreme cold, and fierce Natives. One reported that Roberval "had not the means of founding a settlement, for the men were discontented and unwilling; and if he winters there this year, he believes he will have few men left." Spanish diplomats at the time found the Portuguese even more dismissive of Roberval's chances. According to the ambassador in Lisbon, King John III believed that "where the French had gone the cold is excessive . . . and the sea continuously tempestuous," so "they could not have gone to any place where they could do less damage" to Iberian interests.[47]

There is certainly a whiff of sour grapes in these descriptions. Yet the disillusionment with Canada—and the Canadian climate—comes out clearly. The return of Roberval's colony would provoke even more disappointment in France. As the chronicler André Thevet described, the supposed gold ore from Canada was assayed and turned out to be nothing more than iron pyrite, or fool's gold: "it could not withstand the fire and dissipated and turned to ashes." As for the diamonds, he wrote: "Those who first found them thought to have instant riches, thinking they were real diamonds, of which they brought back an abundance. From this is derived the proverb known to all today: it is [false as] a diamond of Canada."[48]

Thevet was equally disparaging of the country's climate and people. His descriptions misplaced Canada by several degrees of latitudes to the north. He presented it as a land of extreme winters, with rivers and seas frozen for months on end. He devoted a whole chapter of one geography to "these poor Canadians, and how they resist the cold," imagining Indians suffering continual hunger, huddled together in poor huts in a futile search for warmth.[49]

The introduction to Cartier's own account of his second voyage (known as the *Brief récit*) points out that the country he explored lay at "the same climates and parallels" as France. It insisted on Canada's "fertility and richness," if only the French would devote the necessary resources to colonize it.[50] In the same way, cosmographer Jean Alphonse determined that "all the lands of Canada truly deserve to be called New France, because they are at the same latitude. And if they were as well populated as France, my opinion is that they would be just as temperate." The French only needed to come, clear the forests, and cultivate the soil to moderate the country's climate and make it livable.[51]

This kind of argument would become popular during the seventeenth and eighteenth centuries, but there is no evidence it made much impression among Frenchmen at the time.[52] Historians have doubted how much the Cartier expeditions really contributed to French interest in Canada at all. Canadian animal and botanical specimens enjoyed only short-lived interest and novelty, unlike the perennial fascination of tropical exotica. Cartier's *Brief récit* ran through a single edition in 1545, and was scarcely mentioned in French literature for decades. Rumors, imagination, and medieval legends continued to populate French descriptions of the country for generations to come.[53]

Some French maps, particularly the famous decorative charts of Dieppe, incorporated names and geographical features discovered by Cartier. However, their descriptions of the country, when not fanciful, could be brief and unflattering. One such map presented to French king Henry II (r. 1547–1559) noted how Cartier discovered Canada, but added that its "austerité, intemperance, et petit proffit" drove him away. In the following decades, France would devote more attention to colonial expeditions in Florida and even Brazil.[54] "While hopes were entertained of these more southerly enterprises," wrote a later adventurer to Canada, Marc Lescarbot, "the discoveries of Jacques Cartier were forgotten; insomuch that many years passed by during which we French slept, and did nothing memorable at sea."[55]

Lescarbot was not entirely correct. As historian Laurier Turgeon has argued, the Cartier expeditions had begun to incorporate Canada into the maps, the imagination, and the tastes of French men and women—and in particular, their taste for fish. From the 1530s to the 1550s, French fishing fleets at Newfoundland swelled from dozens to hundreds of vessels. By midcentury, salt cod—the major catch of the Newfoundland Banks—was fast becoming a major item of trade and a staple of the French diet.[56]

During the high Middle Ages Europeans had already begun to exploit inland waters to their limits. Elite consumers turned to aquaculture of considerable scale and sophistication in order to supply themselves with the most prestigious freshwater seafood. At the same time, more and more European fishermen ventured out into the open seas, at first mostly for small oily fish such as herring, and then for larger predators, particularly cod. By the 1500s, the general trend of population growth and urbanization, the poor harvests and inflation of the Little Ice Age, and religious strictures on meat during holidays and Lent all drove a new mass demand for seafood, mostly in the form of dried or salted fish.[57]

Thus the discovery of the Newfoundland fisheries had come at an opportune moment. The banks off America's North Atlantic coast provide the perfect waters for cod, especially the Grand Banks off Newfoundland, where the warm Gulf Stream meets the cold nutrient-rich Labrador Current. The abundance of fish became legendary, attracting fishermen every summer from all over Europe, but particularly Frenchmen, who enjoyed better access to the Biscay salt used to cure the catch. With his discoveries, Cartier expanded that fishing into the Gulf of St. Lawrence, too.[58]

The Little Ice Age had a hand in these fisheries, although the connection between climate change and fish catch is by no means straightforward. It would be too simple to say that a given species prefers one temperature or another: factors from salinity to primary productivity, age distribution, and predator and prey abundance all play a part in how climate regulates fish populations. Moreover, reconstruction of historical fish catches remains a work in progress. Nevertheless, the research so far sketches a certain pattern. During warmer periods with a stronger North Atlantic Current, cod and herring tended to do better at the north end of their ranges, in Iceland and Norway respectively. During colder periods, when warm currents weakened, those fish multiplied at the sound end of their ranges, meaning more herring off the coasts of Britain and southern Sweden, and more cod off New England and in the southern part of the North Sea.

The warm North Atlantic Current remained strong throughout most of the 1500s, bringing unusual Arctic warmth by Little Ice Age standards. Taken together, the evidence suggests that the era of Cartier's expeditions would have been a good time for the cod fisheries in Newfoundland and Iceland, and also for herring in the North Sea. Then toward the end of the sixteenth century, the North Atlantic Current weakened and the Arctic abruptly cooled. With that climatic shift, fish populations around Iceland and Norway would have suffered, and those in the English Channel and parts of the Baltic might have benefited (although catches off the British Isles were notably poor during the 1590s). New England cod schools would have become more plentiful and those around the Grand Banks less so.[59]

Compared to already heavily fished waters around Europe, however, those of the New World appeared almost supernaturally prolific, even in bad years. In 1597, the English captain Charles Leigh claimed to have caught one cod per hook per minute from the Gulf of St. Lawrence, while marine birds lined the shores "as thick as stones lie in a paved street."[60] "Whether Northern Europe's once-abundant waters were being depleted," historian W. Jeffrey Bolster writes, "or whether the ecosystem could no longer produce enough to satisfy heightened demand, perhaps because of climate change, is not clear. It is clear, however, that European fishermen wanted more fish than they could catch in home waters."[61]

Europe's exploitation of fisheries overseas was, after all, just one part of a much larger shift in human ecology across the early modern world. Growth, urbanization, commerce, and technology were transforming populations in Europe and Asia into "ecological omnivores" whose consumption of resources reached ever farther beyond their borders. Whalers followed fishermen into new Atlantic and Pacific waters. English and Dutch merchants brought grain and timber from the Baltic. Germans and Italians imported cattle from the plains of Poland and Hungary. Russian Cossacks ventured into Siberia; Turkish and Arab peasants plowed up nomads' grazing lands; and Han settlers carved new farms out of the hillsides of southwestern China. Mughal settlers pressed into the forests of Bengal, and Japanese traders moved into Hokkaido. Sugar plantations would soon draw in millions of enslaved workers and transform the landscapes of whole islands in the Caribbean. All were part of that era's vast expansion into seemingly endless frontiers of new resources.[62]

The fur trade benefited from that same conjuncture of ecological pressures and economic forces. Western Europeans had hunted most of their fur-bearing animals, including the Eurasian beaver, to regional extinction by the early sixteenth century. Yet demand for furs kept rising, driven by growing numbers of urban consumers and particularly their taste for broad-brimmed felt hats of the sort immortalized in Rembrandt paintings. These hats required great quantities of fur to manufacture, and the soft undercoat of beavers worked best of all. (The Little Ice Age cold may have played its part in this trend, too, but we shouldn't assume clothing fashions made any more practical sense in the sixteenth century than they do today.) At first, English and Dutch traders met most demand with furs from Russia, including sable and mink. But in the early 1580s, war between Russia and Sweden temporarily cut off supplies, and North American traders were poised to step in.

During the early 1500s, European fishermen in Canada made little contact with First Nations. By midcentury, as hundreds of ships began to arrive each year and more fishermen came ashore to cure their catch, exchange became frequent and routine. A pidgin language of trade grew up, mixing Basque, Algonquian, and French words. By the 1580s, French merchants were investing in a well-established commerce, sending European manufactures for North American pelts and furs. French clothiers created new fashions out of Canadian supplies and reexported their wares throughout Europe.[63]

Thus the same factors that had made Canada so unappealing in Cartier's time—its vast size, wide bays, thick forests, Native peoples, and cold climate—now made it potentially valuable. Even without gold, diamonds, or a passage to the South Sea, its proven resources of fish and fur would attract another generation of French colonial expeditions at the end of the century.

These expeditions would arrive to find Canada's Native and natural worlds transformed. Archaeologists have discovered that the effects of European contact reached far beyond the coast and deep into the interior of North America. European copper, beads, kettles, ornaments, and axes have turned up hundreds of miles from the original sites of exchange. Some reshaped Native American lives because they were useful. Others served as prestige goods that shook up existing social and political hierarchies. Some traveled along existing trade routes. Others generated new trade routes and new competition to control them. Some nations, such as the Beothuk of Newfoundland, backed away from direct contact with Europeans. Others, such as the Mi'kmaq, stepped boldly into maritime commerce and even piracy. These Native intermediaries often controlled the flow of goods and information, meaning Europeans were unaware how far away and in what ways their commerce brought changes to distant parts of North America and their peoples.[64]

The most profound change, and the most consequential for French Canada, was the disappearance of the St. Lawrence Iroquois. The Iroquoian-speaking peoples encountered by Cartier—and we can identify them as Iroquoian by the vocabularies he recorded—had evidently been horticulturalists, growing corn and beans and living in permanent villages. They survived the winters by dispersing to hunt game and by preserving essential local environmental knowledge, such as the use of *annedda* to cure scurvy. But by the time the French came back to the St. Lawrence valley at the end of the century, they were gone, and their villages, crops, and knowledge had gone with them. Archaeological evidence suggests that sites west of Montreal were abandoned by around 1580 and the rest before 1600, by which time Iroquoian names had also disappeared from French maps of Canada.[65]

Decades ago, scholars blamed the Cartier expeditions for spreading European diseases that supposedly decimated the St. Lawrence Iroquois. Subsequent research has found no evidence that such a brief contact could have brought such catastrophic population loss.[66] Instead, some archaeologists have argued that competition to control European trade goods drove the Iroquois out. Huron invaded from the west, forcing Iroquois to abandon villages such as Hochelaga and Stadacona and to migrate south. These archaeologists point to signs of warfare, cannibalism, and captives taken from the Iroquois, including pottery apparently made by Iroquois women at Huron sites.

Competition over trade, however, was neither the only nor necessarily the most important reason for the disappearance. Another argument implicates the colder phase of the Little Ice Age that began during the late 1500s. Its proponents point out that the evidence for conflict between Huron and Iroquois long predates European trade. Nor do we find many European goods among the St. Lawrence Iroquois that would suggest they ever controlled that trade. Conflict could have arisen from competition over hunting grounds and other natural resources made scarce by a changing climate. Iroquoian pottery could have reached Huron villages by commerce as well as captivity. On the other hand, the long winters characteristic of the late 1500s must have put pressure on Iroquoian food supplies, as suggested by Cartier's account. Or else cold and hunger left the St. Lawrence Iroquois more vulnerable to disease or warfare. The new high-resolution pollen evidence for rapid climate and environmental change lends strong support to this theory. There were indeed changes in the land, and the Iroquois either could not adapt or chose to migrate.[67]

Unlike the English in Virginia and New England, or the Spanish in Florida and New Mexico, the French would find neither the challenges nor the resources of permanent Indian villages when their expeditions returned to the St. Lawrence. They would find a sort of no-man's-land in the midst of a worsening climate and of conflict and competition over commerce.

The fiscal turmoil, political factionalism, and sectarian conflict that beset France in the second half of the sixteenth century delayed the realization of new colonial ambitions in Canada. In 1577, during a brief lull in France's wars of religion, King Henry III (r. 1574–1589) granted the Marquis Troïlus de La Roche de Mesgouez a sweeping commission to conquer the country. The following year, however, the first ships sent by La Roche fell quickly into the hands of the English navy. During a second short break in the conflict in 1582, a coalition of statesmen and merchants organized another Canadian expedition, led by the captain Étienne Bellenger. Based on notes by Richard Hakluyt, it seems they meant to set up a trading post and possibly a colony in Nova Scotia. The expedition departed France early in 1583, made a swift Atlantic crossing, and began to chart the coast of the peninsula that spring, going all the way around to the Bay of Fundy. Bellenger apparently abandoned the enterprise about four months later, when attacks by Indians took the lives of two of his men. Yet furs obtained from exchanges with First Nations brought a handsome profit and attracted new interest to what would later become Canada's maritime provinces. The following year La Roche prepared a second and larger expedition, only to see it collapse when its largest ship was wrecked off the west coast of France.[68]

During the next decade and a half, France plunged even deeper into civil war, as well as famine and disease brought on by the miserable summers and freezing winters of the 1580s and 1590s. Only in 1598, once the new king, Henry IV (r. 1589–1610), had restored foreign and domestic peace, did the French enterprise in Canada resume in earnest. Henry IV appears to have had some long-standing interest in North American colonization, perhaps going back to his association with Huguenot privateers such as Admiral Coligny, of the French Florida expeditions described in Chapter 3. He also had to defend France's long-standing claims to Canada from new English encroachments. During the 1590s Bristol fishermen, whalers, and walrus-hunters had expanded their presence off Newfoundland and moved into the Gulf of St. Lawrence as well. In 1597, a venture led by the Englishman Charles Leigh even planned to overwinter on the Magdalen Islands, but the expedition fell through when one of its ships was captured and the other driven out by French fishermen.[69]

At the Peace of Vervins in 1598, Henry IV failed to get Spain to concede any of its North American claims. By then, however, Spain's statesmen cared even less for Canada than they had in Cartier's time. One of Spain's few natural historians to discuss the country described it as "a land of extreme cold, from which the natives suffer admirably," reminding his readers that "half the company that Jacques Cartier took with him died of cold in this land." When the issue of a French expedition eventually came before the Spanish Council of State in 1604, it concluded that Canada was "very removed from the East and West Indies, and so there does not seem any reason to divert this voyage."[70]

Spain's indifference may have been encouraged by news of France's early failures. In 1598, the ever-unfortunate Marquis de La Roche still held prior rights to colonize Canada, although Henry IV had reduced his title from viceroy of the country to mere lieutenant general. Unable to secure new private or royal funds, he sank his personal fortune into a third Canadian venture, this time bound for Sable Island.

It proved an unfortunate choice. Lying more than 100 miles (165 kilometers) off Nova Scotia, Sable Island is a shifting crescent of sand

peaking just above the sea at the edge of the Sable Banks. João Fagundes is credited with its "discovery" in 1520, when two of his vessels
headed to Newfoundland ran aground on Sable's shores. Hundreds of
captains, including Humphrey Gilbert in 1583, would "rediscover" it
over the following decades, when their ships wrecked on its shallows.
The island measures roughly 30 miles (48 kilometers) long today, but
charts of the late sixteenth century sometimes show it stretching 100
miles (165 kilometers) or more, like some great claw grasping at ships
in the North Atlantic.[71]

La Roche must have seen certain advantages in the place. It was
uninhabited, apart from some castaway livestock; it had fresh water;
and it was strategically positioned to control maritime traffic to and
from Newfoundland. For King Henry IV, it had other benefits. A
colony there would present a challenge to Spain's exclusive claims on
America, but not such a provocation that it might wreck the peace negotiations at Vervins. Sable's isolation also made it an appealing site to
dump unwanted Frenchmen, at a time when some believed France
was impoverished and overpopulated. As Marc Lescarbot would write,
"We see France full of beggars and vagabonds of every kind, not to
mention a vast number who groan in silence." These were the men
(and some women) that he and others expected to populate Canada.[72]
For lack of volunteers, vagrants and prisoners were rounded up for the
expedition.[73]

La Roche sailed to Sable in spring 1598 and deposited about forty-
five to fifty men there. He evidently planned to set up the island as a
temporary base for further exploration and settlement. However, before La Roche could return to Sable from his first reconnaissance of
the mainland, his ship was "overtaken by a wind so strong and violent
that he was compelled to run before it, and in ten or twelve days found
himself in France," as Marc Lescarbot later told the story. Lacking money
or support for a new expedition, La Roche just managed to keep the
island resupplied each year through 1601.

Then in 1602, it all went wrong. Sources differ on exactly what happened, but it appears La Roche ended up a prisoner in France, his estate ruined. Supplies never reached the island. Its colonists murdered
the officers in charge, and then in all likelihood started murdering
each other. When a French ship finally arrived in 1603, it found only
eleven survivors, clad in skins and "reduced to every extremity by the

sterility of the said country," according to testimony in a subsequent court case over the incident.[74]

Whether most had died of violence, hunger, exposure, disease, or some other cause remains unclear. There should have been livestock and marine life enough for the settlers to survive a few years on the island, and there are indications they tried to plant grains as well. However, as the experience of Jamestown demonstrates, the survival of any colony—even one surrounded by natural abundance—depended on local environmental knowledge and settler morale. The exiles at Sable possessed neither. With overuse and without sanitary precautions, the island's shallow aquifer could have been tainted by encroaching salt-water or human waste, poisoning the inhabitants. Located so far out into the ocean, Sable has a mild maritime climate, but the island is still subject to fierce gales and occasional freezing temperatures. There is no mention in the brief surviving accounts of what weather they encountered, but experience elsewhere in Canada suggests those winters were cold indeed.

In the meantime, La Roche's commission under Henry IV left open the possibility that a different enterprise might pursue ventures farther inland. Two merchants invested in the Newfoundland fisheries—Pierre Chauvin, sieur de Tonnetuit, and François Gravé, sieur du Pont—won a limited monopoly of trade along the St. Lawrence valley. Without royal funds, and facing stiff opposition from other French merchants, Chauvin and Gravé sank their own money into a commercial and colonizing venture in early 1600. Their four ships left France that June, sailing around Newfoundland and up the St. Lawrence to Tadoussac. There Chauvin deposited sixteen men to pass the winter, while the remainder of the expedition sought fish and furs.

Sometime during the late sixteenth century, Tadoussac had become a regular summer trading post for French and Montagnais (the First Nations people now usually known as Innu). It had enjoyed a similar role among Native Americans long before that. Tadoussac lies along the north shore of the St. Lawrence at the outflow of the Saguenay River, a deep fjord cut by ice-age glaciers. The cold, nutrient-rich water that flows up from its depths fertilizes a stretch of the St. Lawrence, attracting a bounty of fish and whales. The location formed a natural meeting point for trade (and now a natural destination for summer tourists and whale-watchers). Yet it had never been a place for year-round habitation—and for good reason, as soon became apparent.

The sixteen French colonists crowded into a small wooden trading post for the winter, now reconstructed on the site for the benefit of tourists. It still stood there when Samuel de Champlain first visited Tadoussac two and a half years later. He judged it "the most disagreeable and barren place in the whole country," adding, "If there is an ounce of cold forty leagues up the river [at Quebec], there will be a pound of it here." His account remains our only detailed guide to what happened next that winter of 1600–1601:

> Our men wintering there soon used up their small store, and the winter coming on made them realize the thorough difference between France and Tadoussac. It was like the court of King Petaud, each desiring to be its leader; idleness and laziness along with the diseases which seized them unawares, reduced them to such great straits that they were obliged to entrust themselves to the Indians, who charitably took them in, and they gave up their abode. Some died miserably; the others in great distress awaited the return of the ships.

Based on tree-ring evidence, this year must have been one of the coldest in Tadoussac for centuries—just as it was for Oñate's colonists far away in New Mexico. Champlain blamed the deaths on Saguenay's "dangerous air," but it was almost certainly scurvy that killed them. The Frenchmen succumbed while Chauvin was still trying to scrape together funds for a relief expedition in the spring of 1601.[75]

Arriving late in the season, and stopping first to collect the survivors from Tadoussac, Chauvin had little time left to fish or trade for furs. Failing to yield a profit on his voyage, and still facing resistance from rival merchants, Chauvin had to renegotiate his monopoly in late 1602. When he died early the following year, the enterprise passed to the gentleman officer Aymar de Chaste, whom Henry IV appointed viceroy of New France in expectation of a larger colonizing venture. What revived the enterprise in 1603 were new rumors that the rapids on the way to Saguenay might be navigable after all. Under a new license to "discover and people the country," de Chaste and François

Gravé outfitted three ships that spring. One of them, the *Bonne Renommée,* would carry Champlain on his first voyage to Canada.[76]

As we can gather from his description of the Tadoussac colony, Champlain would find much to learn from these and other early voyages to Canada. Years later, Champlain's history and that of Marc Lescarbot—our two major narrative sources for expeditions of the early seventeenth century—would revisit France's early attempts at North American colonization in an effort to draw lessons.

Both recognized the dangers of the scurvy encountered on Cartier's expeditions, and both equated it with the climate of Canada. "Just as plants exported from their native province, or when often transplanted within the same district, do not give as good results as in their native soil," wrote Lescarbot, "many of Cartier's men, whose bodily dispositions were not in full accord with the temper of the air of that country, were seized by unknown maladies."[77]

Yet Lescarbot and Cartier realized that warmer climates did not ensure success, either. Both criticized the French Florida expedition for failing to bring enough food, and for supposing that "the soil of those countries produces without any sowings," as Champlain put it. Lescarbot concurred: "But in vain does one run and weary oneself in search of havens wherein fate is kind. She is ever the same. It is a good thing to live in a mild climate . . . but death follows us everywhere."[78]

Both also grew critical of those who would judge the climates of distant lands based on their latitude or on a passing impression of the landscape. Champlain declared:

> It is of little use to hurry over to distant lands and go inhabit them, without first of all exploring them and living there at least one whole year, to learn the quality of the country, the diversity of the seasons in order afterwards to lay the foundations for a colony. The majority of the colonizers and travelers do not do this, but are content merely to look at the coasts and latitudes of lands as they pass by without stopping. Others undertake such voyages based on mere relations made to persons who, though well versed in the world's affairs, and with considerable and long experience, nevertheless

being ignorant in these matters, believe that everything
should be governed by the latitude where they are, and it is in
this that they find themselves greatly mistaken.[79]

Lescarbot proposed several explanations for Canada's unexpectedly
cold winters. At one point, he blamed the icebergs carried down by the
Labrador Current. At another, he argued that the east-west rotation of
the heavens around the earth brought cold oceanic vapors to eastern
North America. In still another passage, he blamed Canada's thick
woods and untilled soils for preventing the sun from warming the
ground. His descriptions left no doubt, however, that New France had
colder winters than Europe, and probably colder winters than Canada
has today.[80]

Perceptive though they were, these were judgments written years
after the fact. There is no evidence Champlain had any of these lessons
in mind when he first voyaged to Canada in 1603. At that time, he had
some experience sailing in the Spanish Caribbean, and apparently
some background as a draftsman and cartographer. Some biographies
have pictured him in 1601–1602 working as the official geographer for
King Henry IV, diligently mastering Canada's natural and colonial history. However, there is little evidence to support this assertion.[81]

Champlain too often appears as a visionary leader because he wrote
his own story with the benefit of hindsight. More realistically, we may
say that he was sensible and discerning enough to learn from his own
mistakes and those of his contemporaries. He also had the great good
fortune to survive those mistakes unscathed. Not all of his companions were so lucky.[82]

The 1603 voyage sponsored by Aymar de Chaste, which first carried
Champlain to Canada, sailed to the Lachine rapids past present-day
Montreal, finding them as impassable as had Cartier and Roberval
more than six decades earlier. Along the way, the expedition made new
contacts among the Huron and hunter-gatherer Montagnais who now
occupied the St. Lawrence. Champlain's description of those First Nations and their encounters, along with the arrival of several Indians
accompanying the expedition back to France, earned Champlain credit
at the court of Henry IV and helped draw attention to Canada. Champlain also gathered some basic geographical information, including

the locations of the Great Lakes and Hudson's Bay, which he had the good sense not to mistake for a Northwest Passage.[83]

While the expedition was away, Aymar de Chaste died and La Roche's enterprise at Sable Island collapsed. Pierre Dugua de Monts, a Huguenot officer who had fought in the civil wars alongside Henry IV, and who had earlier joined forces with Chauvin on the Tadoussac colony, used the occasion to press for a new commission in French North America. The king granted most of his requests, appointing him lieutenant general with a monopoly on American furs from Cape Breton to New England, on the condition that de Monts settle sixty Frenchmen in the country each year. Early in 1604, overcoming other merchants' objections to his monopoly, de Monts managed to organize a company to finance the venture, built on the promise of shared profits from the fur trade.[84]

The expedition's two ships, the *Don de Dieu* and *Bonne Renommée*, departed from separate ports in early April 1604. They carried months of supplies and more than a hundred men (not yet any women), including various craftsmen and miners, all in anticipation of finding precious ores rumored to lie somewhere in Nova Scotia. The first ship, carrying Pierre Dugua and Champlain among others, caught favorable winds and made the voyage in exceptional time, dodging the icebergs that still crowded the waters near Sable Island. Off the coast of Nova Scotia, they captured a small French ship, the *Levrette*, discovered trespassing on Dugua's new monopoly. While the lieutenant general waited there for the arrival of the *Bonne Renommée* and its supplies, Champlain and eleven others took a boat to explore the Bay of Fundy, finding what looked to be suitable deposits of iron and silver. With this exciting news, Dugua took the *Don de Dieu* and *Levrette* up to the prospective mines, which turned out to be unworkable. That voyage brought the French past a bay that Champlain named Port-Royal, which seemed a promising site for an eventual colony.

Dugua's hopes, however, were to find a more southerly location to settle, and presumably one with a more temperate climate. That June, the ships explored the New Brunswick and Maine coasts in search of a place to spend the winter—if not as a permanent colony, then at least as a trading post and base for further exploration.

They soon located a site that looked almost ideal. A short way up a wide navigable river, they found a small wooded island that de Monts named St. Croix. Cliffs on three sides made it easy to fortify and suitable to command traffic along the river. "We judged it the best," recalled Champlain, "as much for the location and the good country, as for the commerce we meant to have with the savages of the coast and the interior." The latter were Native Americans whom the French called Etchemin—today's Passamaquoddy nation—and whom they had already contacted during the previous year's expedition up the St. Lawrence. The rivers were full of fish, and once cleared and planted, the land brought up grains quickly. "It was difficult to understand the country without having wintered there, since on arriving in summer everything is so agreeable, on account of the woods, pretty countryside, and fisheries of several sorts that we found there."[85]

Today, those tourists still curious about the place can approach St. Croix (also known as Dochet Island) on either the north or south shore of the St. Croix River—that is, on either the Canadian or U.S. side of the border. Either approach takes them to public monuments inviting visitors to contemplate the hard lessons those first colonists learned there during the harsh winter of 1604–1605. Their story still makes an unsettling contrast with the pleasant prospect of the wide river and small, wooded island in the middle. However, incidental documents and recent excavations of the site only confirm the nightmare described in Champlain's personal narrative.[86]

Late that summer and autumn, the Frenchmen at St. Croix took on extra supplies and settlers from the *Bonne Renommée,* and Dugua sent his two larger ships back to France with the furs obtained from trade. In September, while most of the French were still at work building the settlement, Champlain set out with a dozen sailors and two Etchemin guides on a further exploration of the New England coast. Recalling the same legends that inspired English expeditions to Norumbega, they soon returned disappointed. The Indians along the Maine coast were accustomed to trading only with Mi'kmaq intermediaries. In mid-September, a short way past Kennebec Bay, they ran into bad weather amid a difficult rocky coastline and decided to turn back. "I believe," Champlain concluded, "that this place is just as disagreeable in winter as that of our settlement, of which we were well disappointed."[87]

The colony at St. Croix that had taken shape in the meantime resembled a French town in miniature. Little defended, except by some cannon to control the river, its small separate dwellings and tidy gardens were clustered around a central square.[88] The Frenchmen realized they had little to fear from the neighboring Etchemins, with whom they got on well, but their colony's design also failed to protect them from the cold and wind. Worse still, as archaeologists have discovered, they built their storehouse without any cellar to insulate their food and drink from the cold.[89]

"The winter surprised us sooner than we expected," Champlain recalled. The snows began on October 6, and a deep freeze set in on December 3. Winds from the north and northwest brought a bitter chill. "The cold is harsher and more excessive than in France," Champlain recalled, "and much longer." The ground lay covered in three to four feet (90–120 centimeters) of snow until the end of April, making it an exceptional winter even for that part of the world. The cider froze in its casks, "and each man was given his portion by weight," according to Lescarbot. The little wine strong enough to stay liquid was strictly rationed among the men. Without adequate woods or a spring on the island, the settlers had to venture out across the river daily for firewood and fresh water. Once the frozen river became too treacherous, the Frenchmen drank melted snow and went short of fuel for their fires.[90]

Written and archaeological evidence gives some clues to their diet that autumn and winter. They almost certainly brought aboard the usual ship fare of biscuit and salted meat, but "without any delicacies," as Lescarbot described. The French had set up a small mill on the island, but Champlain complained that most of their grain dried out in the sandy soil. Both he and Lescarbot mentioned gardens with cabbages and other greens, but it is not clear whether or how much they harvested before the early onset of winter. Archaeologists have found jars for olives and glassware for wines. There were hooks, nets, and shot for fishing and hunting. They had few livestock, to judge from the animal bones found on the island, but caught plenty of fowl; Champlain's account mentions mussels and other shellfish taken from the river before it froze. The distribution of archaeological finds suggests the most privileged ate apart from the others and enjoyed a more varied diet, which likely accounts for Champlain's survival and that of Pierre Dugua. Yet with the possible exception of those garden plants, nothing in their diet contained enough vitamin C to ward off scurvy for long.[91]

In some respects, European understanding of that disease had come a long way in the decades since the Cartier expeditions. Many more ships had made the long ocean voyages to the East and West Indies, and many more sailors grew acquainted with the ravages of vitamin C deficiency. The trouble now was not that they had too few explanations or cures for scurvy, but too many. Richard Hawkins, son of the famous privateer, had recommended oranges as an antidote as early as the 1590s. Henry Hudson's men survived their winter in James Bay by eating a plant they recognized as "scurvy grass."

Yet these genuine cures had to compete with any number of bogus recommendations. Hugh Plat, mentioned in Chapter 4 for his helpful advice on famine foods, also wrote a tract titled *Certaine Philosophical Preparations of Foode and Beverage for Sea-men, in Their Long Voyages.* To his credit, he included lemon juice, but he listed it alongside dozens of other foodstuffs with little or no curative power. Marc Lescarbot was typical of learned opinion at the time. Sometimes he argued for a change of diet, recommending "two or three doses of chicken broth" or "good wine" to cure the disease, or more accurately that "the tender herbs of springtime make a perfect remedy." At other times he remained convinced that scurvy emerged from malevolent terrestrial vapors and could be cured with exercise and hot baths, or else warded off by burning down trees. Champlain's less educated guess about their illness came closer to the mark: he later blamed their monotonous diet for causing "bad blood."[92]

The disease began to take hold in December or January, with its characteristic symptoms of rotting gums, loose teeth, and putrid sores. The victims soon suffered lassitude and sharp pains in their stomachs and sides. Champlain recognized the signs of scurvy, but no one could find a cure that worked. Lescarbot later identified it as the same disease that had struck Jacques Cartier's men, but he regretted, "As for the tree called *annedda*, of which the said Cartier speaks, the savages of these regions know it not. So that it was most pitiful to see all save a very few in misery, and the poor patients, who could in no way be relieved, die all full of life."[93]

French Catholics were customarily buried facing east. When excavations of St. Croix first discovered the settlers' graveyard during the 1960s, they found most skeletons aligned toward the sun as it would have risen in March—a date of death that matches our written accounts of that winter. Examination of the bones (now respectfully reinterred) has confirmed scurvy aggravated by general malnutrition as the likely cause of death. Some thirty-five or thirty-six men died, and most of the others were stricken before recovering in April. By then Etchemins had begun to return, offering to barter food. The heavy snows had made for easy tracking of game and a successful hunting season.[94]

By May, the settlers on St. Croix began to doubt their fellow Frenchmen would ever come to relieve them, as promised the previous autumn. In mid-June, just as they were about to crowd onto their small vessels and set off looking for help, the *Bonne Renommée* finally arrived with fresh supplies and more men.

At that point Champlain took a small ship on a second exploration of the New England coast, this time venturing as far south as today's Nauset Harbor, Massachusetts. The voyage is particularly remarkable for an encounter with the Indians of Cape Cod, which Champlain later recorded in his *Voyages:*

> We asked them if they had their permanent residence in this place, and whether it snowed much, which we could not easily learn since we did not understand their language, even though they made an effort at it using signs, taking up sand in their hand, then spreading it on the ground, and indicating that the snow was the same color as our collars, and fell to the depth of a foot, and others indicated that it was less, giving us also to understand that the harbor never froze over; but we were unable to find out whether the snow lasted a long time.[95]

There should be nothing remarkable about European explorers simply asking Native Americans about the local climate—except that there is no indication Champlain had ever done the like before. In fact, this encounter has almost no parallel anywhere in the scores of colonial narratives up to the foundings of Jamestown, Santa Fe, and Quebec. While absence of evidence cannot be taken as evidence of absence, the fact that Europeans almost never recorded actual conversations with Native Americans about local climates and weather may tell us a lot about the miscommunication, mistrust, or overconfidence that marred early encounters.

From this exchange, and from his view of the landscape, Champlain judged that winters in southern New England must be a little more

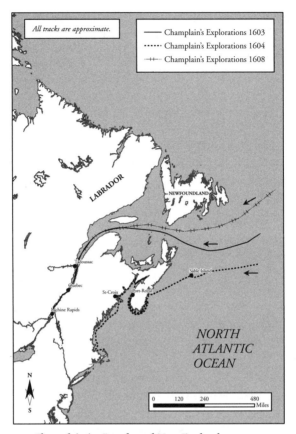

All tracks are approximate.

—— Champlain's Explorations 1603
······ Champlain's Explorations 1604
⁻⁺⁻ Champlain's Explorations 1608

LABRADOR

NEWFOUNDLAND

Tadoussac

Quebec

St-Croix   Port-Royal

Sible Island

Lachine Rapids

NORTH
ATLANTIC
OCEAN

N

S

0   120   240   480
                      Miles

9.2 Champlain in Canada and New England

temperate than the one he had just experienced on St. Croix. However, that summer of 1605 in Massachusetts was evidently no summer at all. "During the time we were there it blew a gale from the northeast, with the sky so overcast that the sun was hardly visible at all. It was very cold, so that we were obliged to put on our overcoats, which we had laid aside." Making little headway in the foul weather and facing a shortage of provisions, Champlain and his detachment were forced to turn back.[96]

By the time Champlain returned to the colony in early August, "the Sieur de Monts decided to remove elsewhere, and to build another settlement to escape the cold and the dreadful winter we had experienced at St. Croix island." Since another winter was coming on fast, and their

New England explorations had turned up no suitable site, Pierre Dugua opted to pack up their dwellings and transfer to Port-Royal, "where we judged the location much more mild and temperate." The *Bonne Renommée* took most of the survivors across the Bay of Fundy. Anxious to defend his monopoly at court, Pierre Dugua joined the ships returning to France in September 1605, leaving François Gravé in charge of the new settlement.[97]

Their preparations this time show they had learned some lessons from their winter at St. Croix. Champlain and Gravé selected a location with wood and fresh water close at hand, and one sheltered from the northwest wind—a wind "that we dreaded," Champlain recalled, "having been tortured by it" at St. Croix. Their settlement this time resembled less an open village than a compact manor, with buildings packed about a central courtyard. Their storehouse, built to minimize exposure to the winter winds, now boasted a proper cellar to protect their food and drink from freezing. They cleared some ground, dug gardens, and planted grains, "which fared well." Another search for copper mines, however, met without success.[98]

"The winter was not as harsh as it had been the year before," Champlain recalled, "nor the snows as great, nor of such long duration." The weather remained cool and rainy, without notable snow or ice, until late December. Nevertheless, the winter lasted long enough to turn deadly. In all likelihood, the settlers had again planted their gardens and grains too late for an adequate harvest. They were short of wine and unfamiliar with Nova Scotia's wild plants and fruits, which would have offered something in the way of vitamin C. They had made contact with neighboring Mi'kmaq peoples, but there is no mention of trade for fresh food. Of the forty-five men left at Port-Royal almost half came down with scurvy and twelve died, including the expedition's chief miner.[99]

A February storm brought gales strong enough to uproot trees. Twice, in March and in April, François Gravé tried to put to sea and explore down the coast. Both times, strong winds and poor piloting nearly wrecked his ship. By mid-June, the settlers again despaired of timely relief from France and again started to patch up their pinnace in preparation to go search for help. A month later, leaving only two men behind, the expedition set off around the coast of Nova Scotia hoping to find rescue.

After a week of difficult sailing, they encountered a French shallop bearing the welcome news that Dugua's relief expedition was already on its way. By the time they returned to Port-Royal at the end of July, the *Jonas* had already unloaded its supplies and its fifty new settlers. Those included Marc Lescarbot as well as Jean de Biencourt Baron de Poutrincourt et de Saint-Just, a nobleman to whom Dugua had granted the seigneury of Port-Royal the year before. Their voyage had been held up by troubles back in their port of departure, La Rochelle, and by storms and contrary winds at sea. Now, realizing they had arrived too late to find a more southerly site before winter, Poutrincourt ordered the roughly 100 Frenchmen to plant their gardens, set up their homes, and prepare the colony at Port-Royal again.[100]

Gravé returned to France on the *Jonas* late that August. In early September Poutrincourt, with Champlain, set out for a third and final reconnaissance of the New England coast. They started by gathering what supplies and plants they could recover from the remains of their settlement at St. Croix. Continuing down the shores of Maine and Massachusetts, they again faced bad weather and difficult sailing. Worse, they fell into hostilities with Indians for the first time, leaving two Frenchmen dead. Demoralized and fearing the onset of winter, they turned back toward Port-Royal in mid-October, having accomplished little. On their return journey, they found the islands off southern Maine already covered in ice "two inches thick, which had clearly been frozen some eight or ten days."[101]

Progress back at Port-Royal was more encouraging. In Lescarbot's rather rosy portrait of the colony, the gardens progressed well and the grain sprouted within a week. Port-Royal's numerous craftsmen devoted a few hours each day to enlarging their habitation and the rest to hunting or gathering shellfish. Welcoming Mi'kmaq gave freely of their food and wares.[102]

The cold came unusually late that year. The day after Christmas, a strong storm came in from the southeast, and on New Year's Eve it began to snow. By mid-January the rivers had frozen up, and by mid-February ice floes began to drift into their harbor. Nevertheless, the air was warmer, and the colonists could often go out in the winter sunshine in nothing more than their doublets. "The winter was not so long as in the preceding years," Champlain concluded, "nor did the snow remain so late upon the ground." March and April were mild enough that they could return to gardening. Lescarbot claimed, perhaps with some exaggeration, "We had fine weather nearly all winter." The spring brought some unusual cold: a late heavy snow came on May 10, and "several heavy white frosts which lasted as late as the tenth and twelfth of June, when all the trees were covered with leaves, except the oaks, which do not put out theirs until about the fifteenth." Based on his experience, Champlain judged the growing season of Nova Scotia to start a month and a half later than in France.[103]

"There was some scurvy among our men," according to Champlain, "but not as violent as it had been in the previous years." He put the death toll at seven, with eight or ten seriously ill. Lescarbot claimed only four died that February and March, and only "those who were downcast or slothful." He credited their good health to an innovation of Champlain's: an "Order of Good Cheer" in which the colony's elite, who dined with Poutrincourt, took turns organizing hunts and banquets for each other. Fresh game contains very little vitamin C, except the raw organ meat. But the "Order" could have contributed to health in other ways. Certainly it was good for morale. It also improved relations with the Mi'kmaq, and it forced the settlers outdoors, where they discovered other fresh foods, including some small red fruits that may have been cranberries. At Poutrincourt's table at least, the wine flowed freely. Nevertheless, the better progress on their gardens and the relatively mild winter must have made the greatest difference of all in their survival.[104]

This modest success in Canada was undone by failures back in France. Dugua had made a strong impression lobbying for his claims at the court of Henry IV, even presenting the king with exotic animals from the New World. French merchants, however, remained implacably opposed to the fur trade monopoly, and Dugua's seizure of French ships caught trespassing on his claims had only stoked more resentment. In any case, the long coastline of Cape Breton and Nova Scotia had proven impossible to patrol for contraband, which undercut the price of furs.[105]

Larger political forces were at work as well. Merchants in Holland were eager to move into the North American fur trade. In 1606, two of Dugua's ships had challenged two Dutch vessels caught trespassing on his claims. In the fight that ensued, the Frenchmen came out the worse, and Dugua's two ships were captured. Rather than pressing for compensation, Henry IV's chief minister, the duc de Sully, tried to use the fur trade as a bargaining chip in diplomacy with the Netherlands. Sully had always dismissed Canadian colonies as of "great expense and little use" and "ill-suited to the nature and brains of the French." By offering to open the Canadian fur trade to their merchants, he hoped to entice the Dutch back into war against their common enemy, Spain. This incident, while not fatal in itself, helped tip Dugua's tottering company into financial ruin by early 1607.[106]

The settlers at Port-Royal received the bad news in late May that year, when Dugua's secretary arrived aboard the *Jonas* with orders to evacuate the colony. After gathering their supplies, the last of the settlers finally left in August. In the meantime, citing "complaints made to the King in his council," and acting "for the good and utility of his people," Henry IV officially revoked Dugua's and other colonial monopolies on July 17, 1607.[107]

The end of the monopoly and the retreat from Port-Royal have often been presented as a setback, even a disaster, for French colonization in North America. "Great was our grief thus to abandon a soil which had produced for us such goodly wheat and so many fairly adorned gardens," lamented Lescarbot; "to abandon the enterprise was in truth to lack courage."[108] In fact, it may have been the greatest stroke of luck the enterprise ever enjoyed. The colony had made some progress in dealing with the Canadian climate and the threat of scurvy. Yet there is no sign it could have withstood what would soon become one of the longest, coldest North Atlantic winters for centuries. Back in France, Lescarbot wrote, "For just as on this side of the ocean that winter [of 1606–1607] was mild, so also the last winter of 1607–8, the most severe we have ever seen, was the same across the ocean, so that many savages died from the severity of the weather, as in France did many poor people and travelers."[109] In all likelihood, scurvy would have dealt both the colony and Canada's reputation a blow that would have taken years to overcome. In the early months of 1608, while settlers at Jamestown died miserably of cold and disease and the Popham colony saw its "hopes . . . frozen to death," the hundred settlers from Port-Royal were fortunate to be back in France, and Dugua and Champlain lucky they were still alive to plan another expedition.

In January 1608, Dugua managed to secure a new monopoly on the Canadian fur trade, this one to last only a single year. His next venture would have to find a way not only to survive the Canadian climate but also to defend his claims and turn a quick profit. Tadoussac, although a ready site for trade, had proven too cold. Island settlements, such as those at Sable and St. Croix, had been death traps. The New England coast, although farther south, did not seem much warmer, and its Indians made for uncertain partners in the fur trade. The shores of Nova Scotia were impossible to patrol. By process of elimination, plans circled back to the same location that had first attracted Cartier in 1535, and Champlain in 1603: Quebec.[110]

The expedition Dugua planned that spring combined commerce and colonization. On April 5, 1608, François Gravé departed France on the 80-ton *Lièvre* to launch the company's fur trade and earn a profit as quickly as possible. Champlain departed on a second ship with the expedition's settlers. Gravé arrived first at Tadoussac in late May, only to lose a fight with a larger Basque ship he discovered encroaching on the company's new monopoly. Champlain, who arrived ten days later, just managed to negotiate the *Lièvre*'s release. During June he explored up the Saguenay River, returning after a few days in disappointment: "All the land that I saw was nothing but mountains and rocky outcrops, mostly covered with pine and birch; a very unpleasant land, as much on one side as the other." Even the hunting proved poor, yielding nothing more than a few small birds, "for the excessive cold there." To the north there was rumored to be a saltwater sea, which Champlain later correctly identified as Hudson's Bay. However, the land to the north was reputed to be even colder, covered by snow for most of the year.[111]

Finally, in early July, Champlain took his settlers to their intended destination. Its name, Quebec, derived from the narrowing of the St. Lawrence River just past the Île d'Orléans. It was a natural site to control the fur trade coming from the upper St. Lawrence valley and beyond. Most of the modern city is located on a high promontory looking far down the river. However, what Champlain had in mind was the low ground near the river's edge: wooded, with streams of fresh water and good soil, and protected by high cliffs from the north wind. Since the disappearance of the St. Lawrence Iroquois, the location had been uninhabited.[112]

Days after arriving, Champlain discovered and defused a mutiny and plot on his life. Its four ringleaders were captured, one executed, and the other three sent back to France with François Gravé that September. Then the remaining settlers set to work in earnest building the colony.

From documents collected over the centuries since, historians have identified at least eighteen of those first twenty-eight colonists of French Canada. They included a baker, a tailor, carpenters, masons, a gardener, a surgeon, and some salaried laborers. They came from all around the country, but mostly northern France. About half could sign their names. In these respects, they would prove remarkably representative of the few thousand men and women who would settle in Quebec during the next century and a half of French rule, and who would become the ancestors of millions of living French Canadians. Although the bad harvests, inflation, and turmoil of the Little Ice Age may have made emigration more attractive, these emigrants did not come predominantly from among France's millions of poor and marginalized. Most came from geographically and economically mobile classes with links to the growing commerce of North Atlantic ports. Most would have been drawn by a sense of opportunity or adventure rather than desperation.[113]

The settlement at Quebec continued to build on the lessons learned at St. Croix. Work began by excavating the colony's storehouse. Then they constructed dwellings around a compact central courtyard for defense against attackers and against the elements. Even more than Port-Royal, the colony resembled a fortress rather than a village. With construction under way, colonists set to work as soon as possible clearing the land for gardens. But there could not have been much time between the sprouting of the first greens and the onslaught of another early winter.[114]

By the time the French sowed their wheat and barley in early October, the weather had already turned cold. "On the third of the month there was a white frost," Champlain recounted, "and on the fifteenth the leaves of the trees began to fall." In mid-November they had their first heavy snowfall, although it did not stay on the ground long. Around that time, the colony also began to suffer its first casualties.[115]

In previous French colonies, trade with First Nations had been one of the few ways settlers found fresh food and survived the winters. That first year at Quebec, however, relations with the Montagnais, although friendly, may have worked against the colonists' survival. The Montagnais, unlike the Iroquoians who once lived in Stadacona, were a seminomadic hunter-gatherer people. They grew no gardens, and as Champlain discovered, they survived the first months of winter from the early autumn run of eels, which they smoked in preparation for the weeks ahead. That autumn, the Montagnais left their dried eels in the care of the French for a month while they themselves went out hunting. They returned in mid-November and shared their catch. Whether because the eels had been poorly preserved or because of unsanitary conditions at the colony, both the Montagnais and French came down with dysentery, and several died.[116]

By January, the ground was covered in snow and ice. Early February brought strong winds and intense cold. On the twentieth of that month, the Montagnais returned to the colony a second time, making a harrowing crossing over the ice floes that choked the St. Lawrence. Their winter hunt had gone badly.

"Hunger so sorely pressed these poor wretches that, not knowing what to do, every man, woman, and child resolved to die or to cross the river, in the hope that I would help them in their dire need," Champlain described. "They came to our settlement so thin and emaciated that they looked like skeletons, most of them being unable to stand." Champlain shared out some of the colony's bread and beans. Then he found to his disgust that the Indians had devoured, half-raw, even the stinking carrion that Frenchmen had left outside the settlement.[117]

This act, which so revolted the French, may have been more than just a sign of hunger. The Montagnais, like many indigenous peoples of the north, ate raw organ meat because, whether they knew it or not, it provided one of their only sources of ascorbic acid during the winter. Meanwhile, the provisions back in the French settlement offered almost nothing in the way of vitamin C.[118]

"The scurvy began to take hold very late, that is in February until mid-April," Champlain recalled:

> Eighteen were struck down and ten died of it, and five others of dysentery. I had some of them opened to see if they were affected like those I had seen in the other settlements. We found the same thing. Some time after, our surgeon died. All this gave us much trouble, on account of the difficulty we had in nursing the sick.

Even one of the Montagnais who stayed on with the French that winter died of the illness.

Recognizing scurvy, Champlain again blamed it on their salted food. "The winter too is partly the cause," he added, "for it blocks the natural heat which causes greater corruption of the blood, and also the earth, when it is opened: certain vapors enclosed in come out and infect the air." He pointed out that the illness affected travelers to all parts of the East and West Indies, and again he lamented the absence of a cure. Lescarbot later wrote that Champlain made "a diligent search" for the *annedda* described by Cartier, but again without success.[119]

Excavations of that first settlement in Quebec have uncovered no improvements in the colonists' diet since the time of the St. Croix settlement, or even the time of Jacques Cartier. They appear to have lived on the same shipboard fare of biscuit, salted meat, cider, and a little wine. Most of the animal bones found from the first years of the settlement came from the few livestock they had brought with them, indicating the colonists lacked the skill or inclination for hunting and fishing. The first arrivals, unaccustomed to the long journey by ship and the primitive conditions of the colony, could have succumbed to multiple ailments and deficiencies, compounding the effects of scurvy.[120]

Recovery began only in mid-April, once the snows melted. In May, the trees broke back into leaf. From these observations, Champlain concluded that despite the fatal winter, the climate of Quebec must be warmer than that of Newfoundland and the lower St. Lawrence valley, where "one finds ice and snow in most places up to the end of May, and the whole great river is blocked with ice."[121]

In early June, a French ship finally arrived with fresh provisions and new instructions for Champlain. He was first to rendezvous with Gravé at Tadoussac, and then proceed back to France, in order to defend the company's fragile monopoly. Champlain, however, insisted on remaining in Canada at least through the summer of 1609.

By staying on, Champlain committed himself to make good on promises of friendship and alliance, first forged back in 1603 and renewed the previous year, with the St. Lawrence Montagnais and Algonquins. These nations now planned a campaign against the Mohawk, an Iroquoian people, "against whom they waged a mortal war, sparing nothing that belonged to them." Champlain saw an opportunity to intervene decisively on the side of French allies, to bring peace to the region and to expand the reach of the fur trade.[122]

That June and July, Champlain led a contingent of twenty Frenchmen, who accompanied the Montagnais war party into Mohawk country, in today's upstate New York. The journey took them around the rapids of what the French would later name the Richelieu River and then down a long, narrow lake that Champlain named for himself, being perhaps the first European ever to set eyes on it. Looking past its eastern shores to the Green Mountains of present-day Vermont, Champlain recalled seeing snow still on their summits in early July. Depending on how much snow he actually saw, this indicates that the snowfall must have been exceptionally heavy or late, even by the standards of the Little Ice Age. Perhaps recalling his experience on the New England coast, Champlain enquired of his Indian allies whether the land was inhabited, and he was assured that indeed "it had beautiful valleys and fertile fields" planted by the local Iroquois.[123]

On July 29, they finally encountered the Mohawk warriors at Ticonderoga, where Lake George empties into the south end of Lake Champlain. Champlain himself remains our only source for the details of what followed. All that night, the two war parties danced and sang, hurling insults at each other. The next morning, the Montagnais advanced to join battle with some 200 of the enemy, with three Iroquois chiefs at their head. The twenty Frenchmen, arquebuses at the ready, hid within the Montagnais ranks. Champlain loaded his gun with four shot, in preparation for a close fight.

"As soon as we landed," Champlain recounted, "they [the Montagnais] began to run around two hundred paces toward their enemy, who stood firm and hadn't yet spotted my companions, who had gone into the woods with some savages."

> Our men began to cry out to me, and to make way for me, they parted in two. I put myself at the front, marching some 20 paces ahead, until I was some 30 paces from the enemy, where just then they perceived me and stopped to consider me, and I them. Once I saw them break ranks to shoot at us, I set my arquebus at my cheek and fired straight at one of the three chiefs. Suddenly, two of them fell to the ground, and one of their companions was wounded.

As the Montagnais gave out a shout, the other French stepped out of the woods and fired into the Mohawk ranks. After a brief stand, the latter took to their heels, and the battle turned into a rout.[124]

Champlain returned to France that autumn of 1609. In his absence, the small colony at Quebec was blessed with a mild winter and few deaths. In 1610, he would again go on campaign with the Montagnais against their Iroquois enemies, cementing the Franco-Algonquian alliance.

With these steps Champlain helped ensure that Quebec, however remote and tenuous its existence during those first years, had allies and trading partners among Canada's First Nations. In this way, he also ensured the colony's commercial purpose and eventual survival. At the same time, his actions committed New France to decades of conflict with Iroquois. The Montagnais, Algonquin, and Huron had also committed themselves, unknowingly, to the arrival of missionaries, alcohol, and European diseases.[125]

After decades of failures and false starts, the French succeeded in planting an enduring colony in North America by learning from mistakes and scaling back ambitions. The colony at Quebec was strategically located and relatively sheltered from the elements; its few inhabitants remained highly dependent on trade and good relations with neighboring nations. In this way, it helped set a pattern for future generations of settlement in what would become New France. At the same time, in Virginia, English colonists chose a path of confrontation and conflict with Native Americans, and consequently dependence on new investment and settlers from Europe. That choice would prove just as fateful in the long run, and catastrophic in the short term.

10

Such Wonders of Afflictions

Throughout 1607 and 1608, the Virginia Company had been reading the private letters of Jamestown's colonists and gathering information from returning sailors and settlers. At first, the company's directors expected the situation to improve on its own, once the colony had survived its early challenges. They blamed its problems on the poor quality of the common settlers, consumed by "idleness and bestial slouth." Once Captain Newport returned from the second supply in January 1609, bringing his negative report and John Smith's angry letter, they realized the colony's problems ran much deeper than that. It was time to put the whole venture on a new foundation.[1]

Under the leadership of the influential London merchant Sir Thomas Smythe, the Virginia Company secured a second royal charter that spring. This charter expanded the company's territory to include all land 200 miles (322 kilometers) north and south of Point Comfort, from the Atlantic coast to the Pacific Ocean (assumed to be much closer than it really was). The company officially closed its New England venture, and called on investors in the defunct Popham colony to join the Jamestown project.[2]

The new charter reacted to the failure of the colony's old governing council and presidency by appointing new officers with "full and abso-lute power and authority, to correct, punish, pardon, govern and rule."[3] Elaborated and later published in 1612, the colony's "Lawes Divine, Morall and Martiall" give a sense of Jamestown's chaos and the com-pany's fears during 1609. The law code ordered death for heresy, blas-phemy, "traiterous words," sodomy, adultery, sacrilege, false witness, embezzlement, stealing food, and mutiny, among other crimes. It strictly forbade sailors from charging settlers more than the fixed rate for provisions, and settlers from conducting any unauthorized trade with Indians. Thomas West, Baron De La Warr, a veteran officer and nobleman, would be the colony's new lord governor and captain general. Sir Thomas Gates, also selected for his military experience, was appointed lieutenant general and would lead the next supply to Virginia.[4]

Gates's instructions contained a mix of new and old ideas for the venture. The first innovation was to find a more direct route across the Atlantic. The usual course by way of the trade winds and the Caribbean was taking too long, wasting too many supplies, and getting too dangerous, "lest you fall into the hand of the Spaniard," as the company warned Gates. A captain named Samuel Argall was appointed to pioneer a new course that would bear southwest to the latitude of the Canaries and then more or less due west to America; he departed in mid-May 1609. In addition, Gates was to find a better location for Virginia's principal colony, keeping Jamestown only as its seaport "because the place is unwholesome." While insisting that the care and conversion of the Indians remained "the most pious and noble end of this plantacion," the directors instructed Gates to overthrow Wahunsenacawh and collect tribute from the Natives. The colony should exploit local commodities and trade, but the company still hoped for profits in "the discouvery either of the southe seas or Royall mines."[5]

The venture would be reorganized on a grander scale. Thus far Jamestown's "week and feeble endeavours consisting of so few persons" had accomplished little but "providing for the necessities of life." As the company directors confided to their agent in Amsterdam, they were preparing at least eight ships and 600 men that summer "to undertake more roundly, the plantation." Even that number might be "too few to defend themselves against an enemy that daily threatens, or to send back a present return, that may answer the expectation of such a business." So the new governor general, De La Warr, would follow with a thousand more colonists in 1610.[6]

An enterprise of that scale needed far more volunteers and investors. In March 1609, the Virginia Company wrote to the lord mayor of London and to the city's guilds and aldermen inviting them to invest in the reorganized company and "to ease the city and suburbs of a swarme of unnecessary inmates, as a continual cause of dearth and famine, and the very originall cause of all the Plagues that happen in this Kingdome." Settlers were promised "meate, drinke and clothing, with an howse, orchard and garden, for the meanest family, and possession of lands to them and their posterity": perhaps the earliest

version of the American dream.[7] At the same time, the company reduced share prices to £12 10s. to attract a wider range of stockholders.
Capitalizing on the recent success of the East India Company, the
Virginia venture attracted some 650 private investors and fifty London
companies, making their 1609 stock offering the most lucrative of any
company's for decades.[8]

The directors realized that the uncertain promise of profit in a risky
enterprise would not suffice to raise enough funding and support.
They needed good public relations, too—or more baldly, propaganda.
In April 1609, Richard Hakluyt finished a translation of one of the
main narratives of the ill-fated Soto expedition, dedicating it to the
councilors and investors of the Virginia Company. He gave it the improbable title *Virginia Richly Valued,* and framed it as an advertisement
for all the commodities and possibilities to be found in the region. At
the same time, Hakluyt helped rush into print a translation of Marc
Lescarbot's newly published history of French expeditions in New
England and Nova Scotia. Lescarbot's accounts of cold winters, scurvy,
and eventual failure could not offer much hope for new colonies in
North America; however, the translator's introduction promised the
Virginia colony "must be far better, by reason it stands more Southerly,
neerer to the Sunne," and so "greater encouragement may be given to
prosecute that generous and godly action."[9]

The Virginia Company itself published broadsides and pamphlets
defending the enterprise against detractors, and rebranding it with a
sense of mission and higher purpose. "The *Principall* and *Maine
Ends*" of the colony were to spread the gospel and "recover out of the
armes of the Divell, a number of poore and miserable soules, wrapt up
unto death." The colony would transplant "the rancknesse and multitude of increase in our people" that threatened to "infest" the kingdom.
It would "add our mighte to the treasury of Heaven" and secondarily to
the public treasury by providing commodities that England was "now
enforced to buy, and receive at the curtesie of other Princes"—that is,
Mediterranean commodities expected to come from the Mediterranean
latitudes of Virginia.[10]

To reach a wider audience, the company sponsored sermons lauding
the colony as part of England's mission to spread the gospel and counter
Catholic Spain.[11] Now more than ever, promoters played upon the
national fear that England overflowed with dirty, dangerous men—a

fear stoked by recurring harvest failures, famine, and plague during recent decades of the Little Ice Age. "There is nothing more daungerous for the estate of common-wealths," warned one, "then when the people do increase to a greater multitude and number then may justly paralell with the largenesse of the place and countrey." Another proffered Virginia as the solution for England's "idle persons" who "swarme in lewd and naughtie practises . . . pestering the land with pestilence and penury, and infecting one another with vice and villanie."[12]

Mere profit-seeking would be "too brutish" for such a noble enter-prise, but the propaganda hinted that investors might do well while doing good.[13] One tract promised "hills and mountaines making a sensible proffer of hidden treasure, never yet searched." The country would provide timber, sassafras, dyes, wines, spices, pearls, silk "com-parable to that of Persia, Turkey, or any other," and "a soyle so rich, fertill, and fruitefull" that Virginia would soon be sending food back to England.[14] Reports of Virginia's harsh climate and poor beginnings were no excuse. Was not Virginia on the same parallel as Spain? And had not the ancestors of the English endured hardships to conquer and civilize Great Britain? "Nay they exposed themselves to frost and colde, snow and heate, raine and tempests, hunger and thirst, and cared not what hardnesse, what extremitie, what pinching miseries they endured."[15]

The frenzy of excitement for the Virginia colony and its sense of religious mission only deepened a conviction that providence, or di-vine purpose, must have been behind the extraordinary events that followed. In less than a year, thanks in large part to accidents of cli-mate and weather, the venture would go from wreck to redemption in the most improbable ways.

The third supply bound for Jamestown first set out in late May 1609. Losing time to take on new provisions and await favorable winds, it only got under way from England on June 18. Less than a week out at sea, one pinnace was separated from the fleet, leaving eight ships in all. The flagship, *Sea Venture*, carried around 150 passengers and all of the colony's intended leaders: Sir Thomas Gates, Admiral George Somers, and Captain Christopher Newport. It was accompanied by six smaller ships and the pinnace *Virginia*, which the settlers of the Po-pham colony had built the year before. Altogether these carried roughly

450 more passengers, including George Percy, John Ratcliffe, Gabriel Archer, and other gentlemen adventurers hoping to strike it rich in the New World.

As per company instructions, they followed the new transatlantic course that Samuel Argall had gone to explore the month before—although nothing had been heard from Argall yet. The fleet sailed southwest to about 30°N latitude, where, as hoped, they found the winds to steer westward toward America. In the "fervent heat" of late July, fevers spread through two of the ships, and thirty-two passengers died by the end of the month. That was only the beginning of their troubles.

Late on August 2, the clouds grew thick and the night turned black. The wind began to howl, "singing and whistling most unusually," as one passenger remembered.[16] The next day they met with what another described as "a most sharpe and cruell storme." Other witnesses, with longer experience at sea, called it by the new name that Spaniards had learned from Natives of the Caribbean: a "hurricane."[17]

"For foure and twenty houres the storme in a restlesse tumult, had blowne so exceedingly, as we could not apprehend in our imaginations any possibility of greater violence," wrote William Strachey; "yet did wee still finde it, not onely more terrible, but more constant, fury added to fury, and one storme urging a second more outrageous then the former." The rain poured and the sea tossed, so "waters like whole Rivers did flood in the aire." Waves crashed over the decks. The wind grew "so violent that men could scarce stand." Its roar drowned out shouts and prayers from the men, women, and children aboard the ships. The day "like an hell of darkenesse turned blacke upon us."[18]

For the flagship, *Sea Venture,* the worst was yet to come. The hull sprang a leak, and in the chaos of the storm, the water in the hold rose 9 feet (almost 3 meters) before anyone realized the danger. By then the oakum caulking spewed from the joints, and it was too late to stop the leaking. "We almost drowned within, whilest we sat looking when to perish from above," Strachey recalled.

All the men on the ship—even gentlemen "such as in all their life times had never done houres worke before"—divided into three crews and two shifts to clear the water in a desperate "labour for life." They threw overboard everything that could weigh down the ship (saving the ship's dog), and "our men stood up to the middles, with buckets, barcios, and kettles, to baile out the water, and continually pumped for three dayes and three nights together, without any intermission; and yet the water seemed rather to encrease, then to diminish." On the third night, St. Elmo's fire danced about the masts, "an apparition of a little round light, like a faint Starre, trembling and streaming along with a sparkeling blaze," and they "observed it with much wonder."[19]

The Little Ice Age had brought Europe some extraordinary weather, including some serious meteorological disasters. However, not every event aroused the same amazement and anxiety nor the same sense of religious wonder. Freezing winters may have been deadly for crops, livestock, and even people. Yet freezing winters were only an extreme case of the regular cycle of seasons. They were not prodigies; they did not mean that the balance of nature had been disturbed. Many would write of the winter of 1607–1608 as the coldest and cruelest in living memory, but few called it a sign of cosmic disorder or a warning from God.[20]

Extraordinary floods and storms, on the other hand, were a sure sign of God's wrath. In early 1607, while the first settlers had made their way to Jamestown, southwestern England was struck by a flood so sudden and powerful that, until recently, meteorologists believed it must have been a tsunami. A southwesterly gale blew hard up the Bristol Channel as the waters rose; then the River Severn burst its banks for miles on either side, "to the ruine of all creatures and places which lay within." Dozens of parishes were inundated, flocks of animals and fields of grain were lost, and hundreds or possibly thousands of people drowned as their houses were swept away. The event produced an outpouring of pamphlets, some even translated into Dutch and French. Their titles carried the conviction of divine judgment: *God's Warning to His People of England* was one, *Miracle upon Miracle* another. "Albeit these swelings up and overflowings of waters proceed from natural causes," wrote one anonymous author, "yet are they the very diseases and monstrous births of nature, sent into the world to terrifie it, and to put it in mind, that the great God, (who holdeth stormes in the prison of the Cloudes at his pleasure, and can enlarge them to breed disorder on the Earth when he growes angry) can as well now drowne all mankind as he did at the first." For some Englishmen, it was a final warning to fly from sin. It probably factored into the decision of one group of English Puritans to escape the following year to

Leiden, from where, thirteen years later, some would take the much longer voyage to Massachusetts as America's original Pilgrims.[21]

It had appeared even more providential when Europeans first encountered hurricanes in the Americas. It was only to be expected that the still unfamiliar, untamed, heathen lands of the New World would be subject to demonic storms unknown in Europe. During the first decades of Spanish colonization, some hoped that the arrival of Christianity had altered the weather. "The natives call these furious winds, which formerly tore up great trees by the roots and destroyed houses, hurricanes," wrote Pietro Martire. "Hispaniola was formerly ravaged with these storms . . . when they believed infernal demons were seen to appear. This terrible curse, it appears, ceased since the sacrifice of the Eucharist has spread to the island; and the demons, which formerly loved to show themselves to the ancient inhabitants, have no longer been seen."[22]

Over time, the observant naturalists and chroniclers of the Spanish Empire gathered real information on the workings of hurricanes and tropical storms, just as they did for the workings of other weather and seasons in the Americas. They identified the signs of a hurricane's approach, and some of their theories about these storms even presaged modern discoveries about their pathways and rotational motion. By the seventeenth century, the Catholic Church had adapted the rites used to combat storms in Spain in order to face hurricanes in the New World; meanwhile, immigrants and creoles modified traditional folk rituals or else adopted new ones from Indians and Africans. Nevertheless, as hurricanes became more familiar to Spanish colonists, they ceased to be awe-inspiring and providential, instead turning into practical challenges for adaptation and administration. As historian Stuart Schwartz has argued, "They were simply too frequent and too random to fit into that 'moral cosmos' of destructive tempests, calves with two heads, deformed babies, epidemics, and recurrent catastrophes as divine punishments by which early modern societies sought to explain their world."[23]

All of which raises a vital question: Just how frequent were hurricanes and other tropical storms during the Little Ice Age? Untimely storms have appeared seemingly at every turn in our story: during the expeditions of Pánfilo de Narváez, Tristán de Luna, and Pedro Menéndez de

Avilés; in the Roanoke colony and in Spanish Florida; and now as the third supply sailed to Jamestown.

Modern instrumental records of these storms reach back only a century. Within that period, Atlantic hurricanes have tended to become larger during warmer years.[24] That does not mean storms were any less frequent or less damaging when the world was cooling. Many factors influence the occurrence of hurricanes and other tropical storms, including the relative distribution of temperatures, the presence or absence of wind shear, and large-scale atmospheric patterns including the El Niño Southern Oscillation and the North Atlantic Oscillation.

Historians and climatologists have turned to two kinds of sources to reconstruct the frequency of storms in past centuries. Some have painstakingly compiled historical descriptions, particularly from Caribbean islands, and have catalogued rates of shipwrecks. Others have identified sedimentary records that preserve traces of storm-force winds. The research so far finds no single long-term trend. The number of hurricanes and other tropical storms has fluctuated from year to year and decade to decade, and different locations have experienced different patterns in storm occurrence. Some periods of the Little Ice Age, particularly the late seventeenth century, appear to have brought fewer hurricanes altogether.

Some of the most recent and innovative studies, however, have found that the mid-1500s to early 1600s actually witnessed the highest average levels of hurricane activity for at least the past 500 years. In particular, counts of all storms, storms identified as hurricanes, total shipwrecks, and rates of shipwrecks found in Spanish colonial records all peaked during the late 1500s. Evidence from physical proxies in Jamaica, the Bahamas, and Bermuda all find increased storm or hurricane activity during the Little Ice Age, and particularly during this period. With respect to storms, as well as droughts and freezing winters, the expeditions in our story were not just individually unlucky. They may have been terribly unfortunate in their timing.[25]

Among the English at this time, such storms were still something novel and terrifying. "Seventeenth-century English accounts are full of the same awe and dread in the face of hurricanes that are found in the early Spanish reports and chronicles," comments Schwartz.[26] William Strachey's description of the hurricane encountered by the *Sea Venture* would prove so sensational back in London that it famously inspired Shakespeare's play *The Tempest.*[27]

Yet it was not only their power and devastation that made these storms, and similar natural disasters, a sign of divine providence. Readers marveled at stories of destruction from winds, rain, and floods. But they marveled even more at amazing stories of survival—at the evidence of God's mercy in saving a few from the general ruin. Shipwrecks upon a stormy sea were a favorite parable for divine judgment. But shipwrecks whose survivors miraculously escaped, like the biblical Jonah, were truly a cause for wonder.[28]

By the morning after the third night of the storm, according to survivors, the passengers aboard the *Sea Venture* were ready to surrender. "All our men, being bitterly spent, tired, and disabled for longer labour, were even resolved, without any hope of their lives, to shut up the hatches, and to have committed themselves to the mercie of the sea, (which is said to be mercilesse) or rather to the mercie of their mightie God and Redeemer," recalled Sylvester Jourdain. They supposedly fetched the last of the ship's alcohol "and drunke one to the other, taking their last leave one of the other, until their more joyfull and happy meeting in a more blessed world."[29]

Then providence intervened. "See the goodnesse and sweet introduction of better hope, by our mercifull God given unto us," wrote Strachey. "Sir George Summers, when no man dreamed of such happinesse, had discovered, and cried Land."[30] Unable to anchor or reach the shore, Admiral Somers ran aground as close as possible, lodging the ship between two rocks about a quarter mile from the beach.[31] With their longboat and skiff, they ferried every passenger to land, and "afterwards had time and fortune to save some good part of our goods and provision," Jourdain recalled.[32]

The location proved equally providential. "We found it to be the dangerous and dreaded Iland, or rather Ilands of the Bermuda," wrote Strachey. And it only completed their sense of wonder that "it pleased our mercifull God, to make even this hideous and hated place, both the place of our safetie, and meanes of our deliverance." "For the Ilands of the Barmudas, as every man knoweth that hath heard or read of them," explained Jourdain, "were never inhabited by any Christian or heathen people, but were ever esteemed, and reputed, a most prodigious and inchanted place, affording nothing but gusts, stormes, and foule weather."[33]

The castaways of the *Sea Venture* were hardly the first Europeans to reach Bermuda. By the time the chronicler Gonzalo Fernández de Oviedo visited those remote islands in 1515, they were already well known. Within a decade, they had become a stopping point for ships returning from the Indies, and they had acquired their reputation for shipwrecks and storms. A 1527 Portuguese plan to settle Bermuda hoped that ships returning from the Caribbean "could find some rest and relief for so long a voyage, and that by settling it they could remedy the storms that are bred in it." However, as the Spanish chronicler Antonio de Herrera y Tordesillas wrote in 1601, "thus far no one has seen or made a settlement in the said island, where sailors visit only with great caution, because of the bad weather."[34] Samuel de Champlain, too, passed by the islands on his first Atlantic crossing, and he shared the common experience and common wisdom of sailors. He described it as "a mountainous island difficult to approach on account of the dangers that surround it. It almost always rains there, and thunders so frequently, that it seems as if heaven and earth must come together."[35]

However, the *Sea Venture* survivors soon found this reputed "isle of devils" to be a terrestrial paradise. Its wildlife was harmless and still naive to the dangers of humans. In half an hour, the castaways could catch enough fish for a day. In an afternoon, they could gather thousands of fowl or eggs. Hogs set loose from a prior shipwreck teemed on the main island, and the settlers rounded up dozens each week to fatten on Bermuda's wild fruits and berries. When those ran out in late winter, they feasted on sea turtles, "of which wee daily turned up great store."[36]

They also found one of the mildest maritime climates in the world. Bermuda is perpetually humid but balmy. Temperatures almost never fall below freezing and rarely top 32°C (90°F). Strachey described the summer as "very hot and pleasant," with intermittent thunderstorms. Apart from the occasional northeasterly winds, the castaways were little troubled by winter cold, even without adequate clothing or shelter.[37] "Wherefore my opinion sincerely of this Iland is, that whereas it hath beene, and is still accounted, the most dangerous, infortunate, and most forlorne place of the world, it is in truth the richest, health-fullest, and pleasing land . . . and meerely naturall, as ever man set foote upon," wrote Jourdain.[38]

Colonial promoters had once written of Virginia as a thinly populated and pristine country that called out for settlement and cultivation: "a lande, even as God made it," in the words of John Smith. Now the contrast between Virginia and Bermuda revealed just how wrong they had been. Virginia, like the rest of the Americas, was not truly a new world but another old one. Its landscape, flora, and fauna had long been shaped by human action—by hunting, fire, clearance, and cultivation. Much of its territory and natural bounty were already spoken for, however sparse its inhabitants may have appeared to some European explorers. Real natural abundance in this world of climate change, ecological pressures, and expanding frontiers was to be found where humans had scarcely or never set foot before. Bermuda, in 1609, was still one of those rare places.[39]

As the expedition's leaders, Thomas Gates and George Somers, quickly realized, the equipment salvaged from the *Sea Venture* and the natural bounty of the islands offered them everything they needed to outfit new vessels and finish the voyage to America. First, they quickly converted their longboat into a pinnace and sent it onward to Virginia, hoping to contact the Jamestown colonists and arrange a rescue. That boat was never heard from again. By December, they set about building two small ships with which to take all the settlers to Virginia.

Yet many castaways found Bermuda so wonderful they resolved never to leave. As early as September, Gates and Somers uncovered a "conspiracy" to stop the preparation of the pinnace. As construction of the two ships advanced that winter, more passengers, some incited by Puritan religious leanings, plotted to break from the Virginia colony and stay in Bermuda instead. Lastly, in late March 1610, during the final preparations for departure, a handful of conspirators tried to seize the ships' provisions. The ringleader of this plot was caught and executed, but three men hid out of reach in the woods, determined to stay behind at all costs. All the other *Sea Venture* castaways departed for Jamestown that May aboard their two new ships, the *Patience* and *Deliverance.*[40]

Of course, this island paradise would not long survive the ravages of invading people and their pests and pets. Within a few years, Bermuda's primitive abundance was already in decline. New colonists, focused on growing tobacco and preparing for a possible Spanish invasion, failed to plant enough food to feed themselves. Then in 1614 the main island was invaded by rats, "multiplyinge themselves by an infinite encrease," wrote one colonist, who compared them to Pharaoh's plague. "They spared not the fruites of plants or trees, neither the plants themselves, but ate them up," recalled another. Cats and dogs brought to kill the rats wreaked more havoc on the island's ecosystem. With their crops destroyed, Bermuda's colonists returned to devouring the islands' wildlife with abandon. As one described: "Monstrous was it to see, how greedily every thing was swallowed downe; how incredible to speake, how many dozen of those poore silly creatures, that even offered themselves to the slaughter, were tumbled downe into their bottomelesse mawes." As early as 1620, the Bermuda colony tried passing laws to protect its declining populations of sea turtles and marine birds.[41]

As he wrote his 1624 *General History*, John Smith continued to marvel at the waves of invading worms, insects, and rats. He concluded: "A man would thinke it a tabernacle of miracles, and the worlds wonder, that from such a Paradise of admiration . . . should spring such wonders of afflictions as are onely fit to be sacrificed upon the highest altars of sorrow, thus to be set upon the highest Pinacles of content, and presently throwne downe to the lowest degree of extremity, as you see have beene the yeerely succeedings of those Plantations."[42]

Smith no doubt wrote from the bitterness of his own experiences in Virginia. During early 1609—according to his version of the story and that of his partisans—he managed to steer Jamestown through extraordinary difficulties. He survived assassination attempts by allies of Wahunsenacawh, and through calculated shows of violence and goodwill he managed to patch up the colony's peace with the nearby Paspahegh and Chickahominy Indians. At the same time, the president put the lazy colonists back to work gathering timber and naval stores and digging Jamestown's first well "of excellent sweete water." Their hogs and chickens multiplied, and the colonists, too, looked set to survive this summer better than the previous two.[43]

Disaster struck again that May. They discovered their corn taken the winter before had turned "halfe rotten, the rest so consumed with the many thousand rats (increased first from the ships) that we knewe not how to keepe that little wee had." At that time of year—and even in a good year—Virginia's Natives had little or nothing to trade while they lived off the land and waited for the next corn to ripen. Relying on Indian guidance, Smith dispersed his colonists to fish, hunt, and gather. Most went to the shores to collect oysters; "others would gather as much *Tockwough* roots in a day, as would make them bread a weeke, so that of those wilde fruites, fish, and berries, these lived very well." However, "such was the most strange condition of some 150, that had they not beene forced . . . to gather and prepare their victuall they would all have starved, and have eaten one another." They would have sold anything to the Indians, complained Smith, guns or their own souls, for a basket of corn.[44]

Smith's complaints had more than a grain of truth. But the problem was more than mere idleness. Jamestown's colonists were not survivalists or even modern campers. They had little or no experience living off the land, and no expectation of or preparation for this kind of hard labor and unfamiliar, monotonous diet. Just as in the summer of 1607, they suffered from shock, exhaustion, and malnutrition. Survivors later described a strange sickness that "caused all our skinns to peele off, from head to foote, as if we had beene flayed."[45]

Such was the situation when Captain Samuel Argall arrived in July. His voyage to chart a new route across the North Atlantic had taken nine weeks altogether, two of those weeks becalmed at sea, but otherwise free from storms or other mishaps.[46] Although not a supply expedition, the captain fished the teeming waters for fresh sturgeon and shared what biscuit, wine, and beer he could spare, "what relieved us for the space of a month."[47] After that, the colony's real troubles began.

While the leaking *Sea Venture* made its desperate landing in Bermuda, the remaining ships were scattered in the hurricane. The stormy weather, winds, and rain continued off and on for five or six more days. Waves crashing over the decks and water pouring into the holds ruined their provisions. "Some lost their Masts, some their Sailes blowne from their Yards," recalled one passenger. Sickness continued to spread, claiming more lives, "and in this miserable estate we arrived in Virginia."[48]

The *Blessing, Unity, Lion,* and *Falcon* rediscovered each other and reached Jamestown together on August 21. The *Diamond* arrived a few days later, followed by the *Swallow* and, more than a month later, the *Virginia*. Even without the *Sea Venture* and the lives lost from storm and sickness, the third supply more than quadrupled the number of settlers at Jamestown from roughly 90 to 400. Bringing little or no food of their own, the new arrivals "fell uppon that small quantitye of corne, not beinge above seaven acres, which we with great penury and sufferance had formerly planted," the colony's survivors testified, "and in three days, at the most, [they] wholly devoured it."[49] Smith shared out what little else they had to the ships' sailors for their return voyage to England, an action that roused the ire of Gabriel Archer, George Percy, and others.

"Now we did all lament much the absence of our Governor," wrote Archer, "for contentions began to grow, and factions, and partakings." With Gates, Somers, Newport, and the colony's official instructions all cast away with the *Sea Venture,* Jamestown fell into a crisis of leadership. Smith despised the aristocrats aboard the third supply as idle dilettantes with ridiculous expectations and "little or no care of any thing, but to pamper their bellies."[50] They despised him, in turn, as an "ambitious, unworthy, and vainglorious fellow" who had led the colony into chaos "and gave not any due respect to many worthy Gentlemen."[51] With his official term as president coming to an end, the gentlemen in the third supply rejected Smith's leadership and nominated Francis West, a kinsman of Baron De La Warr, to succeed him.

Fearing starvation and conflict, Smith sent most of the new arrivals away to plant new settlements. About 120 of them, led by the gentleman John Martin, traveled to the Indian village of Nansemond. Francis West took as many upriver to the falls, and John Ratcliffe took a smaller number to Point Comfort, at the mouth of the James.

Whatever might be said of Smith's abrasive style and his dealings with Virginia's Indians, the fate of these outlying settlements demonstrates just how much worse he might have done. Arriving among the Nansemond Indians, Martin sent two messengers to negotiate the purchase of an island for their settlement. When those messengers failed to return promptly, some of the Englishmen "beat the savages out of the island, burned their houses, ransacked their temples, [and] took down the corpses of their dead kings from their tombs," George Percy recalled. Meanwhile, Martin and others on the mainland seized the chief's son and "accidentally" shot him. Facing hunger and reprisals by Nansemond warriors, Martin and most of his followers retreated to Jamestown, leaving a smaller detachment under his lieutenant, Michael Sicklemore, to hold their fort.[52]

In the meantime, Francis West's company arrived at the falls and set up camp in a place that John Smith derided as flood-prone and "round invironed with many intollerable inconveniences." Smith came and negotiated with the nearby Powhatan Indians to buy a village for the English settlers to live in, and offered in return that the English would help defend them against the Monacans, the Powhatans' traditional enemy beyond the falls. In Smith's version of events, West and his company refused his advice and "so tormented" neighboring Indians with theft and violence "that they dailie complained to Captaine Smith he had brought them for protectors worse enimies then the Monocans themselves." Days later, Powhatan warriors launched an assault on the English camp, and Indian archers began to pick off stragglers. At first, the settlers agreed to relocate to the village that Smith claimed to have purchased from the Powhatans. However, West apparently blamed Smith for Indian hostilities and insisted on returning his company to their original campsite.[53]

On his way back to Jamestown, Smith suffered a life-threatening injury when a bag of gunpowder exploded in his lap. Although he survived the accident, Smith had to board the last ship bound for England that October. The leadership of Jamestown passed instead to the sickly George Percy, who remains our principal source for the events that followed.

After Smith's departure, the situation in Virginia fell apart. Lieutenant Sicklemore and other Englishmen at Nansemond "were found also slain with their mouths stopped full of bread, being done as it seemeth with contempt and scorn, that others might expect the like when they should come to seek for bread and relief," recalled Percy. The rest of the Englishmen at Nansemond soon retreated to Jamestown "to feed upon the poor store we had left us." Not long after, Francis West and his followers returned from the falls, having lost another eleven men to Indian attacks.[54] The passengers aboard the recently arrived *Virginia* and a few other colonists were sent to live at Point Comfort, but most of the frightened and hungry colonists now crowded for safety into Jamestown's fort.

Smith's partisans later claimed that his departure set off a general uprising among the Indians: "For the Savages no sooner understood of Captain Smiths losse, but they all revolted, and did murder and spoile all they could incounter."[55] However, the disaster had been some time in the making. With their incessant demands for food, the colonists had long since worn out their welcome. Smith may have managed to patch up relations with nearby Indian villages, but Jamestown's relations with the Powhatans and their allies had never recovered from the violent encounters of the previous winter. The arrival of more than 200 hungry, aggressive colonists at Nansemond and the falls tipped this conflict into open warfare. If Smith really believed that he could preserve the peace between Englishmen and Powhatans by offering a little copper and military assistance, then he had gambled too heavily on his powers of persuasion and intimidation.

From Wahunsenacawh's point of view, Smith's proposals must have seemed ridiculous. Trade and smuggling had seriously devalued English metals, manufactures, and even weapons. The newcomers could hardly serve as useful allies when they could not even feed themselves. The chief now had more to gain by eliminating the colonists than by trading with them.[56]

There was a further consideration that must have weighed heavily with Wahunsenacawh and his allies in the autumn of 1609. The drought had persisted for yet another year. The corn harvest had been poor again, and even wild plants and game animals must have suffered. Eyewitnesses to the colony's troubles, who disagreed about so much else, were later unanimous about one point: the Indians had little food to spare even if they had been willing to trade it.[57]

That did not keep Wahunsenacawh from using food as a tool of war. However bad the situation among Virginia's Indians, he knew it was far worse among the English colonists. Once the settlers had returned from Nansemond and the falls, the stores at Jamestown dwindled, by Percy's reckoning, to "a poor allowance of half a can of meal for a man a day." The Powhatan chief delivered a message that he would gladly barter corn if only the Englishmen would come to him to receive it. Had the colonists not been so desperate, they probably would have seen through the ruse. Instead John Ratcliffe led one pinnace carrying about thirty men, who carelessly entered a Powhatan village expecting to trade. Taken by surprise, only one lived to tell of the fatal ambush.

Ratcliffe was captured, tortured, and skinned alive with mussel shells, "and so for want of circumspection miserably perished," wrote Percy.[58] Meanwhile, Francis West took the *Swallow* and thirty-six men up the Potomac, hoping to find more willing trade partners outside of Wahunsenacawh's sphere of influence. To ensure those Indians delivered enough food, West resorted to "some harsh and cruel dealing," which apparently involved chopping off limbs and heads. Once they had their ship loaded with corn, West and his followers deserted Jamestown and sailed for England instead, bearing grim news for the Virginia Company.[59]

These disasters left the settlers in "misery and want," as Percy recalled, "all of us at Jamestown beginning to feel the sharp prick of hunger which no man truly describe but he which hath tasted the bitterness thereof." Wahunsenacawh now closed the trap on the colony, cutting it off from all trade and employing Indian warriors to pick off any settlers who ventured beyond the fort. The food inside gave out quickly, and starvation followed. "And it is true, the Indian killed as fast without . . . as Famine and Pestilence did within," wrote Strachey.[60]

Jamestown's "starving time" that winter may be the sole episode in this book that has truly endured in America's national memory. Yet the full story remains to be told. Archaeological investigations continue to confirm and elaborate its grim details.

Contemporaries may have exaggerated the number of victims, but not by much. Carefully counting English arrivals and casualties, roughly 240 people must have been in Jamestown at the start of winter in late 1609. By the following summer, just 60 were left alive.[61] A burial ground has been discovered just outside the fort; its occupants, many only in their twenties at their time of death, lie together hastily interred. It is now believed to be the final resting place of most "starving time" victims.[62]

It seems those victims actually died of starvation, too. William Strachey tried to blame the colony's well, "fed by the brackish River owzing into it, from whence I verily beleeve, the chiefe causes have proceeded of many diseases and sicknesses." However, recent excavation and testing reveals that the well dug in 1609 would have drawn clean fresh water. Nor is there other clear evidence of infectious disease that winter that could not have arisen from hunger itself. This time, the leading cause of death really was "meere famine."[63]

Details from both written accounts and archaeology shed further light on Jamestown's desperate struggle for food. The first victims were livestock and other animals. In 1609, the colony's free-ranging, half-wild hogs had multiplied to 500 or 600, so settlers had to move most of them to nearby Hog Island. As the starving time commenced that winter, the English in Jamestown slaughtered all those they could reach, and Indians killed off the rest. Soon the hens, chicks, goats, sheep, and even horses followed.[64]

But that was only the beginning. Zooarchaeological evidence has confirmed Percy's testimony that they lived on "vermin as dogs, cats, rats, and mice" and even searched the woods "to feed upon serpents." Excavations have indeed turned up butchered horse bones, as well as the skeletons of dogs, cats, rats, turtles, and even venomous snakes.[65]

As the last animals ran out, colonists turned to more desperate measures. "Some to satisfy their hunger have robbed the [common] store for the which I caused them to be executed," recalled Percy. "Many of our men this starving time did run away unto the Salvages whom we never heard of after." Others sought in vain to barter anything they could for food from local Indians. Excavations have even turned up English silver coins refashioned as trade beads.[66]

Another freezing Little Ice Age winter thwarted attempts to gather wild foods. Survivors later testified to "the depth of winter, when by reason of the colde, it was not possible for us to endure to wade in the water (as formerly) to gather oysters to satisfie our hungry stomacks." Instead, they were "constrained to digge in the grounde for unwhole-some rootes whereof we were not able to get so many as would suffice us, in respect of the frost at that season and our poverty and weak-ness." The colonists soon turned to true famine foods, picking mush-rooms, toadstools, "or what els we founde growing upon the grounde that would fill either mouth or belly." Some reportedly ate excrement or gnawed on the leather of their boots.[67]

"And now famine beginning to look ghastly and pale in every face, that nothing was spared to maintain life and to do those things which seem incredible." According to George Percy's account and the testi-mony of survivors, colonists began to dig up corpses and eat the dead. "Some adventuringe to seeke releife in the woods, dyed as they sought it, and weare eaten by others who found them dead." Finally, one man was discovered to have murdered and eaten his wife. He confessed under torture, and Percy ordered him executed.[68]

These stories of eating vermin and other humans are sometimes dismissed as mere clichés, typical of the famine narratives of medieval and early modern Europe. Tragically, in the Jamestown famine, as in so many other famines throughout history, people really did go to every extreme in the struggle for food—even cannibalism.[69] In 2012, researchers discovered a human skeleton while excavating a cellar in Jamestown's fort. The context, including animal bones and pottery fragments, dated the remains to the starving time. Forensic examination revealed the skeleton belonged to a teenage English girl. Breaks and chops indicated her head and limbs had been removed from her body. Her skull had been sawed with a sharp knife to remove flesh, and punctured to reach her brain.[70]

The Powhatans and their allies evidently relaxed their siege of Jamestown around mid-May 1610, probably to attend to spring planting.[71] George Percy recovered from his illness and took a boat to reach the outlying settlement at Point Comfort. He found its thirty or so colonists not only in good health but "so well stored" with crabs and other shellfish that they had fed them to their hogs. Percy came to the obvious conclusion that it was time to evacuate Jamestown and move its survivors downriver.[72]

Then providence seemed to intervene once more. "The next morning we espied a boat coming," wrote Percy; "we hailed them and understood that Sir Thomas Gates and Sir George Somers were come in these pinnaces which by their great industry they had built in the Bermudas with the remainder of their wracked ship and other wood they found in the country. Upon which news we received no small joy."

That joy soon vanished. As the *Deliverance* and *Patience* came ashore at Point Comfort, "a mightie storme of Thunder, Lightning, and Raine, gave us a shrewd and fearefull welcome," Strachey recalled. George Somers learned from Percy of the famine at Jamestown, "whereupon wee hastened up and found it true." They reached the fort at the beginning of June, only to see its palisades torn down, gates torn from the hinges, "and emptie houses (which Owners death had taken from them) rent up and burnt, rather then the dwellers would step into the Woods a stones cast off from them, to fetch other fire-wood." The sixty skeletal survivors reportedly cried out to the arriving ships: "We are starved, we are starved!"[73]

Thomas Gates promptly took charge of the colony and weighed his options. The *Deliverance* and *Patience* had come with barely enough provisions to reach Virginia. Colonists at Jamestown continued to die from Indian attacks and the lingering effects of famine. "All which considered, it pleased our Governour to make a Speech unto the Company," wrote Strachey. The ablest men would do their best to find food for the colony. Failing that, Gates promised, he would bring them all home—"at which there was a general acclamation, and shoute of joy on both sides."[74]

In ordinary times, Gates might have found a way to save Jamestown by living off the land. After all, John Smith had managed to keep most of the colonists alive the summer before through hunting and gathering. Yet Gates faced added challenges and misfortunes. Many of the colonists were starved, and all were demoralized. Powhatan attacks made it still harder to forage for food. Moreover, the drought continued, leaving little to eat even for Virginia's Indians.

Adding to their misfortunes, the fish never came upriver that year. The quantity of fish bones discovered at Jamestown suggests the colonists had depended heavily on fish during the first years of the colony. In particular, they enjoyed a bounty of sturgeon every May to October, when those great fish—up to 14 feet (3.5 meters) and over 700 pounds (320 kilograms)—swam up the rivers of the Chesapeake to spawn. In the colder waters of the Little Ice Age, they may have grown older and even larger than those of today. During the summer of 1609, John Smith boasted of "more Sturgeon then could be devoured by dogge and man." Captain Samuel Argall also fished for sturgeon during his stay at Jamestown, catching so many "that he could have loaded the ship with them." By astonishing misfortune, however, the only person aboard Argall's vessel who knew how to salt or pickle the fish had died on the voyage over.

When Gates sent out the ships in early June 1610, they caught next to nothing. "Albeit we labored, and hold our Net twenty times day and night, yet we tooke not so much as would content halfe the Fishermen," Strachey recalled. They sailed downriver as far as Cape Comfort, "scarse getting so much Fish as served their owne Company."

There is no way to know for certain why the sturgeon never arrived that summer. It could well be that the drought had rendered the lower James River so salty that those and other fish were deterred from spawning nearby.[75]

With barely two weeks' provisions left on hand, Gates and the colony's other leaders gave up. They outfitted and loaded their four remaining vessels: the *Deliverance, Patience, Discovery,* and *Virginia.* Around June 13, they sent one of those downriver to Point Comfort to deliver instructions to the settlers still there. Then on the morning of June 17, Gates announced his decision to the colonists at Jamestown. They would board the ships and "with all speede convenient" sail for Newfoundland, in the hope that fishermen might take them in and deliver them back to England. They set sail at noon the same day. Gates was reportedly the last man to board, to ensure that no one carried out a threat to burn down the settlement that had been the cause of so much misery.

The ships leaving Jamestown made slow progress downriver. On the morning of June 18 they had only reached Mulberry Island, a few miles away. There, lying at anchor and waiting for the outgoing tide, they sighted an English longboat working its way up the river, rushing to bring them news.

It was the last and most providential turn of events in the colony's painful founding. Baron De La Warr's fleet from England, carrying new colonists and enough provisions to last 400 settlers for a year, had arrived at the mouth of the James River just two days before. The lord governor had sent the longboat just in time to intercept the departing ships and bring them back to Jamestown. As John Smith later wrote, "God would not have it so abandoned."[76] Jamestown joined those first few European outposts—St. Augustine, Quebec, and Santa Fe—that had survived the confusions, perils, and disappointments of colonizing North America in the Little Ice Age.

Conclusion

Jamestown's improbable rescue in the summer of 1610 brings our story to a fitting close. After so many misfortunes on land and at sea, a chance encounter at just the right moment brought the colony back from the brink. The Little Ice Age did not end in 1610. Nor did the challenge of adapting to the climate and environment of North America. Jamestown's survival nevertheless marked a turning point in colonial history. The English now had their first enduring colony in Virginia, and the French had theirs in Quebec. The Spanish Empire had chosen to hold on to its outposts in Florida and New Mexico. The Little Ice Age would still trouble Europe's colonization of North America, but it would delay it no longer.

Our story of Europe's first century of encounter with North America has been a story of such accidents, contingencies, and, above all, misfortunes of weather. The first Spanish expeditions into the Southeast, one after another, fell victim to storms, hunger, disease, and freezing winters. The French Huguenot attempts to settle Florida suffered famine and ended with a hurricane. England's first colony at Roanoke disappeared after drought, hunger, and storms. Spain's outpost in Florida fared so poorly in its environment that Philip III debated whether it was even worth saving. The Popham colony ended after a single winter of extreme cold in Maine. Arctic expeditions foundered in freezing weather, storms, and sea ice. The Spanish Empire's expeditions to New Mexico faced icy winters, drought, or both. Several of the first French colonies in Canada and New England were decimated by scurvy during long winters. And of course there was the Jamestown colony, whose repeated misfortunes of drought, disease, famine, freezing winters, and shipwrecks would culminate in the "starving time" of 1610, just before its seemingly miraculous rescue.

It is worthwhile to remember how differently history would have turned out but for these accidents. It is not just that individual expeditions might have succeeded where they failed, but that the whole geography and chronology of colonization would have unfolded differently. There was nothing inevitable, or even probable, about the Spanish Empire's failure to colonize eastern North America before England, and nothing inevitable about its acceptance of English designs on Virginia. It was only during a window of climate-driven crisis starting in the 1590s that France and England were able to break into Spanish claims in North America and establish their first enduring colonies. But for some extreme Little Ice Age winters, New England and California might have been colonized years or decades sooner, and Quebec might have seemed much more inviting to colonists. Given the very high birth rates of settlers in these lands, the whole demography of North America would have been altered as a result. The path that history actually took was not the only or even most likely one.

Yet we can also step back and see larger patterns and processes at work in this history. Viewed one at a time, it may appear that each European expedition to North America fell victim to adverse *weather*—to unexpected storms, exceptional droughts, or outbreaks of unusual cold. But when we consider all of these events together, all this weather becomes a manifestation of a larger phenomenon: climate change. The Little Ice Age helps explain how so many expeditions, all across the continent and over the course of almost a century, would so often face disaster.

This is especially true when we understand the Little Ice Age in all its complexity. There were at least four major elements within this climatic shift, all of which played important parts in our story.

First, the Little Ice Age involved a very long-term cooling of the Northern Hemisphere, which ended only in the nineteenth century. This cooling meant that throughout the colonial period, North America was subject to cold that most of us have rarely or never experienced in our own lifetimes. Growing seasons were shorter and winter weather more extreme; explorers and settlers faced potentially deadly cold in regions where it would be unusual or unheard of today. This included a widespread cold spell around 1540, felt in expeditions as far apart as Soto's in the Southeast, Coronado's in the Southwest, and Cabrillo's along the California coast.

Second, the Little Ice Age included a period of stronger global cooling during the late sixteenth and early seventeenth centuries. This cooling meant that, even as Europeans might have adapted to North America's freezing winters, those winters became still more extreme. Native Americans may have been forced to change their ways of life or to migrate when faced with shorter growing seasons and changing environments. At the same time, Europe faced recurring harvest failures, famines, and epidemics—a crisis that shaped Spanish, French, and English views of colonies and emigration to North America. In the Arctic, cooling occurred more abruptly right around the year 1600, posing a new challenge for voyages in search of a Northwest Passage.

Third, the Little Ice Age included years of anomalous climate in the wake of large tropical volcanic eruptions. One of the most important clusters of eruptions began in the 1580s and culminated in the Huaynaputina eruption of 1600, which brought some of the coldest years in one of the coldest decades in more than a millennium. The timing of this volcanic weather could not have been worse for European expeditions in North America. Anomalies of cold and drought brought hardship and death to explorers and settlers as far apart as New Mexico and New England, and left lasting impressions of extreme cold in regions as diverse as California, Florida, and Maine.

Fourth, this phase of the Little Ice Age during the late sixteenth and early seventeenth centuries meant more than just lower temperatures. Oceanic currents and atmospheric circulation shifted in significant ways. Sailors in the North Atlantic faced more sea ice. Fish may have migrated to new waters. Hurricanes and other storms became more frequent, wrecking a greater share of ships around the Caribbean, Bahamas, and Bermuda. Extraordinary droughts seized the American Southwest and mid-Atlantic during the 1580s to 1600s, withering maize and other plants.

When it came to the weather, therefore, each set of expeditions faced particular difficulties. For some the problem was storms, for others droughts, and for others still it was freezing winters. Each presented different vulnerabilities. Many were at risk of starvation if winds and rough seas disrupted supplies. Some depended on successful foraging, or on exchanging with Indians, or (more rarely) on crops they had planted themselves. Long voyages in the Arctic risked encounters with sea ice. Long voyages in the tropics risked encounters with hurricanes. In warmer climates, weather could be a factor in water quality and the spread of disease. In colder climates, even the best-provisioned expeditions could succumb to scurvy without fresh food during a long winter.

The common challenge of every expedition was to adjust expectations and preparations in line with realities. Virginia, New England, Florida, New Mexico, and Quebec are all fine places to live today. Personal preferences aside, none has an intrinsically bad climate. Within limits, it is hard to say that any climate is absolutely bad. Heat and cold, rain and drought: each can bring benefit or harm depending on circumstances. What matters is what people know and what they have adapted to. With the exception of Arctic expeditions, the first explorers and settlers did not abandon their missions or die in these lands simply because their climates were too harsh. They did not fail only because it was the Little Ice Age—although Little Ice Age climates did pose very real challenges. They failed mainly because they did not know what to expect and how to adapt.

Even without the Little Ice Age, Europeans of the time would have struggled to understand and adjust to the unfamiliar climate of North America. Popular notions equating climates with latitudes proved a poor guide to weather and seasons in the New World. Classical meteorology offered as many ways to explain away evident differences in climate as to really understand them. If North America had simply been hotter or colder than Europe at the same parallels, then the process of learning and adapting might have been faster. But the new continent presented both hotter summers and colder winters. Its seasons were more variable, its storms stronger, and its patterns of precipitation suited to different crops.

The process of learning took time, and it did not happen automatically. The Spanish Empire's repeated failures in La Florida and New Mexico illustrate how easily false hopes and rumors could obscure the accumulation and transmission of reliable information about the climates and environments of new lands, especially when there was little continuity between one expedition and the next. In England, the process of information gathering began later but worked more systematically, thanks in part to the work of Richard Hakluyt and to the close personal association among London's early planners and promoters of American colonies. England's colonial joint-stock companies may have had the benefit of pooling experience and information as well as capital. Nevertheless, expeditions to Roanoke, Maine, and Virginia still came woefully unprepared for the seasons they encountered. Arguably, French Canadian expeditions provide the clearest demonstration of actual learning and adaptation. The continued presence of Samuel de Champlain, and the fact both he and the whole French enterprise survived the disaster of St. Croix, contributed to decisions that would help preserve colonies at Port-Royal and then Quebec. Then again, even these colonies still faced high death rates from scurvy, since both French and First Nations had lost the knowledge of its cure.

The complexities of Little Ice Age climate change made the process of learning and adaptation harder still. Even as Europeans tried to comprehend North America's climate, that climate was changing. It was hard to set expectations of average seasons when early explorers and colonists encountered so many extremes. Too often, these extremes were dismissed as mere anomalies, unrepresentative of the "true" climate of a region. Or else they generated an exaggerated fear and disappointment toward North America's climates, sapping enthusiasm and investment for new ventures. Extreme weather might even be taken as evidence of sorcery or divine wrath—a perception with powerful consequences for how Europeans viewed their expeditions and how Europeans and Indians may have viewed each other.

In hindsight, it is obvious that European explorers and settlers should have made more use of local indigenous knowledge about environments and climates in North America. Populations who successfully adapted to the Little Ice Age cold, including the Wabanaki, demonstrate how much explorers and settlers might have learned: fashioning suitable clothes and footwear; finding the right places to plant, hunt, and fish; and learning the signs of impending weather. Descriptions of local indigenous knowledge in early colonial narratives are mostly conspicuous by their absence. Climatic changes and extremes seem only to have exacerbated the almost inevitable mistrust and conflict between Natives and newcomers in early colonies—above all, conflict over scarce food, fuel, and warm clothing. Archaeological and historical research will no doubt continue to reveal the ways in which the Little Ice Age put pressure on Native American communities, and the ways in which those communities responded. What emerges from the story of early colonial encounters are those fatal consequences that arose from combined pressures of climatic extremes and European invasions.

A final pattern in this story has been the way that the Little Ice Age back in Europe shaped the nature and timing of colonies in North America. It was bad luck—but not only bad luck—that so many colonial expeditions took place during some of the coldest years of the past two millennia. Climatic extremes during the 1590s opened a window for those ventures in the first place. Years of famine and epidemic disease in Castile critically weakened the Spanish Empire and added to the urgency to make peace with France and England. Spain's sense of crisis at home, as well as disappointment with wars and colonies abroad, left officials unwilling to risk more lives and money defending what seemed like worthless territory in North America. At the same time, climate-driven subsistence crises in France and England left some in those countries looking for ways to dispose of hungry, poor, and vagrant subjects. The misfortune of these French and English colonies was to arrive just as the climate of North America turned even colder, in the wake of the Huaynaputina eruption of 1600.

Modern technologies and infrastructures, not to mention the pace and direction of climatic change, all make global warming a very different prospect than the cooling of the Little Ice Age. Yet if there is one lesson from our story relevant to today, it is this: that it takes time to understand new climates, and until that understanding has set in, it is hard to begin adapting.

Europeans confronting North America in the Little Ice Age had to make sense of a novel and changing climate by gathering local observations, comparing them to past experiences, and searching for patterns. Every encounter with some unexpected extreme would have raised troubling questions. Was this a single anomaly, or a sign that the continent really had a different climate? Was it just mischance, or something to be feared on every voyage to this part of the world from now on? The answers could determine whether the next expedition succeeded or failed, and, just as important, whether it took place at all.

Despite tremendous advances in meteorology and climate modeling, most individuals today are in a similar position. We have a much better sense of where the world is headed in terms of average temperatures and sea levels. However, we often have little sense of what these large-scale changes mean in terms of local, practical experiences. Every time we face extreme weather, similar questions arise: Is this global warming, or just natural variability? Is it a sign of things to come, or just an anomaly?

Models and statistics can sometimes give us answers. At the least, they provide indications of which kind of weather may be probable under which projected climatic conditions. However, the sort of answers we get from models and statistics are not what most people experience personally or feel intuitively. Human psychology may be both too quick to grasp at false patterns and yet too slow to let go of familiar expectations in order to make the constant calculations necessary to understand and adjust to a changed or changing climate. Much has altered between the world of the Little Ice Age and our world today, but that psychology has not. History may at least offer concrete examples and intuitive insights where models and statistics fall short.

Finally, this story reminds us how difficult and yet how vital it is to set the events of the past within the context of past climates. We do not usually think of climate reconstruction as part of the work of historians. Nor do we think of past climates as part of historical preservation, in the same way we preserve traditional architecture and landscapes or place historical objects in museums. However, with global warming we are losing a part of our heritage, just as surely as when monuments are destroyed or artifacts lost. Climates have always changed—but not with the speed they are changing now. It is not only that we risk losing historical sites, and even whole islands and neighborhoods, to rising sea levels. We risk becoming cut off from the rhythms of past seasons, their patterns of weather, the timing of their freezes and thaws, frosts and flowerings, and the feel of their extremes. We can excavate archaeological sites and rebuild old buildings. We can even recreate historical villages and have actors wear the clothes and speak the parts. We cannot, however, turn back the clock on global warming to recover the climates of colonial times. What we can do is make the best use of physical and written evidence, of scientific and historical research, to understand the past and contemplate our choices for the future. We can recognize the Little Ice Age and all it meant for history, including Europe's first encounters with North America.

Notes

Acknowledgments

Index

Notes

Abbreviations

AGI   Archivo General de Indias, Seville, Spain.

ASV   Archivio di Stato di Venezia, Venice, Italy.

CDIE   Martín Fernández de Navarrete, ed., *Colección de documentos inéditos para la historia de España,* 113 vols. (Madrid: Academia de la Historia, 1842).

CWCJS   Philip L. Barbour, ed., *The Complete Works of Captain John Smith (1580–1631),* 3 vols. (Chapel Hill, NC: University of North Carolina Press, 1986).

DCE   Richard Flint and Shirley Cushing Flint, eds., *Documents of the Coronado Expedition, 1539–42* (Dallas: Southern Methodist University Press, 2005).

ENEV   David B. Quinn and Alison M. Quinn, eds., *The English New England Voyages, 1602–1608* (London: Hakluyt Society, 1983).

HGIO   Antonio de Herrera y Tordesillas, *Historia general de las Indias Ocidentales, o de los hechos de los castellanos en las islas y tierra firme del mar oceano* (Amberes: Juan Bautista Verdussen, 1728).

HGNI   Gonzalo Fernández de Oviedo y Valdés, *Historia general y natural de las Indias,* ed. Juan Pérez de Tudela y Bueso [1535] (Madrid: Atlas, 1992).

JVFC   Philip L. Barbour, ed., *The Jamestown Voyages under the First Charter, 1606–1609,* 2 vols. (Cambridge: Cambridge University Press, 1969).

LP   Herbert Ingram Priestley, ed., *The Luna Papers: Documents Relating to the Expedition of Don Tristán de Luna y Arellano for the Conquest of La Florida in 1559–1561,* 2 vols. (Deland: Florida State Historical Society, 1928).

NAW   David B. Quinn and Alison M. Quinn, eds., *New American World: A Documentary History of North America to 1612* (London: Arno, 1979).

PN   Richard Hakluyt, ed., *The Principal Navigations, Voyages, Traffiques and Discoveries of the English Nation* (New York: Macmillan, 1904).

RV   David B. Quinn, ed., *The Roanoke Voyages, 1584–1590*, 2 vols.
          (London: Hakluyt Society, 1955).
WMQ   *William and Mary Quarterly.*

Additional Sources

A Cold Welcome was written to be accessible to a general audience. Specialists interested in issues of historiography and methodology may be interested in several academic publications written in preparation for this book:

Sam White, "Cold, Drought, and Disaster: The Little Ice Age and the Spanish Conquest of New Mexico," *New Mexico Historical Review* 89 (2014): 425–58.

Sam White, "The Real Little Ice Age," *Journal of Interdisciplinary History* 44 (2014): 327–352.

Sam White, "'Shewing the Difference betweene Their Conjuration, and Our Invocation on the Name of God for Rayne': Weather, Prayer, and Magic in Early American Encounters," *WMQ* 72 (2015): 33–56.

Sam White, "Unpuzzling American Climate: New World Experience and the Foundations of a New Science," *Isis* 106 (2015): 544–566.

Sam White, Christian Pfister, and Franz Mauelshagen, eds., *The Palgrave Handbook of Climate History* (London: Palgrave, forthcoming).

Sam White, Richard Tucker, and Kenneth Sylvester, "North American Climate History," in *Cultural Dynamics of Climate Change and the Environment in Northern America,* ed. Bernd Sommer (Leiden: Brill, 2015), 109–136.

Readers interested in learning more about historical climate reconstruction and climate history may also wish to consult the website and database maintained by the author and colleagues at http://climatehistory .net.

Introduction

1. Quoted in Vivian Etting, "The Rediscovery of Greenland during the Reign of Christian IV," *Journal of the North Atlantic* 2 (2009): 156.
2. James Hall, "A Report to King Christian IV of Denmark on the Danish Expedition to Greenland, under the Command of Captain John Cunningham, in 1605," in *Danish Arctic Expeditions, 1605–1620,* ed. Christian Carl August Gosch (London: Hakluyt Society, 1897), 1:1–19, quote from 4–5; Etting, "Rediscovery of Greenland."

3. James Hall, "An Account of the Danish Expedition to Greenland, under the Command of Captain Godske Lindenow, in 1606," in *Danish Arctic Expeditions, 1605–1620*, ed. Christian Carl August Gosch (London: Hakluyt Society, 1897), 1:54–81, quotes from 73, 64.

4. Isaac de La Peyrère, *Relation du Groenland* (Paris: Augustin Courbe, 1647), n.p. I am assuming that de La Peyrère, a Frenchman, was using the Gregorian calendar, but he could have been using Julian dates, then current in Denmark. The original information from this voyage all comes from Claus Christoffersen Lyschander, *Den Grønlandske Chronica* (Copenhagen: Benedicte Laurentz, 1608).

5. For various proxy measurements of North Atlantic paleotemperature and the onset of the Little Ice Age, see J. M. Grove, "The Initiation of the 'Little Ice Age' in Regions round the North Atlantic," *Climatic Change* 48 (2001): 53–82; J. Guiot et al., "Last-Millennium Summer-Temperature Variations in Western Europe Based on Proxy Data," *Holocene* 15 (2005): 489–500; William Patterson et al., "Two Millennia of North Atlantic Seasonality and Implications for Norse Colonies," *Proceedings of the National Academy of Sciences* 107 (2010): 5306–5310; P. D. Jones, C. Harpham, and B. M. Vinther, "Winter-Responding Proxy Temperature Reconstructions and the North Atlantic Oscillation," *Journal of Geophysical Research: Atmospheres* 119 (2014): 6497–6505; Hilmar A. Holland et al., "Decadal Climate Variability of the North Sea during the Last Millennium Reconstructed from Bivalve Shells (*Arctica islandica*)," *Holocene* 24 (2014): 771–786.

6. For general histories of the Greenland Vikings, see, e.g., Kirsten Seaver, *The Frozen Echo: Greenland and the Exploration of North America ca. AD 1000–1500* (Stanford, CA: Stanford University Press, 1996).

7. Maanasa Raghavan et al., "The Genetic Prehistory of the New World Arctic," *Science* 345 (2014): 1255832.

8. The reason for the collapse or abandonment of the Greenland Viking settlements has become a contentious scholarly topic. Jared Diamond, in *Collapse: How Societies Choose to Fail or Succeed* (New York: Viking, 2005), has held out the fate of the Greenland Vikings as a sort of cautionary tale for failures of cultural adaptability in the face of environmental stress and climatic change. Other scholars have sharply disagreed with this interpretation; see especially Joel Berglund, "Did the Medieval Norse Society in Greenland Really Fail?," in *Questioning Collapse: Human Resilience, Ecological Vulnerability, and the Aftermath of Empire*, ed. Patricia A. McAnany and Norman Yoffee (New York: Cambridge University Press, 2010), 45–70. The balance of recent scholarship has found a very close coincidence between Little Ice Age cooling and the abandonment of the eastern and especially western settlements; it also emphasizes that Viking settlements were marginal and particularly vulnerable to long winters. See especially L. K. Barlow et al., "Interdisciplinary Investigations of the End of the Norse Western Settlement in Greenland," *Holocene* 7 (1997): 489–499; William J. D'Andrea et al., "Abrupt Holocene Climate Change as an Important Factor for Human Migration in West Greenland," *Proceedings of the National Academy of Sciences* 108 (2011): 9765–9769;

Andrew J. Dugmore et al., "Cultural Adaptation, Compounding Vulnerabilities
and Conjunctures in Norse Greenland," *Proceedings of the National Academy
of Sciences* 109 (2012): 3658–3663; Sofia Ribeiro et al., "Climate Variability in
West Greenland during the Past 1500 Years: Evidence from a High-Resolution
Marine Palynological Record from Disko Bay," *Boreas* 41 (2012): 68–83. The
state of discussion at the time of writing is summarized in Eli Kintisch, "Why
Did Greenland's Vikings Disappear?," *Science* 354 (2016): 696–701.

9. A. E. J. Ogilvie, "Documentary Evidence for Changes in the Climate of
Iceland, A.D. 1500 to 1800," in *Climate since A.D. 1500*, ed. R. S. Bradley and
P. D. Jones, rev. ed. (London: Routledge, 1995), 92–117, quote from 108. The
evidence for global cooling during the late sixteenth century is discussed in
more depth in Chapter 4. Changes in North Atlantic sea ice are examined
in more depth in Chapter 5.

10. E.g., Joseph Judge, "Exploring Our Forgotten Century," *National Geographic*,
March 1988, 331–362; Charles M. Hudson and Carmen Chaves Tesser, eds.,
The Forgotten Centuries: Indians and Europeans in the American South, 1521–1704
(Athens: University of Georgia Press, 1994). For recent developments in the
historiography of the sixteenth century, see Karen Ordahl Kupperman, "Before
1607," *WMQ* 72 (2015): 3–24; Karen Halttunen, "Grounded Histories: Land
and Landscape in Early America," *WMQ* 68 (2011): 513–532; Eric Hinderaker
and Rebecca Horn, "Territorial Crossings: Histories and Historiographies of
the Early Americas," *WMQ* 67 (2010): 395–432.

11. Bernard Bailyn, *The Barbarous Years: The Peopling of British North America: The
Conflict of Civilizations, 1600–1675* (New York: Alfred A. Knopf, 2012); Karen
Kupperman, *The Jamestown Project* (Cambridge, MA: Harvard University
Press, 2007), 1. See also Gary B. Nash, *Red, White, and Black: The Peoples of
Early North America*, 6th ed. (Upper Saddle River, NJ: Prentice Hall, 2009);
Alan Taylor, *American Colonies*, Penguin History of the United States (New
York: Viking, 2001); Kathleen Donegan, *Seasons of Misery: Catastrophe and
Colonial Settlement in Early America* (Philadelphia: University of Pennsylvania
Press, 2014).

12. Among many critical new works of early colonial history relevant to this
project, see Kupperman, *Jamestown Project*; Karen Kupperman, *Roanoke: The
Abandoned Colony* (Lanham, MD: Rowman and Littlefield, 2007); Daniel K.
Richter, *Before the Revolution: America's Ancient Past* (Cambridge, MA: Harvard
University Press, 2011); Robbie Franklyn Ethridge, *From Chicaza to Chickasaw:
The European Invasion and the Transformation of the Mississippian World,
1540–1715* (Chapel Hill: University of North Carolina Press, 2010); Richard
Flint, *No Settlement, No Conquest: A History of the Coronado Entrada* (Albu-
querque: University of New Mexico Press, 2008). Many new publications
came out to coincide with the 400th anniversary of Jamestown; see, e.g.,
Peter C. Mancall, ed., *The Atlantic World and Virginia, 1550–1624* (Chapel Hill:
University of North Carolina Press, 2007); David Armitage, ed., *The World
of 1607* (Williamsburg, VA: Jamestown-Yorktown Foundation, 2007).

13. E.g., Richard Flint, ed., *Great Cruelties Have Been Reported: The 1544 Investigation of the Coronado Expedition* (Dallas, TX: Southern Methodist University

Press, 2002); Richard Flint and Shirley Cushing Flint, eds., *Documents of the Coronado Expedition, 1539–42* (Dallas, TX: Southern Methodist University Press, 2005); Conrad E. Heidenreich and K. Janet Ritch, eds., *Samuel de Champlain before 1604: Des Sauvages and Other Documents Related to the Period* (Toronto: Champlain Society, 2010); and the documents transcribed and published by the University of California's Cibola Project, http://escholarship .org/uc/rcrs_ias_ucb_cibola (last accessed March 17, 2016).

14. E.g., William M. Kelso, *Jamestown, the Buried Truth* (Charlottesville: University of Virginia Press, 2006); Jeffrey P. Brain, Peter Morrison, and Pamela Crane, *Fort St. George: Archaeological Investigation of the 1607–1608 Popham Colony on the Kennebec River in Maine* (Augusta: Maine State Museum, 2007); Hélène Côté, "The Archaeological Collection from the Cartier-Roberval Site (1541–43): A Remarkable Testimony to French Colonization Efforts in the Americas," *Post-Medieval Archaeology* 43 (2009): 71–86; Steven R. Pendery and H. W. Borns, eds., *Saint Croix Island, Maine: History, Archaeology, and Interpretation* (Augusta: Maine Historic Preservation Commission and the Maine Archaeological Society, 2012); J. Michael Francis, Kathleen M. Kole, and David Hurst Thomas, "Murder and Martyrdom in Spanish Florida: Don Juan and the Guale Uprising of 1597," *Anthropological Papers of the American Museum of Natural History* 95 (2011): 1–154.

15. For the state of the field of climate history, see Sam White, Christian Pfister, and Franz Mauelshagen, eds., *The Palgrave Handbook of Climate History* (London: Palgrave, forthcoming). For the historiography of North American climate history, see also Sam White, Richard Tucker, and Kenneth Sylvester, "North American Climate History," in *Cultural Dynamics of Climate Change and the Environment in Northern America,* ed. Bernd Sommer (Leiden: Brill, 2015), 109–136. For examples of North American climate history covering the colonial period, see, e.g., Lesley-Ann Dupigny-Giroux and Cary J. Mock, eds., *Historical Climate Variability and Impacts in North America* (Berlin: Springer, 2009); Georgina Endfield, *Climate and Society in Colonial Mexico* (London: Blackwell, 2008); Katherine A. Grandjean, "New World Tempests: Environment, Scarcity, and the Coming of the Pequot War," *WMQ* 68 (2011): 75–100; Stuart B. Schwartz, *Sea of Storms: A History of Hurricanes in the Greater Caribbean from Columbus to Katrina* (Princeton, NJ: Princeton University Press, 2015).

1. Where Everything Must Be Burning

1. Nicolás Wey Gómez, *The Tropics of Empire: Why Columbus Sailed South to the Indies* (Cambridge, MA: MIT Press, 2008).

2. Pietro Martire d'Anghiera, *De Orbe Novo: The Eight Decades of Peter Martyr d'Anghera,* ed. Francis Augustus MacNutt (New York: G. P. Putnam's Sons, 1912), 1:133. Other examples include, e.g., Fracanzio da Montalboddo, *Paesi Novamente Retrovati et Novo Mondo da Alberico Vesputio Florentino Intitulato* (Vicenza: Henrico Vicentino, 1507), ccxviii. See also Craig Martin, "Experience of the New World and Aristotelian Revisions of the Earth's Climates during the Renaissance," *History of Meteorology* 3 (2006): 1–15.

3. Karen Kupperman, "The Puzzle of the American Climate in the Early Colonial Period," *American Historical Review* 87 (1982): 1262–1289; Karen Kupperman, "Climate and Mastery of the Wilderness in Seventeenth-Century New England," in *Seventeenth-Century New England,* ed. David Hall and David Allen (Boston: Colonial Society of Massachusetts, 1984), 3–37.

4. To be more precise, "climates" (*klimata*) in classical geography referred to the north-south coordinates of certain familiar locations, rather than bands of latitude or zones of heat and cold. In practice, however, early modern discussions of geography often collapsed these different concepts. The classical definition may have been significant in the way that "climate" eventually came to refer to the characteristic weather of a given location. See Franz Mauelshagen, "Climate as a Scientific Paradigm—Early History of Climatology to 1800," in *The Palgrave Handbook of Climate History,* ed. Sam White, Christian Pfister, and Franz Mauelshagen (London: Palgrave, 2018).

5. Roger G. Barry, "A Brief History of the Terms Climate and Climatology," *International Journal of Climatology* 33 (2013): 1317–1320.

6. Sam White, "Unpuzzling American Climate: New World Experience and the Foundations of a New Science," *Isis* 106 (2015): 544–566.

7. For an overview of geographical confusions, see Seymour I. Schwartz, *The Mismapping of America* (Rochester, NY: University of Rochester Press, 2003).

8. Dava Sobel, *Longitude: The True Story of a Lone Genius Who Solved the Greatest Scientific Problem of His Time* (London: Fourth Estate, 1998).

9. On Columbus and his navigation, see Douglas T. Peck, *Cristoforo Colombo: God's Navigator* (Columbus, WI: Columbian Publishers, 1993), 27–30. For a detailed examination of the practice of navigation in the first Jamestown voyages, see Robert D. Hicks, *Voyage to Jamestown: Practical Navigation in the Age of Discovery* (Annapolis, MD: Naval Institute Press, 2011). For the development of navigational instruments in Atlantic seafaring during the early seventeenth century, see Sara J. Schechner, "New Worlds, New Scientific Instruments: Cosmology, Mathematics, and Power in the Time of Jamestown," in *The World of 1607,* ed. David Armitage (Williamsburg, VA: Jamestown-Yorktown Foundation, 2007), 230–239. For the general history of early modern European navigation, see also E. G. R. Taylor, *The Haven-Finding Art: A History of Navigation from Odysseus to Captain Cook* (New York: Abelard-Schuman, 1957) and David Watkin Waters, *The Art of Navigation in England in Elizabethan and Early Stuart Times* (New Haven: Yale University Press, 1958).

10. José de Acosta, *Natural and Moral History of the Indies,* ed. Jane E. Mangan, trans. Frances López-Morillas (Durham, NC: Duke University Press, 2002), 89. Cf. Juan de Cárdenas, *Primera parte de los problemas y secretos maravillosos de las indias* [1591], ed. Xavier Lozoya, 5th ed. (Mexico City: Academia Nacional de Medicina, 1980), 67–71; Gonzalo Fernández de Oviedo y Valdés, *Historia general y natural de las Indias,* ed. Juan Pérez de Tudela y Bueso (Madrid: Atlas, 1992), 2:318–319, 4:330, 336–337.

11. *Obras del Padre Bernabé Cobo* (Madrid: Biblioteca de Autores Españoles, 1964), 55.

12. On the blend of old and new ideas, and debates over the impact of New World discoveries, see J. H. Elliott, *The Old World and the New, 1492–1650* (Cambridge: Cambridge University Press, 1970); Anthony Grafton, *New Worlds, Ancient Texts: The Power of Tradition and the Shock of Discovery* (Cambridge, MA: Harvard University Press, 1992); Karen Kupperman, ed., *America in European Consciousness 1493–1750* (Chapel Hill: University of North Carolina Press, 1995); Klaus Vogel, "European Expansion and Self-Definition," in *The Cambridge History of Science,* ed. Katherine Park and Lorraine Daston (Cambridge: Cambridge University Press, 2006), 3:818–839; John Gascoigne, "Crossing the Pillars of Hercules: Francis Bacon, the Scientific Revolution and the New World," in *Science in the Age of the Baroque,* ed. Ofer Gal and Raz Chen-Morris (Dordrecht: Springer Netherlands, 2012), 217–237.

13. For recent research on academic meteorology in the sixteenth century, see Craig Martin, *Renaissance Meteorology: Pomponazzi to Descartes* (Baltimore: Johns Hopkins University Press, 2011). On Aristotle's legacy, see H. Howard Frisinger, "Aristotle's Legacy in Meteorology," *Bulletin of the American Meteorological Society* 54 (1973): 198–204. For discussion of popular works, see S. K. Heninger, *A Handbook of Renaissance Meteorology, with Particular Reference to Elizabethan and Jacobean Literature* (Durham, NC: Duke University Press, 1960).

14. Though first published in 1543, the Copernican heliocentric view would not catch on with the educated public until well into the seventeenth century. All of the sixteenth-century writers about American geography and colonies discussed in this volume evidently assumed a geocentric universe.

15. Matthias Heymann, "The Evolution of Climate Ideas and Knowledge," *Wiley Interdisciplinary Reviews: Climate Change* 1 (2010): 581–597.

16. Jorge Cañizares-Esguerra, "New World, New Stars: Patriotic Astrology and the Invention of Amerindian Creole Bodies in Colonial Spanish America, 1600–1650," in *Nature, Empire, and Nation: Explorations of the History of Science in the Iberian World* (Stanford, CA: Stanford University Press, 2006), 64–95; Joyce E. Chaplin, *Subject Matter: Technology, the Body, and Science on the Anglo-American Frontier, 1500–1676* (Cambridge, MA: Harvard University Press, 2001); Rebecca Earle, *The Body of the Conquistador: Food, Race, and the Colonial Experience in Spanish America, 1492–1700* (Cambridge: Cambridge University Press, 2012).

17. Cárdenas, *Problemas y secretos,* 74–84, 88.

18. Bernardo de Vargas Machuca, *Milicia y descripción de las Indias* [1599] (Bogota: Fondo de Promoción de la Cultura, 2003). Cf. Juan de Torquemada, *Monarquía Indiana* [1615] (Mexico City: Editorial Porrúa, 1969), 7.

19. Martín Fernández de Enciso, *Suma de geografía,* ed. Mariano Cuesta Domingo (Madrid: Museo Naval, 1987), 85–86. A contemporary English translation included the same passage: Martín Fernández de Enciso, *A Brief Summe of Geographie,* ed. E. G. R. Taylor, trans. Roger Barlow (London: Hakluyt Society, 1932), 13. For the influence of Fernández de Enciso's idea, see Cárdenas, *Problemas y secretos,* 85–87; Cobo, *Obras,* 62; Jean Alfonse, *Cosmographie, avec l'espère et régime du soleil et du nord,* ed. Georges Musset (Paris: Leroux, 1904), 86–87.

20. José de Acosta, *Historia Natural y Moral de las Indias*, ed. Fermín del Pino (Madrid: Consejo Superior de Investigaciones Científicas, 2008), 115–116, and book III, passim.

21. White, "Unpuzzling."

22. For popular history, see especially Brian Fagan, *The Little Ice Age: How Climate Made History 1300–1850* (New York: Basic Books, 2000). For an overview of evidence and debates, see Sam White, "The Real Little Ice Age," *Journal of Interdisciplinary History* 44 (2014): 327–352.

23. The term "little ice-age" was coined in François Matthes, "Report of the Committee on Glaciers," *Transactions—American Geophysical Union* 20 (1939): 53–82. On the glaciology of the Little Ice Age, see Jean Grove, *The Little Ice Age* (London: Meuthen, 1988). On the development of European climate reconstruction techniques, see Emmanuel Le Roy Ladurie, *Times of Feast, Times of Famine: A History of Climate since the Year 1000*, trans. Barbara Bray (New York: Noonday Press, 1971); H. H. Lamb, *Climate, History and the Modern World*, 2nd ed. (London: Routledge, 1995); Rudolf Brázdil et al., "Historical Climatology in Europe—The State of the Art," *Climatic Change* 70 (2005): 363–430; Franz Mauelshagen, *Klimageschichte der Neuzeit, 1500–1900* (Darmstadt: Wissenschaftliche Buchgesellschaft, 2010); Emmanuel Le Roy Ladurie, *Naissance de l'histoire du climat* (Paris: Hermann, 2013). For an overview of historical climatology in China, see Q.-S. Ge et al., "Coherence of Climatic Reconstruction from Historical Documents in China by Different Studies," *International Journal of Climatology* 28 (2008): 1007–1024.

24. For overviews of proxy climate records for recent millennia and centuries, see, e.g., P. D. Jones et al., "High-Resolution Palaeoclimatology of the Last Millennium: A Review of Current Status and Future Prospects," *Holocene* 19 (2009): 3–49; Raymond S. Bradley, *Paleoclimatology: Reconstructing Climates of the Quaternary*, 3rd ed. (Amsterdam: Elsevier, 2015).

25. See especially the work of the PAGES 2k project, as in Moinuddin Ahmed et al., "Continental-Scale Temperature Variability during the Past Two Millennia," *Nature Geoscience* 6 (2013): 339–346.

26. See discussion and examples in Ulf Büntgen and Lena Hellmann, "The Little Ice Age in Scientific Perspective: Cold Spells and Caveats," *Journal of Interdisciplinary History* 44 (2013): 353–368.

27. Heinz Wanner et al., "Mid- to Late Holocene Climate Change: An Overview," *Quaternary Science Reviews* 27 (2008): 1791–1828; Jan Esper et al., "Orbital Forcing of Tree-Ring Data," *Nature Climate Change* 2 (2012): 862–866; Hui Jiang et al., "Solar Forcing of Holocene Summer Sea-Surface Temperatures in the Northern North Atlantic," *Geology* 43 (2015): 203–206; Zhengyu Liu et al., "The Holocene Temperature Conundrum," *Proceedings of the National Academy of Sciences* 111 (2014): E3501–E3505; Belén Martrat et al., "Multi-Decadal Temperature Changes off Iberia over the Last Two Deglaciations and Interglacials and Their Connection with the Polar Climate," *Past Global Changes* 23 (2015): 10–11; C. F. Schleussner and G. Feulner, "A Volcanically Triggered Regime Shift in the Subpolar North Atlantic Ocean as a Possible Origin of the Little Ice Age," *Climate of the Past* 9 (2013): 1321–1330.

28. At the time of writing, the state of the field on volcanic forcing is discussed in M. Sigl et al., "Timing and Climate Forcing of Volcanic Eruptions for the Past 2,500 Years," *Nature* 523 (2015): 543–549. On solar forcing, the classic study is John A. Eddy, "Climate and the Role of the Sun," *Journal of Interdisciplinary History* 10 (1980): 725–747. For analysis supporting a modest role for solar forcing, see, e.g., Petra Breitenmoser et al., "Solar and Volcanic Fingerprints in Tree-Ring Chronologies over the Past 2000 Years," *Palaeogeography, Palaeoclimatology, Palaeoecology* 313–314 (2012): 127–139. Recent analysis based on climate models suggests, however, that solar forcing cannot have contributed as much as volcanic forcing to Little Ice Age cooling; see, e.g., Andrew P. Schurer, Simon F. B. Tett, and Gabriele C. Hegerl, "Small Influence of Solar Variability on Climate over the Past Millennium," *Nature Geoscience* 7 (2014): 104–108; A. R. Atwood et al., "Quantifying Climate Forcings and Feedbacks over the Last Millennium in the CMIP5-PMIP3 Models," *Journal of Climate* 29 (2016): 1161–1178. Other studies have questioned whether apparently cyclical patterns in temperature could be due to random variability or internal climatic processes rather than solar cycles: e.g., Danny McCarroll, " 'Study the Past, if You Would Divine the Future': A Retrospective on Measuring and Understanding Quaternary Climate Change," *Journal of Quaternary Science* 30 (2015): 154–187; T. Edward Turner et al., "Solar Cycles or Random Processes? Evaluating Solar Variability in Holocene Climate Records," *Scientific Reports* 6 (2016): 23961.

29. Emmanuel Le Roy Ladurie, Daniel Rousseau, and Anouchka Vasak, *Les fluctuations du climat, de l'an mil à nos jours* (Paris: Fayard, 2011).

30. E.g., Martin P. Tingley and Peter Huybers, "Recent Temperature Extremes at High Northern Latitudes Unprecedented in the Past 600 Years," *Nature* 496 (2013): 201–205. See Chapter 4 for more details.

31. Christian Pfister, "Weeping in the Snow: The Second Period of Little Ice Age–Type Impacts, 1570–1630," in *Kulturelle Konsequenzen der Kleine Eiszeit,* ed. Wolfgang Behringer, Hartmut Lehmann, and Christian Pfister (Göttingen: Vandenhoeck und Ruprecht, 2005), 31–86, quotation on 78; Martin Hille, "Mensch und Klima in der frühen Neuzeit: Die Anfänge regelmässiger Wetterbeobachtung, 'Kleine Eiszeit,' und ihre Wahrnehmung bei Renward Cysat (1545–1613)," *Archiv für Kulturgeschichte* 83 (2001): 63–91.

32. John Stow, *The Annales, or, Generall Chronicle of England* (London: Thomas Adams, 1615), 656–658, 661, 667, 673, 677–679, 681, 685–686, 695–697, 741–742, 750, 764, 766, 768–769, 782, 784, 787–797, 812, 827, 844, 883; William Camden and Robert Norton, *Annals, or the Historie of the Most Renowned and Victorious Princesse Elizabeth, Late Queene of England,* 3rd ed. (London: Benjamin Fisher, 1635), 52–53, 180, 242, 450, 529; Todd Gray, ed., *The Lost Chronicle of Barnstaple 1586–1611* (Exeter: Devonshire Association, 1998), 61–63, 80–91; William Adams, *Adams's Chronicle of Bristol,* ed. Francis F. Fox (Bristol: J. W. Arrowsmith, 1910), 120, 149, 153, 178. For a season-by-season summary of weather conditions in this period in England, see John Kington, *Climate and Weather* (London: Collins, 2010), 248–258.

33. Frank Oberholzner, "From an Act of God to an Insurable Risk: The Change in the Perception of Hailstorms and Thunderstorms since the Early Modern Period," *Environment and History* 17 (2011): 133–152.

34. William F. Ruddiman, *Plows, Plagues, and Petroleum: How Humans Took Control of Climate* (Princeton, NJ: Princeton University Press, 2005).

35. Key studies include Z. Wang et al., "Large Variations in Southern Hemisphere Biomass Burning during the Last 650 Years," *Science* 330 (2010): 1663–1666; Robert A. Dull et al., "The Columbian Encounter and the Little Ice Age: Abrupt Land Use Change, Fire, and Greenhouse Forcing," *Annals of the Association of American Geographers* 100 (2010): 755–771; R. J. Nevle et al., "Neotropical Human–Landscape Interactions, Fire, and Atmospheric CO_2 during European Conquest," *Holocene* 21 (2011): 853–864; Logan Mitchell et al., "Constraints on the Late Holocene Anthropogenic Contribution to the Atmospheric Methane Budget," *Science* 342 (2013): 964–966; Daniel Pasteris et al., "Acidity Decline in Antarctic Ice Cores during the Little Ice Age Linked to Changes in Atmospheric Nitrate and Sea Salt Concentrations," *Journal of Geophysical Research: Atmospheres* 119 (2014): 5640–5652; William Ruddiman et al., "Does Pre-Industrial Warming Double the Anthropogenic Total?," *Anthropocene Review* 1 (2014): 147–153; Richard Hunter and Andrew Sluyter, "Sixteenth-Century Soil Carbon Sequestration Rates Based on Mexican Land-Grant Documents," *Holocene* 25 (2015): 880–885; Jed O. Kaplan, "Holocene Carbon Cycle: Climate or Humans?," *Nature Geoscience* 8 (2015): 335–336.

36. M. Rubino et al., "Low Atmospheric CO_2 Levels during the Little Ice Age due to Cooling-Induced Terrestrial Uptake," *Nature Geoscience* 9 (2016): 691–694.

37. Alfred W. Crosby, *The Columbian Exchange: Biological and Cultural Consequences of 1492* (Westport, CT: Greenwood, 1972); Alfred W. Crosby, *Ecological Imperialism* (New York: Cambridge University Press, 1986); William Cronon, *Changes in the Land: Indians, Colonists, and the Ecology of New England* (New York: Hill and Wang, 1983); Elinor Melville, *A Plague of Sheep: Environmental Consequences of the Conquest of Mexico* (Cambridge: Cambridge University Press, 1994); Jared M. Diamond, *Guns, Germs, and Steel: The Fates of Human Societies* (New York: W. W. Norton, 2005); Charles C. Mann, *1491: New Revelations of the Americas before Columbus* (New York: Alfred A. Knopf, 2006).

38. Quoted in David Watts, *The West Indies: Patterns of Development, Culture, and Environmental Change since 1492* (Cambridge: Cambridge University Press, 1987), 78.

39. For a critical review of earlier works on the Columbian Exchange, see David S. Jones, "Virgin Soils Revisited," *William and Mary Quarterly* 60 (2003): 703–742, and David S. Jones, *Rationalizing Epidemics: Meanings and Uses of American Indian Mortality since 1600* (Cambridge, MA: Harvard University Press, 2004). On the timing and causes of epidemics in Spanish America, see David Noble Cook, *Born to Die: Disease and the New World Conquest, 1492–1650* (New York: Cambridge University Press, 1998), and Massimo Livi Bacci, *Conquest: The Destruction of the American Indios* (Cambridge: Polity, 2008). For contemporary understandings of disease and their social and cultural dimensions in Spanish America, see Sherry Lee Fields, *Pestilence and Head-*

colds: Encountering Illness in Colonial Mexico (New York: Columbia University Press, 2008).

40. Camilla Townsend, ed., *Here in This Year: Seventeenth-Century Nahuatl Annals of the Tlaxcala-Puebla Valley* (Stanford, CA: Stanford University Press, 2010), 169–175. For a detailed analysis of climate and impacts in colonial Tlaxcala, see Bradley Skopyk, "Undercurrents of Conquest: The Shifting Terrain of Indigenous Agriculture in Colonial Tlaxcala, Mexico," Ph.D. diss., York University, 2010. Matthew Therrell et al., "Tree-Ring Reconstructed Maize Yield in Central Mexico: 1474–2001," *Climatic Change* 74 (2006): 493–504, also finds physical evidence for droughts and frosts in 1594, 1597, 1598, 1599, and 1604.

41. See Jones, *Rationalizing Epidemics,* and William Denevan, "The Pristine Myth: The Landscape of the Americas in 1492," *Annals of the Association of American Geographers* 82 (1992): 369–385.

42. R. Acuna-Soto et al., "Megadrought and Megadeath in 16th-Century Mexico," *Emerging Infectious Diseases* 8 (2002): 360–362; R. Acuna-Soto et al., "When Half of the Population Died: The Epidemic of Hemorrhagic Fevers of 1576 in Mexico," *FEMS Microbiology Letters* 240 (2004): 1–5.

43. Felipe Guaman Poma de Ayala, *The First New Chronicle and Good Government: On the History of the World and the Incas up to 1615,* ed. Roland Hamilton (Austin: University of Texas Press, 2009), 70–71. I have here translated *estado* as "fathom."

44. For New England: Dean R. Snow and Kim M. Lanphear, "European Contact and Indian Depopulation in the Northeast: The Timing of the First Epidemics," *Ethnohistory* 35 (1988): 15–33, and Brenda J. Baker, "Pilgrim's Progress and Praying Indians: The Biocultural Consequences of Contact in Southern New England," in *In the Wake of Contact: Biological Responses to Conquest,* ed. Clark Spencer Larsen and George R. Milner (New York: Wiley-Liss, 1994), 35–45. For Canada: Gary Warrick, "European Infectious Disease and Depopulation of the Wendat-Tionontate (Huron-Petun)," *World Archaeology* 35 (2003): 258–275. For the Spanish Southeast: Cameron B. Wesson, "Prestige Goods, Symbolic Capital, and Social Power in the Protohistoric Southeast," in *Between Contacts and Colonies: Archaeological Perspectives on the Protohistoric Southeast,* ed. Cameron B. Wesson and Mark A. Rees (Tuscaloosa: University of Alabama Press, 2002), 110–125; Rebecca Saunders, "Seasonality, Sedentism, Subsistence, and Disease in the Protohistoric," in *Between Contacts and Colonies: Archaeological Perspectives on the Protohistoric Southeast,* ed. Cameron B. Wesson and Mark A. Rees (Tuscaloosa: University of Alabama Press, 2002), 32–48; and Dale L. Hutchinson, "Entradas and Epidemics in the Sixteenth-Century Southeast," in *Native and Spanish New Worlds: Sixteenth-Century Entradas in the American Southwest and Southeast,* ed. Clay Mathers, Jeffrey M. Mitchem, and Charles M. Haecker (Tucson: University of Arizona Press, 2013), 140–151. For the Southwest: Sam White, "Cold, Drought, and Disaster: The Little Ice Age and the Spanish Conquest of New Mexico," *New Mexico Historical Review* 89 (2014): 431; Matthew J. Liebmann et al., "Native American Depopulation, Reforestation, and Fire Regimes in the Southwest United States, 1492–1900 CE," *Proceedings of the National Academy of Sciences* 113 (2016): E696–E704. For Virginia: Stephen R. Potter, *Commoners, Tribute, and Chiefs: The Development of*

Algonquian Culture in the Potomac Valley (Charlottesville: University Press of Virginia, 1993), 165–166, and Martin Gallivan, "The Archaeology of Native Societies in the Chesapeake: New Investigations and Interpretations," *Journal of Archaeological Research* 19 (2011): 305.

45. See examples from Jesuit missionaries in Florida (Chapter 2) and the English colony in Roanoke (Chapter 3). Other possible examples appear in Álvar Núñez Cabeza de Vaca, "Relación [1542]," in Rolena Adorno and Patrick Charles Pautz, eds., *Álvar Núñez Cabeza de Vaca: His Account, His Life, and the Expedition of Pánfilo de Narváez* (Lincoln: University of Nebraska Press, 1999), 1:217–219, and "The *Primrose* Journal," in *RV,* 1:306.

46. See especially Suzanne Alchon, *A Pest in the Land: New World Epidemics in Global Perspective* (Albuquerque: University of New Mexico Press, 2003); Paul Kelton, *Epidemics and Enslavement: Biological Catastrophe in the Native Southeast* (Lincoln: University of Nebraska Press, 2007); and Catherine M. Cameron, Paul Kelton, and Alan C. Swedlund, eds., *Beyond Germs: Native Depopulation in North America* (Tucson: University of Arizona Press, 2015).

47. Marvin T. Smith, "Aboriginal Population in the Postcontact Southeast," in *The Forgotten Centuries: Indians and Europeans in the American South, 1521–1704,* ed. Charles M. Hudson and Carmen Chaves Tesser (Athens: University of Georgia Press, 1994), 257–275. The peak of missionization and the first epidemic are described in John Tate Lanning, *The Spanish Missions of Georgia* (Chapel Hill: University of North Carolina Press, 1935), 161–166; the destruction of Timucua is recounted in Jerald Milanich and William Sturtevant, eds., *Francisco Pareja's 1613 Confessionario: A Documentary Source for Timucuan Ethnography* (Tallahassee: Florida Division of Archives History and Records Management, 1972), 1–3. On the timing of droughts and epidemics, see Paul E. Hoffman, *Florida's Frontiers* (Bloomington: Indiana University Press, 2002), 109–110, 126–128, 147.

2. Such Great Snows We Thought We Were Dead Men

1. *HGIO,* 1:207–209. For efforts to trace the route of the voyage, see Samuel Eliot Morison, *The European Voyages of Discovery: The Southern Voyages* (New York: Oxford University Press, 1974), 502–516; Douglas T. Peck, "Reconstruction and Analysis of the 1513 Discovery Voyage of Juan Ponce de León," *Florida Historical Quarterly* 71 (1992): 133–154; Jerald T. Milanich, "Charting Juan Ponce de León's 1513 Voyage to Florida: The Calusa Indians amid Latitudes of Controversy," in *La Florida: Five Hundred Years of Hispanic Presence,* ed. Viviana Díaz Balsera and Rachel A. May (Gainesville: University Press of Florida, 2014), 49–68.

2. *HGIO,* 2:24: "pasado muchos trabajos en la Navegación."

3. *HGNI,* 4:320. Oviedo uses the term *temple* to mean "climate."

4. This misimpression, and the experience of Spanish expeditions in La Florida up to ca. 1590, is the subject of Paul E. Hoffman, *A New Andalucia and a Way to the Orient: The American Southeast during the Sixteenth Century* (Baton Rouge: Louisiana State University Press, 1990).

5. See Chapter 10 for further discussion of hurricanes in the early colonial Americas.

6. On Mediterranean climate patterns and agriculture, see, e.g., A. Grove and O. Rackham, *The Nature of Mediterranean Europe* (New Haven: Yale University Press, 2001) and John Wainwright and John B. Thornes, *Environmental Issues in the Mediterranean: Processes and Perspectives from the Past and Present* (London: Routledge, 2004). On the naturalization of Mediterranean crops in Spanish America, see William W. Dunmire, *Gardens of New Spain: How Mediterranean Plants and Foods Changed America* (Austin: University of Texas Press, 2004).

7. For efforts to uncover their history and experience, see, e.g., Richard Flint, "Armas de La Tierra: The Mexican Indian Component of Coronado Expedition Material Culture," in *The Coronado Expedition to Tierra Nueva,* ed. Richard Flint and Shirley Cushing Flint (Niwot: University Press of Colorado, 1997), 57–72.

8. For examples of these studies, see Eberhard Gischler et al., "A 1500-Year Holocene Caribbean Climate Archive from the Blue Hole, Lighthouse Reef, Belize," *Journal of Coastal Research* 24 (2008): 1495–1505; J. N. Richey et al., "1400 Yr Multiproxy Record of Climate Variability from the Northern Gulf of Mexico," *Geology* 35 (2007): 423–426; Casey Saenger et al., "Regional Climate Variability in the Western Subtropical North Atlantic during the Past Two Millennia," *Paleoceanography* 26 (2011): PA2206; Casey Saenger et al., "Tropical Atlantic Climate Response to Low-Latitude and Extratropical Sea-Surface Temperature: A Little Ice Age Perspective," *Geophysical Research Letters* 36 (2009): L11703; J. N. Richey et al., "Regionally Coherent Little Ice Age Cooling in the Atlantic Warm Pool," *Geophysical Research Letters* 36 (2009): L21703; T. M. Cronin et al., "The Medieval Climate Anomaly and Little Ice Age in Chesapeake Bay and the North Atlantic Ocean," *Palaeogeography, Palaeoclimatology, Palaeoecology* 297 (2010): 299–310. The evidence suggests stronger and more consistent cooling in the Gulf of Mexico than off the Atlantic coast. The most recent high-resolution, long-term data come from oxygen isotope ratios in calcites left by *G. ruber,* a species of planktonic foraminifera. These findings confirm the circumstantial evidence for cold anomalies, based on wider ocean-atmospheric circulation patterns, recently proposed in Dennis B. Blanton, "The Factors of Climate and Weather in Sixteenth-Century La Florida," in *Native and Spanish New Worlds: Sixteenth-Century Entradas in the American Southwest and Southeast,* ed. Clay Mathers, Jeffrey M. Mitchem, and Charles M. Haecker (Tucson: University of Arizona Press, 2013), 99–121.

9. Broxton W. Bird et al., "Midcontinental Native American Population Dynamics and Late Holocene Hydroclimate Extremes," *Scientific Reports* 7 (2017): 41628.

10. For the state of the art in tree ring and other proxy climate reconstruction at the time of writing, see Raymond S. Bradley, *Paleoclimatology: Reconstructing Climates of the Quaternary,* 3rd ed. (Amsterdam: Elsevier, 2015). The tree-ring-based North American Drought Atlas can be viewed either by year, http://iridl.ldeo.columbia.edu/SOURCES/.LDEO/.TRL/.NADA2004/.pdsi -atlas.html, or by geographical grid point, http://www.ncdc.noaa.gov/paleo /newpdsi.html (accessed March 15, 2017). The pioneering studies of

tree-ring-reconstructed drought in the southeastern United States have been the work of David Stahle and colleagues; see especially D. Stahle et al., "North Carolina Climate Changes Reconstructed from Tree Rings: A.D. 372 to 1985," *Science* 240 (1988): 1517–1519; D. Stahle and Malcolm Cleaveland, "Tree-Ring Reconstructed Rainfall over the Southeastern USA during the Medieval Warm Period and Little Ice Age," *Climatic Change* 26 (1994): 199–212; D. Stahle et al., "Tree-Ring Reconstructed Megadroughts over North America since AD 1300," *Climatic Change* 83 (2007): 133–149; and D. Stahle et al., "The Lost Colony and Jamestown Droughts," *Science* 280 (1998): 564–567. Several lines of evidence also indicate that Florida became drier during colder periods; see Emmy I. Lammertsma et al., "Sensitivity of Wetland Hydrology to External Climate Forcing in Central Florida," *Quaternary Research* 84 (2015): 287–300.

11. Paul Kocin, Alan Weiss, and Joseph Wagner, "The Great Arctic Outbreak and East Coast Blizzard of February 1899," *Weather and Forecasting* 3 (1988): 305–318.

12. For overviews of protohistoric Mississippian cultures, see, e.g., James Axtell, *The Indians' New South: Cultural Change in the Colonial Southeast* (Baton Rouge: Louisiana State University Press, 1997); John F. Scarry, ed., *Political Structure and Change in the Prehistoric Southeastern United States* (Gainesville: University Press of Florida, 1996); Cameron B. Wesson and Mark A. Rees, eds., *Between Contacts and Colonies: Archaeological Perspectives on the Protohistoric Period Southeast* (Tuscaloosa: University of Alabama Press, 2002); Charles M. Hudson, Thomas J. Pluckhahn, and Robbie Franklyn Ethridge, eds., *Light on the Path: The Anthropology and History of the Southeastern Indians* (Tuscaloosa: University of Alabama Press, 2006). Here and throughout this book, I use the terms *chiefdom* and *community* in much the same way and for the same reasons discussed in Robin Beck, *Chiefdoms, Collapse, and Coalescence in the Early American South* (New York: Cambridge University Press, 2013), 25–34.

13. David G. Anderson, *The Savannah River Chiefdoms: Political Change in the Late Prehistoric Southeast* (Tuscaloosa: University of Alabama Press, 1994), 286–288; Larry V. Benson, Timothy R. Pauketat, and Edward R. Cook, "Cahokia's Boom and Bust in the Context of Climate Change," *American Antiquity* 74 (2009): 467–483; Bird et al., "Midcontinental Native America Population Dynamics"; William C. Foster, *Climate and Culture Change in North America AD 900–1600* (Austin: University of Texas Press, 2012), 79–80.

14. Foster, *Climate and Culture Change*, 118–119; David H. Dye, "Warfare in the Protohistoric Southeast 1500–1700," in *Between Contacts and Colonies: Archaeological Perspectives on the Protohistoric Southeast*, ed. Cameron B. Wesson and Mark A. Rees (Tuscaloosa: University of Alabama Press, 2002), 126–141; David G. Anderson, David W. Stahle, and Malcolm K. Cleaveland, "Paleoclimate and Potential Food Reserves of Mississippian Societies: A Case Study from the Savannah River Valley," *American Antiquity* 60 (1995): 258–286.

15. The best narratives are in Hoffman, *New Andalucia*; Paul E. Hoffman, "Lucas Vázquez de Ayllón's Discovery and Colony," in *The Forgotten Centuries: Indians and Europeans in the American South, 1521–1704*, ed. Charles M. Hudson and Carmen Chaves Tesser (Athens: University of Georgia Press, 1994), 36–49;

and Douglas T. Peck, "Lucas Vásquez de Ayllón's Doomed Colony of San Miguel de Gualdape," *The Georgia Historical Quarterly* 85 (2001): 183–198.

16. Peck, "Lucas Vásquez de Ayllón's Doomed Colony," argues for the Waccamaw River instead.

17. Paul E. Hoffman, *Florida's Frontiers* (Bloomington: Indiana University Press, 2002), 25.

18. *NAW*, 1:251–252.

19. Martire d'Anghiera, *De Orbe Novo: The Eight Decades of Peter Martyr d'Anghera*, ed. Francis Augustus MacNutt (New York: G. P. Putnam's Sons, 1912), 2:269–270.

20. *HGNI*, 4:325–326.

21. At the time of writing there were ongoing archaeological surveys to locate the site of the wrecked flagship and the colony, but neither had yet made any finds.

22. *HGNI*, 4:326.

23. Hoffman, *New Andalucia*, 74–75. This observation appears valid whether San Miguel de Gualdape was as far south as Sapelo Island (as Hoffman proposes) or as far north as Tybee Island (as Peck proposes).

24. *HGNI*, 4:327, 329.

25. *HGNI*, 4:328–329.

26. Martire d'Anghiera, *Orbe Novo*, 2:419.

27. Scholarly narratives of Narváez's career and expedition are found in Andrés Reséndez, *A Land So Strange: The Epic Journey of Cabeza de Vaca* (New York: Basic Books, 2007) and Robin Varnum, *Álvar Núñez Cabeza de Vaca: American Trailblazer* (Norman: University of Oklahoma Press, 2014). For primary sources and commentary, see Rolena Adorno and Patrick Charles Pautz, eds., *Álvar Núñez Cabeza de Vaca: His Account, His Life, and the Expedition of Pánfilo de Narváez*, 3 vols. (Lincoln: University of Nebraska Press, 1999).

28. For estimates of numbers, see Reséndez, *Land So Strange*, 62; on desertions, see Varnum, *Álvar Núñez*, 43.

29. Cabeza de Vaca, "Relación," in Adorno and Pautz, *Álvar Núñez Cabeza de Vaca*, 1:29–31.

30. On Miruelo as pilot and the expedition's geographical confusion, see Adorno and Pautz, *Álvar Núñez Cabeza de Vaca*, 2:74–83, and Varnum, *Álvar Núñez*, 49–53.

31. Cabeza de Vaca, "Relación," 1:67; *HGNI*, 291; Reséndez, *Land So Strange*, 104.

32. *HGNI*, 4:289, 296; Cabeza de Vaca, "Relación," 1:57, 99, 101, 111–113.

33. "The Account by a Gentleman from Elvas," in *The De Soto Chronicles: The Expedition of Hernando de Soto to North America*, ed. Lawrence A. Clayton, Vernon J. Knight, and Edward C. Moore (Tuscaloosa: University of Alabama Press, 1993), 1:48. For more on the reception of Cabeza de Vaca's story, see Adorno and Pautz, *Álvar Núñez Cabeza de Vaca*, 3:119–130.

34. The narrative of the Soto expedition is drawn primarily from Charles M. Hudson, *Knights of Spain, Warriors of the Sun: Hernando de Soto and the South's Ancient Chiefdoms* (Athens: University of Georgia Press, 1997), still the definitive study of the expedition. Critical editions of primary sources are in Clayton, Knight, and Moore, eds., *De Soto Chronicles*. I have followed the

arguments of Martin Malcolm Elbl and Ivana Elbl, "The Gentleman of Elvas and His Publisher," in *The Hernando de Soto Expedition: History, Historiography, and "Discovery" in the Southeast,* ed. Patricia Galloway (Lincoln: University of Nebraska Press, 1997), 45–97, and Hudson, *Knights of Spain,* 445–447, for the authenticity and independence of the sources, as opposed to the criticisms of Patricia Galloway, "The Incestuous Soto Narratives," in *The Hernando de Soto Expedition: History, Historiography, and "Discovery" in the Southeast,* ed. Patricia Galloway (Lincoln: University of Nebraska Press, 1997), 11–44. In brief, these arguments are (1) the similar descriptions and phrases reflect common Iberian tropes of the Indies rather than actual copying; (2) there are sufficient differences in dates and place names to indicate independent efforts to record the same itinerary; (3) the Rángel account includes moralizing passages characteristic of Oviedo (who retells it in his chronicle) but these are missing from the Elvas account, which is supposedly borrowed from the Oviedo version; and (4) the Rángel narrative breaks off after the third winter of the expedition, while more than a third of the Elvas narrative concerns events after this point.

35. Violence and warfare have been major themes of the recent historiography of early European expeditions into the Southeast. See, e.g., Matthew Jennings, *New Worlds of Violence: Cultures and Conquests in the Early American Southeast* (Knoxville: University of Tennessee Press, 2011) and Robbie Franklyn Ethridge, *From Chicaza to Chickasaw: The European Invasion and the Transformation of the Mississippian World, 1540–1715* (Chapel Hill: University of North Carolina Press, 2010).

36. Robert S. Weddle, "Soto's Problems of Orientation: Maps, Navigation, and Instruments in the Florida Expedition," in *The Hernando de Soto Expedition: History, Historiography, and "Discovery" in the Southeast,* ed. Patricia Galloway (Lincoln: University of Nebraska Press, 1997), 219–233.

37. For a recent multidisciplinary investigation and discussion on the location and events of the Battle of Mabila, see Vernon J. Knight, ed., *The Search for Mabila: The Decisive Battle between Hernando de Soto and Chief Tascalusa* (Tuscaloosa: University of Alabama Press, 2009); Hudson, *Knights of Spain,* 238–245.

38. "Account by a Gentleman from Elvas," 1:74.

39. Rodrigo Rangel, "Account of the Northern Conquest and Discovery of Hernando de Soto," in Clayton, Knight, and Moore, eds., *De Soto Chronicles,* 1:247–306, quote from 297.

40. "Account by a Gentleman from Elvas," 1:108.

41. Luys Hernández de Biedma, "Relation of the Island of Florida," in Clayton, Knight, and Moore, eds., *De Soto Chronicles,* 1:221–246, quote from 243.

42. "Account by a Gentleman from Elvas," 1:130.

43. Hudson, *Knights of Spain,* 376–381.

44. See overview in Foster, *Climate and Culture Change,* 135–144.

45. See the discussion in Hudson, *Knights of Spain,* 279–280.

46. For more on the regional drought and its correspondence to descriptions in the Soto narratives, see Blanton, "Factors of Climate," and the discussion of droughts and rain prayers in Chapter 6.

47. Hoffman, *New Andalucia*, 84–102.
48. Francisco López de Gómara, *Historia general de las Indias y vida de Hernán Cortés*, ed. Jorge Gurría Lacroix (Caracas: Biblioteca Ayacucho, 1979), 65–66.
49. Hoffman, *New Andalucia*, 125–152. Oviedo's account of the Narváez expedition, written during the 1540s, would make fun of Spaniards who "believe that the Indies must be like a kingdom of Portugal or Navarre, or at least a single small territory, where everyone knows one another"; in *HGNI*, 4:300.
50. Francisco Cervantes de Salazar, *Life in the Imperial and Loyal City of Mexico in New Spain, and the Royal and Pontifical University of Mexico: As Described in the Dialogues for the Study of the Latin Language*, ed. Carlos Eduardo Castañeda, trans. Minnie Lee Barrett Shepard (Austin: University of Texas Press, 1953), 79.
51. E.g., Marqués de Pidal and Miguel Salvá, eds., "Carta del Doctor Pedro de Santander á S. M. fecha en Sevilla á 15 de julio de 1557," in *CDIE*, 26:340–365. See also Hoffman, *New Andalucia*, 145–157.
52. Andrés González de Barcía Carballido y Zúñiga, *Chronological History of the Continent of Florida*, ed. Anthony Kerrigan (Gainesville: University of Florida Press, 1951), 33. Although Barcía was writing more than a century after the events, his description is borne out in several instances by Velasco's correspondence with Luna during 1559–1560. See, e.g., *LP*, 1:180 / 181, 184–186 / 185–187, 2:148 / 149, 2:154 / 155. *LP* remains the major documentary collection for the expedition. I have cited both the original Spanish and English page numbers, making changes to Priestley's translations where appropriate.
53. Quoted in Herbert Ingram Priestley, *Tristán de Luna: Conquistador of the Old South* (Glendale, CA: Arthur H. Clark, 1936), 72–74.
54. The size of the expedition is attested in the letter of Tristán de Luna, August 25, 1559, *LP*, 2:210 / 211. Augustín Dávila y Padilla, *Historia de la fundación y discurso de la provincia de Santiago de México*, 2nd ed. (Brussels: Meerbeque, 1625), 194, also mentions "las mil y quinientas personas que alli avia." At the time, Pensacola Bay was known as Ochuse, but Luna officially renamed it the Bahía de Santa María Filipina. By 1560 the Spanish settlement there came to be known as Polonza.
55. Some sources believed there was enough food for a year: see Fray Pedro de Feria in *LP*, 2:326 / 327, and Dávila y Padilla, *Historia*, 194. Upon arrival, in early September, Luna claimed enough for only eighty days, but he may have been trying to hurry the first resupply mission; see *LP*, 2:212 / 213. For recently discovered evidence of the expedition's expenditures and supply, see John E. Worth, "Documenting Tristán de Luna's Fleet and the Storm That Destroyed It," *Florida Anthropologist* 62 (2009): 83–92. Luna's report quoted in *LP*, 2:210 / 211. The viceroy's letter is AGI Patronato 19, r. 9. (*LP*, 2:274 / 275 is a copy of a copy of a copy of this original.)
56. Worth, "Documenting Tristán de Luna's Fleet," 90.
57. Quoted in *LP*, 2:244 / 245. A more dramatic description is in Dávila y Padilla, *Historia*, 193. For archaeology, see Roger C. Smith et al., *The Emanuel Point Ship: Archaeological Investigations, 1997–1998* (Pensacola: Archaeology Institute, University of West Florida, 1998), esp. 64; Worth, "Documenting

Tristán de Luna's Fleet." Further analysis of the shipwreck is in James Daniel Collis, "Empire's Reach: A Structural and Historical Analysis of the Emanuel Point Shipwreck," M.A. thesis, University of West Florida, 2008, and Gregory D. Cook, "Luna's Ships: Current Excavation on Emanuel Point II and Preliminary Comparisons with the First Emanuel Point Shipwreck," *Florida Anthropologist* 62 (2009): 93–100.

58. *LP*, 2:244 / 245.

59. Dávila y Padilla, *Historia*, 193.

60. The most detailed account of this part of the expedition appears in Dávila y Padilla, *Historia*, 193–202. Although based on eyewitness accounts from friars accompanying the expedition, it is not an entirely reliable source. However, in this case the outline of events is confirmed by a soldier's official testimony, reproduced in *LP*, 2:288 / 289.

61. *LP*, 1:56–65, 92–129.

62. Quoted in *LP*, 1:180 / 181. The inadequacy of the resupply is also attested by officers of the expedition in *LP*, 2:120 / 121. The delay was due to the need to resupply the Spanish imperial *flota* first; see *LP*, 1:183.

63. Based on the declaration of the *maestre de campo* in *LP*, 1:152 / 153.

64. Dávila y Padilla, *Historia*, 200.

65. *LP*, 1:84 / 85, 94 / 95. The *maestre de campo* gave further testimony of the Indians' scorched-earth tactics in *LP*, 1:154 / 155.

66. *LP*, 1:132–145, quotes from 138–143, 142–145.

67. See Philip II, December 18, 1559, *LP*, 2:14–19, and Luis de Velasco, August 20, 1560, *LP*, 1:180–195.

68. Dávila y Padilla, *Historia*, 206, claims that "the Spanish were almost three hundred, between children and adults, masters and servants," but it is difficult to confirm this figure.

69. Paul E. Hoffman, "Did Coosa Decline between 1541 and 1560?," *Florida Anthropologist* 50 (1997): 25–29.

70. The most detailed account of the Coosa expedition comes from Dávila y Padilla, *Historia*, 202–217; however, he adds an otherwise unconfirmed tale of Spanish soldiers joining Coosa warriors in an attack on a neighboring chiefdom. The letters of the soldiers and friars generally confirm the other elements of Dávila y Padilla's account; they appear in *LP*, 1:218–247 (quotes from 230 / 231 and 246 / 247).

71. *LP*, 1:146–151, 152–177; 2:11, 71, 85, 129.

72. *LP*, 2:136–138 / 137–139.

73. The fate of the expedition is described in Priestley, *Tristán de Luna*, 148–149.

74. These calculations are based on ibid., 157–158, but also take into account the expedition sent to Coosa and alternative estimates of the number remaining in Ochuse and Nanipacana, such as those Priestley himself reproduced in *LP*, 2:132 / 133. The return of the sick, women, and children is attested in *LP*, 2:120 / 121.

75. See descriptions in Velasco's letters to Luna in *LP*, 2:152–177, and AGI Patronato 19, r. 10: "They suffered greatly from hunger. . . . They were reduced to eating the hides of cattle . . . and all the horses they had."

76. AGI Patronato 19, r. 12. For a detailed narrative attempting to reconcile the different testimonies about the expeditions, see Hoffman, *New Andalucia,* 173–181.

77. For "bad climate" (*mal temple*), see Dávila y Padilla, *Historia,* 218. On the winter of 1559–1560, see Luis de Velasco, May 6, 1560, *LP,* 1:92–129, quotes from 108 / 109; and Velasco's admonishment to Luna in summer 1561 that "this winter you ought to take great pains to see that the soldiers are well clothed and lodged," suggesting that had not been so the previous winter, *LP,* 1:188 / 189. Rains, heat, and cold are described in *LP,* 1:162 / 163, 202 / 203. For the report from Coosa, see Fray Domingo and others to Viceroy Velasco, August 1, 1560, *LP,* 1:232–243 (quotes from 238 / 239), and Fray Domingo Salazar, July 31, 1560, *LP,* 1:244–247.

78. Juan López de Velasco, *Geografía y descripción universal de las Indias* (Madrid: Atlas, 1971), 84–94. On his background and sources, see José María López Piñero et al., eds., *Diccionario histórico de la ciencia moderna en España* (Barcelona: Península, 1983), 542–543.

3. The Land Itself Would Wage War

1. Marcel Trudel, *Les vaines tentatives, 1524–1603,* Histoire de la Nouvelle France 1 (Montreal: Fides, 1963), 134.

2. Complete text, translation, and commentary of the voyage in Lawrence C. Wroth, ed., *The Voyages of Giovanni da Verrazzano, 1524–1528* (New Haven: Yale University Press, 1970), quote from 123.

3. Ibid., quote from 129 ("alquanto piu fredda, per accidente et non per natura"). On routes and locations, see Samuel Eliot Morison, *The European Discovery of America: The Northern Voyages* (New York: Oxford University Press, 1971), 287–313.

4. On this diplomatic and political background to French Florida, see John T. McGrath, *The French in Early Florida: In the Eye of the Hurricane* (Gainesville: University Press of Florida, 2000), 9–72, and Éric Thierry, "La Paix de Vervins et les ambitions françaises en Amérique," in *Le Traité de Vervins,* ed. Jean-François Labourdette, Jean-Pierre Poussou, and Marie-Catherine Vignal (Paris: Presses de l'Université de Paris–Sorbonne, 2000), 373–390. On America as a Huguenot refuge, see Frank Lestringant, *L'expérience huguenote au nouveau monde (XVIe siècle)* (Geneva: Librarie Droz, 1996), 29–41.

5. The site was discovered in 1996 and has been undergoing excavations. See Chester B. DePratter, Stanley South, and James Legg, "The Discovery of Charlesfort (1562–1563)," *Transactions of the Huguenot Society of South Carolina* 101 (1996): 39–48; Victor D. Thompson, Chester B. DePratter, and Amanda D. Roberts Thompson, "A Preliminary Exploration of Santa Elena's Sixteenth Century Colonial Landscape Though Shallow Geophysics," *Journal of Archaeological Science: Reports* 9 (2016): 178–190.

6. Jean Ribault, *The Whole and True Discouerye of Terra Florida* (London: Thomas Hacket, 1563), quote from 72.

7. René Goulaine de Laudonnière, *L'histoire notable de la Floride* (Paris: Guillaume Auvray, 1586), fols. 23–32.

8. For the motivations and recruitment of the expedition, see McGrath, *French in Early Florida*, 100–101.

9. Jacques Le Moyne de Morgues, *The Work of Jacques Le Moyne de Morgues: A Huguenot Artist in France, Florida, and England*, ed. Paul H. Hulton (London: British Museum, 1977), 1:121.

10. On the heat wave and illness, see Laudonnière, *Histoire notable*, 58v–59r. On Laudonnière's weak leadership, see Le Moyne de Morgues, *Works*, 1:121.

11. Laudonnière, *Histoire notable*, 62r, 79v–80r, indicates that the last real delivery of food from the Indians came in late November, and that the Indians dispersed from January through March. Cf. Le Moyne de Morgues, *Works*, 1:131, 147, 149. Local archaeology suggests the sandy marine soils there had never supported much indigenous agriculture. See James J. Miller, *An Environmental History of Northeast Florida* (Gainesville: University Press of Florida, 1998), 86.

12. NADA grid point 40: ftp://ftp.ncdc.noaa.gov/pub/data/paleo/drought /pdsi2004/data-by-gridpt/240.txt (accessed March 15, 2017).

13. Le Moyne de Morgues, *Works*, 1:131. The threat of starvation and the role of food shortages in the disorder of the colony are also attested by one of the Spanish who was captured by the mutineers. See the testimony of Ruiz Manso in Charles E. Bennett, *Laudonnière and Fort Caroline: History and Documents* (Gainesville: University of Florida Press, 1964), 103–106.

14. Paul E. Hoffman, *A New Andalucia and a War to the Orient: The American Southeast during the Sixteenth Century* (Baton Rouge: Louisiana State University Press, 1990), 212–215.

15. Ibid., 225–228. A detailed narrative of Pedro Menéndez Avilés's expedition and conquest of Florida is in Eugene Lyon, *The Enterprise of Florida: Pedro Menéndez de Avilés and the Spanish Conquest of 1565–1568* (Gainesville: University Presses of Florida, 1976).

16. For narratives of the race to Florida and the Spanish assault, see Lyon, *Enterprise of Florida*, and McGrath, *French in Early Florida*.

17. Laudonnière, *Histoire*, 105v–106r; Le Moyne de Morgues, *Works*, 133.

18. Laudonnière, *Histoire*, 107v. Witnesses differ slightly on the date of the storm, which may have been the nineteenth or twenty-first of September.

19. Escapes and massacres described in detail from the French perspective in Nicolas Le Challeux, *Discours de l'histoire de la Floride* (Dieppe, 1566), and from the Spanish perspective in the two contemporary biographies of Menéndez de Avilés, Gonzalo Solís de Merás and Bartolomé Barrientos, *Menéndez de Avilés y la Florida: crónicas de sus expediciones*, ed. Juan Carlos Mercado (Lewiston, NY: Edwin Mellen Press, 2006). On the French reprisal, see the anonymously written "Histoire memorable de la reprinse de l'isle de la Floride [1568]," in Suzanne Lussagnet, *Les français en Amérique pendant la deuxième moitié du XVIe siècle* (Paris: Presses Universitaires de France, 1958), 241–249.

20. See his letter to King Philip, October 15, 1565, in Juan Carlos Mercado, ed., *Cartas sobre la Florida, 1555–1574* (Madrid: Iberoamericana, 2002), 139–157; also Jerald T. Milanich, *Laboring in the Fields of the Lord: Spanish Missions and Southeastern Indians* (Washington, DC: Smithsonian Institution Press, 1999), 86–89, and Hoffman, *New Andalucia*, 225–227. Robert S. Weddle, *Spanish Sea:*

The Gulf of Mexico in North American Discovery, 1500–1685 (College Station: Texas A&M University Press, 1985), 328, discusses Menéndez de Avilés's plans for settlements linking Florida and Mexico.

21. Complete records and discussion are in Charles M. Hudson, *The Juan Pardo Expeditions: Exploration of the Carolinas and Tennessee, 1566–1568* (Washington, DC: Smithsonian Institution Press, 1990).

22. Lyon, *Enterprise*, 153.

23. Grant L. Harley et al., "Suwannee River Flow Variability 1550–2005 CE Reconstructed from a Multispecies Tree-Ring Network," *Journal of Hydrology* 544 (2017): 438–451.

24. Karen L. Paar, "Climate in the Historical Record of Sixteenth-Century Spanish Florida: The Case of Santa Elena Re-Examined," in *Historical Climate Variability and Impacts in North America*, ed. Lesley-Ann Dupigny-Giroux and Cary J. Mock (Dordrecht, Netherlands: Springer, 2009), 51.

25. Details in Hoffman, *New Andalucia*, 231–266, and Paul E. Hoffman, *Florida's Frontiers* (Bloomington: Indiana University Press, 2002), 52–56. Detailed analysis of the first years at Santa Elena in Karen Lynn Paar, "To Settle Is to Conquer: Spaniards, Native Americans, and the Colonization of Santa Elena in Sixteenth-Century Florida," Ph.D. diss., University of North Carolina, 1999.

26. See particularly the letter of Antonio Sedeño, March 6, 1570, in Félix Zubillaga, ed., *Monumenta Antiquae Floridae (1566–1572)* (Rome: Monumenta Historica Societatis Iesu, 1946), 421–429. Overview of the Jesuit effort in Félix Zubillaga, *La Florida: La missión Jesuítica (1566–1572) y la colonización Española* (Rome: Institutum Historicum S.I., 1941) and Nicholas P. Cushner, *Why Have You Come Here? The Jesuits and the First Evangelization of Native America* (New York: Oxford University Press, 2006), 31–48.

27. Tree-ring record in David W. Stahle et al., "The Lost Colony and Jamestown Droughts," *Science* 280 (1998): 564–567. Quotations from Clifford Merle Lewis and Albert J. Loomie, eds., *The Spanish Jesuit Mission in Virginia, 1570–1572* (Chapel Hill: University of North Carolina Press, 1953), 85–87. Further mentions of the unusual cold appear in the letter of Juan Rogel, in Lewis and Loomie, eds., *Spanish Jesuit Mission*, 107, and the later relation of Juan de la Carrera in the same volume, 126.

28. Charlotte M. Gradie, "Spanish Jesuits in Virginia: The Mission That Failed," *Virginia Magazine of History and Biography* 96 (1988): 131–156. Other discussions include Karen Kupperman, *The Jamestown Project* (Cambridge, MA: Harvard University Press, 2007), 103–105; James Horn, *Land as God Made It: Jamestown and the Birth of America* (New York: Basic Books, 2005), 1–9; Helen C. Rountree, *Pocahontas's People: The Powhatan Indians of Virginia through Four Centuries* (Norman: University of Oklahoma Press, 1990), 15–20. Anna Brickhouse, *The Unsettlement of America: Translation, Interpretation, and the Story of Don Luis de Velasco, 1560–1945* (New York: Oxford University Press, 2015) discusses the historical life and literary afterlife of Paquiquineo.

29. "Investigation of Licenciate Gamboa," in Jeannette M. Connor, ed., *Colonial Records of Spanish Florida: Letters and Reports of Governors and Secular Persons*

(Deland: Florida State Historical Society, 1925), 1:82–101; and "The Settlers of Florida," in Connor, ed., *Colonial Records*, 1:144–185.

30. "Investigation," 1:99.

31. Elizabeth J. Reitz, "Evidence for Animal Use at the Missions of Spanish Florida," in *The Spanish Missions of La Florida*, ed. Bonnie G. McEwan (Gainesville: University Press of Florida, 1993), 376–398; Elizabeth Jean Reitz and C. Margaret Scarry, *Reconstructing Historic Subsistence with an Example from Sixteenth-Century Spanish Florida* (Glassboro, NJ: Society for Historical Archaeology, 1985).

32. Antonio Sedeño, February 8, 1572, in Zubillaga, ed., *Monumenta*, 492–506; Paar, "To Settle," 65–66.

33. "Settlers of Florida," 1:146–148. On the conditions and expectations of settlers from Spain in Florida, see Eugene Lyon, "Spain's Sixteenth-Century North American Settlement Attempts: A Neglected Aspect," *Florida Historical Quarterly* 59 (1981): 275–291.

34. On the background to the uprising, see Eugene Lyon, "Santa Elena: A Brief History of the Colony, 1566–1587," in *Pedro Menéndez Avilés*, Spanish Borderlands Sourcebooks 24 (New York: Garland, 1995), 481–505, and Paar, "To Settle," 129–188. See also the testimony of Don Cristóbal de Erraso in Connor, ed., *Spanish Documents*, 1:192–203.

35. Hoffman, *New Andalucia*, 270–274. For primary descriptions of Pedro Menéndez Marqués's campaign of "fire and blood" against the Guale, see Pedro Menéndez Marqués to the Audiencia of Santo Domingo, April 2, 1579, in Connor, ed., *Spanish Documents*, 2:224–227, and Luis Jerónimo de Oré, excerpted in *RV*, 2:802–803. For the role of drought, see Paar, "Climate in the Historical Record" and Paar, "To Settle," 229–230.

36. On the international political background to the Roanoke colony, Karen Kupperman, *Roanoke: The Abandoned Colony*, 2nd ed. (Lanham, MD: Rowman and Littlefield, 2007), 1–12. On the rise of English overseas expeditions and privateering, see Kenneth Andrews, *Trade, Plunder and Settlement: Maritime Enterprise and the Genesis of the British Empire, 1480–1630* (Cambridge: Cambridge University Press, 1984) and Mark G. Hanna, *Pirate Nests and the Rise of the British Empire, 1570–1740* (Chapel Hill: University of North Carolina Press, 2015), 21–101.

37. Kupperman, *Roanoke*, 68–72.

38. The details here are obscure and the sources are contradictory. The best attempt to make sense of them is in James P. Horn, *A Kingdom Strange: The Brief and Tragic History of the Lost Colony of Roanoke* (New York: Basic Books, 2010), 54–55.

39. "Arthur Barlowe's Discourse," in *RV*, 1:91–116, quotes from 106, 108.

40. Francisco Marqués de Villalobos to Philip II, Jamaica, June 27, 1586, in Irene Wright, ed., *Further English Voyages to Spanish America, 1583–1594: Documents from the Archives of the Indies at Seville Illustrating English Voyages to the Caribbean, the Spanish Main, Florida, and Virginia* (London: Hakluyt Society, 1951), 174–176.

41. David B. Quinn, *Set Fair for Roanoke: Voyages and Colonies, 1584–1606* (Chapel Hill: University of North Carolina Press, 1985), 62.

42. Letter to F. Walsingham, August 12, 1585, in *RV*, 1:199–204, quote from 201.

43. Recent archaeology has found signs of English activity there dating from the 1580s, but not conclusive traces of the fort itself. See Ivor Noël Hume, *The Virginia Adventure, Roanoke to James Towne: An Archaeological and Historical Odyssey* (New York: Alfred A. Knopf, 1994); National Park Service, *Secrets in the Sand: Archeology at Fort Raleigh, 1990–2010, Manteo, North Carolina: Archeological Resource Study* (Manteo, NC: National Park Service, 2011); and ongoing reports at http://www.firstcolonyfoundation.org (accessed March 15, 2017).

44. "Briefe and True Report," in *RV*, 1:317–387, quote from 378. The fear of English sorcery and the descent into conflict are examined in detail in Michael Leroy Oberg, *The Head in Edward Nugent's Hand: Roanoke's Forgotten Indians* (Philadelphia: University of Pennsylvania Press, 2008).

45. See Spanish dispatches in *RV*, 2:726–742, and Wright, *Further English Voyages*, xxii n. 3, 13–15, 175–176.

46. Documents in *RV*, 2:744–751. On the workings of Spanish intelligence and decision-making regarding England's American ventures, see David B. Quinn, *England and the Discovery of America, 1481–1620* (New York: Albert A. Knopf, 1974), 264–281.

47. See Alonso Suárez de Toledo, June 27, 1586, in Wright, *Further English Voyages*, 171–174, quote from 173. On Caribbean diseases and defense against European invaders, see J. R. McNeill, *Mosquito Empires: Ecology and War in the Greater Caribbean, 1620–1914* (New York: Cambridge University Press, 2010).

48. E.g., Pedro Menéndez Marqués, June 17, 1586, in Wright, *Further English Voyages*, 163–164, and Luis Oré, in *RV*, 2:802–816, quote from 804. Cf. "The *Primrose* Journal," in *RV*, 1:303–308, quote from 304, and "Sir Francis Drake's Expedition," in *RV*, 1:294–303, quote from 296. For another narrative of the raid, see James W. Covington, "Drake Destroys St. Augustine: 1586," *Florida Historical Quarterly* 44 (1965): 81–93.

49. "The *Primrose* Journal," in *RV*, 1:303–308 ("wee understoode that the Hyabans had burn[ed] there towne themselves" [305–306]); Gabriel de Luxán and Diego Fernández de Quiñones, July 1, 1586, in Wright, *Further English Voyages*, 184–185 ("for as soon as the English came down upon the fort, the Indians began to burn the town"); Alonso Suárez de Toledo, July 3, 1586, in Wright, *Further English Voyages*, 449 ("finding himself surrounded by the enemy and Indians, Pedro Menéndez Marqués withdrew").

50. *RV*, 1:305; Alonso Sancho Saez, August 12, 1586, in Wright, *Further English Voyages*, 200; Gabriel de Luxán and Diego Fernández de Quiñones, July 1, 1586, in Wright, *Further English Voyages*, 184–185.

51. Paar, "To Settle," 279–280.

52. In Wright, *Further English Voyages*, 187.

53. Quinn, *Set Fair*, 165; Stahle et al., "Lost Colony"; Dennis B. Blanton, "If It's Not One Thing It's Another: The Added Challenges of Weather and Climate for the Roanoke Colony," in *Searching for the Roanoke Colonies: An Interdisciplinary Collection*, ed. E. Thomson Shields and Charles Robin Ewen (Raleigh: North Carolina Department of Cultural Resources, Division of Archives and History,

2003), 169–176; and "Briefe and True Report," in *RV*, 1:317–387, quote from 377.

54. "Ralph Lane's Discourse," in *RV*, 1:255–294, quote from 283. For more interpretations of the food supply and Indian motivations, see Quinn, *Set Fair*, 214–215; Kupperman, *Roanoke*, 83; Horn, *Kingdom Strange*, 89–92; and Oberg, *Head in Edward Nugent's Hand*.

55. *RV*, 1:291, and "The *Primrose* Journal," quote from 1:307–308.

56. *RV*, 1:322, 336, 383, 325, 335–336, 383.

57. *RV*, 1:257, quote from 272–273; Horn, *Kingdom Strange*, 111–113, 120; Kupperman, *Roanoke*, 105–106.

58. "John White's Narrative of His Voyage," in *RV*, 2:523. The reasons are not entirely clear, since we have only White's version of the story. Horn, *Kingdom Strange*, 149–150, even speculates that White preferred to stay at Roanoke and colluded with Fernandes.

59. *RV*, 2:525–526.

60. "John White's Narrative," 2:528–530.

61. *RV*, 2:535.

62. See documents in *RV*, 2:761–771, 778–816; Wright, *Further English Voyages*, 230–234; and Paul E. Hoffman, *Spain and the Roanoke Voyages* (Raleigh: North Carolina Department of Cultural Resources, 1987), which draws on the same sources as those in this chapter.

63. Wright, *Further English Voyages*, 232–233; Luis Jerónimo de Oré, excerpted in *RV*, 2:803.

64. *RV*, 2:779–281; Oré, in *RV*, 2:809. A newly discovered document from 1611 claims that Indians told the Spanish where they could find English colonists in the Chesapeake, but that the voyage had to turn back due to short supplies and a quarrel that resulted in the killing of an Indian. Unfortunately, this account, written in 1611, cannot be taken as reliable evidence that the English had actually moved to the Chesapeake by 1588. See Joseph Hall, "Glimpses of Roanoke, Visions of New Mexico, and Dreams of Empire in the Mixed-Up Memories of Gerónimo de La Cruz," *WMQ* 72 (2015): 323–350.

65. *RV*, 2:536–537.

66. Their story survives thanks to the deposition of a Spanish prisoner on board the ships, who later escaped. See *RV*, 2:786–795.

67. K. S. Douglas, H. H. Lamb, and C. Loader, *A Meteorological Study of July to October 1588: The Spanish Armada Storms*, Climatic Research Unit Publications 6 (Norwich: University of East Anglia, 1978); Horn, *Kingdom Strange*, 183.

68. "John White's Narrative," 2:608–609.

69. *RV*, quote from 2:611–612.

70. *RV*, 2:613–615, quotes from 617, 618.

71. Quinn, *Set Fair*, 341–377; Kupperman, *Roanoke*, 132–135.

72. Wright, *Further English Voyages*, 234.

73. "The Relation of Pedro Diaz," in *RV*, 2:786–795, quotes from 790, 791.

74. E.g., AGI Ind. Gen. 741, n. 248.

75. Hoffman, *New Andalucia*, 274–290.

76. Wright, *Further English Voyages*, 233.

77. In *RV*, 2:822–825, quote from 824. Vicente González had been playing up the promise and strategic value of the region throughout the 1580s, including a report to the Council of the Indies in 1587. See Paar, "To Settle," 282, and Paul E. Hoffman, "New Light on Vicente Gonzalez's 1588 Voyage in Search of Raleigh's English Colonies," *North Carolina Historical Review* 63 (1986): 199–223. The newly discovered account in Hall, "Glimpses of Roanoke," also describes "the land so good and so temperate, which is neither cold nor hot" (347).

78. Probably the same that had almost wrecked Drake's fleet on the Carolina Banks. See Paar, "Climate in the Historical Record," 55.

79. Wright, *Further English Voyages*, 163, 754–755; Paar, "To Settle," 282.

80. On Spanish incorporation of Native American political hierarchies, see Amy Turner Bushnell, "Ruling 'the Republic of Indians' in Seventeenth-Century Florida," in *Powhatan's Mantle: Indians in the Colonial Southeast*, ed. Gregory A. Waselkov, Peter H. Wood, and M. Thomas Hatley (Lincoln: University of Nebraska Press, 2006), 195–214.

81. The missionary endeavor in Florida has been studied both by Franciscan scholars and by historians and ethnographers of the Guale and Timucua. See John Tate Lanning, *The Spanish Missions of Georgia* (Chapel Hill: University of North Carolina Press, 1935); Maynard J. Geiger, "The Franciscan Conquest of Florida (1573–1618)," Ph.D. diss., Catholic University of America, Washington, DC, 1937; Michael V. Gannon, *The Cross in the Sand: The Early Catholic Church in Florida, 1513–1870* (Gainesville: University of Florida Press, 1965); Bonnie G. McEwan, ed., *The Spanish Missions of La Florida* (Gainesville: University Press of Florida, 1993); John H. Hann, *A History of the Timucua Indians and Missions* (Gainesville: University Press of Florida, 1996); and Milanich, *Laboring in the Fields*.

82. Population estimates in John R. Dunkle, "Population Change as an Element in the Historical Geography of St. Augustine," *Florida Historical Quarterly* 37 (1958): 3–32; Hoffman, *Florida's Frontiers*, 74–76. On the recruitment of families and the role of women in Spanish Florida, see Paar, "To Settle," 96–108. The subsidy is analyzed in Amy Turner Bushnell, *Situado and Saban: Spain's Support System for the Presidio and Mission Provinces of Florida*, Anthropological Papers of the American Museum of Natural History 74 (Washington, DC: American Museum of Natural History, 1994), 43–44. On supplies, see Hoffman, *Florida's Frontiers*, 78. The conflicting evidence on living standards is explained in Michael Gannon, "The New Alliance of History and Archaeology in the Eastern Spanish Borderlands," *WMQ* 49 (1992): 321–334. See also Elizabeth J. Reitz, "Comparison of Spanish and Aboriginal Subsistence on the Atlantic Coastal Plain," *Southeastern Archaeology* 4 (1985): 41–50. At the time of writing, archaeologists were using ground-penetrating radar to locate further Spanish settlement activity in La Florida, but excavations had not yet been done. See Victor D. Thompson, Chester B. DePratter, and Amanda D. Roberts Thompson, "A Preliminary Exploration of Santa Elena's Sixteenth Century Colonial Landscape Though Shallow Geophysics," *Journal of Archaeological Science: Reports* 9 (2016): 178–190. For evidence on crops in Spanish Florida,

see Fray Andrés de San Miguel, "Relación de los trabajos que la gente de una nao padeció," in *Dos antiguas relaciones de la Florida,* ed. Genaro García (Mexico City: J. Aguilar Vera, 1902), 206, and Charles W. Arnade, *Florida on Trial, 1593–1602* (Coral Gables, FL: University of Miami Press, 1959), esp. 27, 30.

83. For precipitation and crop yields, see Hoffman, *Florida's Frontiers,* 88; for recurring problems of drought, see ibid., 109, 120, 126, 169–170. The problem of sandy soil is described by Pedro Menéndez Marqués in AGI Santo Domingo 224, 27 Dec 1583. See also Alonso Gregorio de Escobedo, *Pirates, Indians and Spaniards: Father Escobedo's La Florida,* ed. James W. Covington, trans. A. F. Falcones (St. Petersburg, FL: Great Outdoors, 1963), 23–24, 146.

84. Descriptions of cold are in, e.g., Zubillaga, ed., *Monumenta,* 507, 509; Lewis and Loomie, *Spanish Jesuit Mission,* 126; Escobedo, *Pirates, Indians and Spaniards,* 134–135. For snow, see Luis Oré, *Relación histórica de la Florida, escrita en el siglo XVII.,* ed. Atanasio López (Madrid: Ramona Velasco, 1931), 1:87. André Thevet's *Universal Cosmography* (1575) offers one example of a European impression of Florida's cold as he described how Florida Natives thought hell was "a very cold country, because the greatest unease they suffer is the cold"; *André Thevet's North America: A Sixteenth-Century View,* ed. Roger Schlesinger and Arthur Stabler (Kingston: McGill-Queen's University Press, 1986), 147. The Spanish official chronicler during the 1600s, Antonio de Herrera y Tordesillas, also explained to readers that "the land is cold" and "there are always great storms"; see *HGIO,* decade 4, book 2, chapter 5, 2:65. Frosts and snow are very rare in the region today, and since instrumental records began in the mid-1800s, on only five occasions has it stayed below freezing around St. Augustine for more than one day at a time. See Miller, *Environmental History of Northeast Florida,* 20–21.

85. Kupperman, *Roanoke,* 89 (quote), 135.

4. Bitter Remedies

1. Detailed investigation and primary sources in J. Michael Francis, Kathleen M. Kole, and David Hurst Thomas, "Murder and Martyrdom in Spanish Florida: Don Juan and the Guale Uprising of 1597," *Anthropological Papers of the American Museum of Natural History* 95 (2011): 1–154. For at least one case of Spanish officers using a form of waterboarding, see Luis Oré, *Relación histórica de la Florida, escrita en el siglo XVII,* ed. Luis López (Madrid: Ramona Velasco, 1931), 2:21–22. On the role of the drought, see Andrés González de Barcía Carballido y Zúñiga, *Chronological History of the Continent of Florida,* ed. Anthony Kerrigan (Gainesville: University of Florida Press, 1951), 184, 186–187. The uprising, drought, and negative impression they left of Florida are also described in Juan de Torquemada, *Monarquía Indiana* (Mexico City: UNAM, 1983), 73–79.

2. Paul E. Hoffman, *Florida's Frontiers* (Bloomington: Indiana University Press, 2002), 75 (fire), 88 (drought and harvest); Charles W. Arnade, *Florida on Trial, 1593–1602* (Coral Gables, FL: University of Miami Press, 1959), 15 (storm). The

town's characteristic thatched-roof, wattle-and-daub construction contributed to the risk of fire; see Francis, Kole, and Thomas, "Murder and Martyrdom," 18. On architecture, see Albert C. Manucy, *Sixteenth-Century St. Augustine: The People and Their Homes* (Gainesville: University Press of Florida, 1997).

3. Atanasio López, "Cuatro cartas sobre las misiones de la Florida," *Archivo Ibero-Americano* 1 (1914): 355–368, quote from 358; Arnade, *Florida on Trial,* 1–20, quote from 14.

4. Arnade, *Florida on Trial,* 21–80; John Tate Lanning, *The Spanish Mission of Georgia* (Chapel Hill: University of North Carolina Press, 1935), 120–125.

5. Events summarized in Hoffman, *Florida's Frontiers,* 91–97.

6. AGI Mexico, n. 10, 1602-5-31.

7. AGI Santo Domingo 224, r. 6 n. 48, 1605-3-19; Arnade, *Florida on Trial,* 81–85.

8. For interpretations of ascent and weaknesses of the Spanish Empire in the sixteenth century, see J. H. Elliott, *Imperial Spain: 1469–1716* (London: Penguin, 1963); Henry Kamen, *Empire: How Spain Became a World Power, 1492–1763* (New York: HarperCollins, 2003); Henry Kamen, *Spain, 1469–1714: A Society of Conflict,* 4th ed. (New York: Routledge, 2014); Geoffrey Parker, *The Grand Strategy of Philip II* (New Haven, CT: Yale University Press, 2000). On the cost of Philip II's wars, see Bartolomé Yun Casalilla, *Marte contra Minerva: el precio del imperio español, c. 1450–1600* (Barcelona: Crítica, 2004). On Philip III's diplomacy, see Paul Allen, *Philip III and the Pax Hispanica, 1598–1621: The Failure of Grand Strategy* (New Haven, CT: Yale University Press, 2000).

9. See esp. Gustaf Utterström, "Climatic Fluctuations and Population Problems in Early Modern History," *Scandinavian Economic History Review* 3 (1955): 3–47, and Peter Clark, ed., *The European Crisis of the 1590s: Essays in Comparative History* (London: Allen and Unwin, 1985). For a recent review and new evidence, see Geoffrey Parker, "History and Climate: The Crisis of the 1590s Reconsidered," in *Climate and Cultural Transition in Europe,* ed. Claus Leggewie (Leiden: Brill, forthcoming).

10. Massimo Livi Bacci, *Population and Nutrition: An Essay on European Demographic History* (New York: Cambridge University Press, 1991), 64–67; Massimo Livi Bacci, *The Population of Europe* (London: Blackwell, 2000), 91–116; Robert Woods, "Ancient and Early Modern Mortality: Experience and Understanding," *Economic History Review,* new series, 60 (2007): 373–399; Erik R. Seeman, *Death in the New World: Cross-Cultural Encounters, 1492–1800* (Philadelphia: University of Pennsylvania Press, 2010).

11. For retreat of the plague, see Andrew Appleby, "The Disappearance of Plague: A Continuing Puzzle," *Economic History Review* 33 (1980): 161–173; Paul Slack, "The Response to Plague in Early Modern England: Public Policies and Their Consequences," in *Famine, Disease, and the Social Order in Early Modern Society,* ed. John Walter and Roger Schofield (Cambridge: Cambridge University Press, 1989), 167–187. For mortality trends, see Robert W. Fogel, "Economic Growth, Population Theory, and Physiology: The Bearing of Long-Term Processes on the Making of Economic Policy," *American Economic Review* 84 (1994): 369–395. For population estimates, see Livi Bacci, *Population of Europe,* 8–9. "Four Horsemen" discussed in A. Cunningham and O. Grell, *The Four*

Horsemen of the Apocalypse (New York: Cambridge University Press, 2000). Quotations from Thomas Hobbes, *Leviathan: or, The Matter, Forme and Power of a Commonwealth, Ecclesiasticall and Civill*, ed. A. R. Waller (Cambridge: Cambridge University Press, 1904), 84. For an overview on the "general crisis" and climate, see Geoffrey Parker, *Global Crisis: War, Climate Change and Catastrophe in the Seventeenth Century* (New Haven, CT: Yale University Press, 2013). The demographic and economic turning point of the 1600s is discussed in Anne E. C. McCants, "Historical Demography and the Crisis of the Seventeenth Century," *Journal of Interdisciplinary History* 40 (2009): 195–214, and Jan de Vries, "The Economic Crisis of the Seventeenth Century after Fifty Years," *Journal of Interdisciplinary History* 40 (2009): 151–194.

12. The original study of sixteenth-century inflation is E. H. Phelps Brown and Sheila Hopkins, "Wage-Rates and Prices: Evidence for Population Pressure in the Sixteenth Century," *Economica* 24 (1957): 289–306. For general accounts of silver, population pressure, and inflation, see Ralph Davis, *The Rise of the Atlantic Economies* (Ithaca, NY: Cornell University Press, 1973), 88–124, and Harry Miskimin, *The Economy of Later Renaissance Europe 1460–1600* (New York: Cambridge University Press, 1977), 20–82. For recent studies of productivity and population, see Gregory Clark, "The Condition of the Working Class in England, 1209–2004," *Journal of Political Economy* 113 (2005): 1307–1340, and Gregory Clark, "The Long March of History: Farm Wages, Population, and Economic Growth, England 1209–1861," *Economic History Review* 60 (2007): 97–135. On height, see Koepke Nikola and Baten Joerg, "The Biological Standard of Living in Europe during the Last Two Millennia," *European Review of Economic History* 9 (2005): 61–95.

13. E.g., Emmanuel Le Roy Ladurie, *The Peasants of Languedoc*, trans. John Day (Urbana: University of Illinois Press, 1976), 11–145, esp. 51–83; Victor Skipp, *Crisis and Development: An Ecological Case Study of the Forest of Arden, 1570–1674* (Cambridge: Cambridge University Press, 1978); Sam White, *The Climate of Rebellion in the Early Modern Ottoman Empire* (New York: Cambridge University Press, 2011), 52–77, 104–122.

14. E.g., Geoffrey Parker, *Europe in Crisis, 1598–1648*, 2nd ed. (London: Blackwell, 2001), 11–16; Jack Goldstone, *Revolution and Rebellion in the Early Modern World* (Berkeley: University of California Press, 1991), 31–37; White, *Climate of Rebellion*, 52–77.

15. John Walter, "The Social Economy of Dearth in Early Modern England," in *Famine, Disease and the Social Order in Early Modern Society*, ed. John Walter and Roger Schofield (Cambridge: Cambridge University Press, 1989), 75–128; Paul Slack, "Mortality Crises and Epidemic Disease in England 1485–1610," in *Health, Medicine and Mortality in the Sixteenth Century*, ed. C. Webster (Cambridge: Cambridge University Press, 1979), 9–59; Morgan Kelly and Cormac Ó Gráda, "Living Standards and Mortality since the Middle Ages," *Economic History Review* 67 (2014): 358–381. Kelly and Ó Gráda find that the link between living standards and mortality faded during the mid-seventeenth century.

16. On regions and environments, see Mary J. Dobson, "Contours of Death: Disease, Mortality and the Environment in Early Modern England," *Health*

Transition Review 2 (1992): 77–95; R. W. Hoyle, "Famine as Agricultural Catastrophe: The Crisis of 1622–4 in East Lancashire," *Economic History Review* 63 (2010): 974–1002. On urbanization, see Jan de Vries, *European Urbanization 1500–1800* (Cambridge, MA: Harvard University Press, 1984). On urban mortality, see Chris Galley, "A Model of Early Modern Urban Demography," *Economic History Review* 48 (1995): 448–469; J. Landers, "Mortality and Metropolis: The Case of London 1675–1825," *Population Studies* 41 (1987): 59–76.

17. E.g., Andrew B. Appleby, "Epidemics and Famine in the Little Ice Age," *Journal of Interdisciplinary History* 10 (1980): 643–663; Patrick Galloway, "Annual Variations in Deaths by Age, Deaths by Cause, Prices, and Weather in London 1670 to 1830," *Population Studies* 39 (1985): 487–505; Ronald Lee, "Short-Term Variation: Vital Rates, Prices, and Weather," in *The Population History of England 1541–1871: A Reconstruction*, ed. E. A. Wrigley and R. S. Schofield, 2nd ed. (Cambridge: Cambridge University Press, 1989), 356–401. Regional and seasonal variations in France point to the same conclusions: see Alan Bideau et al., "La mortalité," in *Histoire de la population française*, ed. Jacques Dupâquier (Paris: Presses Universitaires de France, 1988), 2:239–243.

18. Jean Meuvret, "Les crises de subsistances et la démographie de la France d'Ancien Régime," *Population* 1 (1946): 643–650; Hoyle, "Famine"; Bruce M. S. Campbell, "Nature as Historical Protagonist: Environment and Society in Pre-Industrial England," *Economic History Review* 63 (2010): 281–314; Bruce M. S. Campbell and Cormac Ó Gráda, "Harvest Shortfalls, Grain Prices, and Famines in Preindustrial England," *Journal of Economic History* 71 (2011): 859–886. Lee, "Short-Term Variation," 371–374, finds a relationship that is "very weak" but "highly significant statistically" between prices and mortality but also notes "non-linear effects" from especially bad price shocks, and finds that these effects were higher in the period 1548–1640.

19. The connection between hunger and disease has been a complex topic of investigation; see Robert I. Rotberg and Theodore K. Rabb, eds., *Hunger and History: The Impact of Changing Food Production and Consumption Patterns on Society* (Cambridge: Cambridge University Press, 1985) and Livi Bacci, *Population and Nutrition*. Only certain infections can take advantage of chronic or acute malnutrition. On causes of famine mortality, see Patrick Galloway, "Basic Patterns in Annual Variations in Fertility, Nuptiality, Mortality, and Prices in Pre-Industrial Europe," *Population Studies* 42 (1988): 275–302, and Joel Mokyr and Cormac Ó Gráda, "What Do People Die of during Famines: The Great Irish Famine in Comparative Perspective," *European Review of Economic History* 6 (2002): 339–363. On contagion, see Jona Schellekens, "Irish Famines and English Mortality in the Eighteenth Century," *Journal of Interdisciplinary History* 27 (1996): 29–42. On declining marriage and fertility, see Lee, "Short-Term Variation," and Emmanuel Le Roy Ladurie, "L'aménorrhée de famine (XVIIe–XXe siècles)," *Annales* 24 (1969): 1589–1601.

20. See esp. Christian Pfister, "Weeping in the Snow: The Second Period of Little Ice Age–Type Impacts, 1570–1630," in *Kulturelle Konsequenzen der Kleine*

Eiszeit, ed. Wolfgang Behringer, Hartmut Lehmann, and Christian Pfister (Göttingen: Vandenhoeck und Ruprecht, 2005), 62–66.

21. John Florio, *Perpetuall and Naturall Prognostications of the Change of Weather Gathered out of Divers Ancient and Late Writers, and Placed in Order for the Common Good of All Men* (London: John Wolfe, 1591), fol. B8.

22. Gerónimo Cortés, *Lunario nuevo, perpetuo, y general, y pronostico de los tiempos, universal* (Madrid: Pedro Madrigal, 1598), fol. 10.

23. On the Little Ice Age as an age of extremes, see Sam White, "The Real Little Ice Age," *Journal of Interdisciplinary History* 44 (2014): 327–352. On these particular episodes, see Bruce Campbell, *The Great Transition: Climate, Disease and Society in the Late-Medieval World* (New York: Cambridge University Press, 2016); C. Camenisch et al., "The 1430s: A Cold Period of Extraordinary Internal Climate Variability during the Early Spörer Minimum with Social and Economic Impacts in North-Western and Central Europe," *Climate of the Past* 12 (2016): 2107–2126; Marcel Lachiver, *Les années de misère: La famine au temps du Grand Roi, 1680–1720* (Paris: Fayard, 1991); John Post, *Food Shortage, Climatic Variability, and Epidemic Disease in Preindustrial Europe* (Ithaca, NY: Cornell University Press, 1985); John D. Post, *The Last Great Subsistence Crisis in the Western World* (Baltimore: Johns Hopkins University Press, 1977).

24. Michael Sigl et al., "A New Bipolar Ice Core Record of Volcanism from WAIS Divide and NEEM and Implications for Climate Forcing of the Last 2000 Years," *Journal of Geophysical Research: Atmospheres* 118 (2013): 1151–1169.

25. Pedro Simón, *Noticias historiales de las conquistas de Tierra Firme en las Indias occidentales: partes segunda y tercera* (Bogota: Medardo Rivas, 1892), 127–129; J.-C. Thouret et al., "Quaternary Eruptive History of Nevado del Ruiz (Colombia)," *Journal of Volcanology and Geothermal Research* 41 (1990): 225–251.

26. Antonio Vázquez de Espinosa, *Compendio y descripción de las Indias Occidentales,* ed. Charles Upson Clark (Washington, DC: Smithsonian Institution Press, 1948), 468–477, quote from 472; L. A. Jara, J.-C. Thouret, and J. Dávila, "The AD 1600 Eruption of Huaynaputina as Described in Early Spanish Chronicles," *Boletín de la Sociedad Geológica del Perú* 90 (2000): 121–132; J.-C. Thouret et al., "Reconstruction of the AD 1600 Huaynaputina Eruption Based on the Correlation of Geologic Evidence with Early Spanish Chronicles," *Journal of Volcanology and Geothermal Research* 115 (2002): 529–570; Shanaka L. De Silva and Gregory A. Zielinski, "Global Influence of the AD 1600 Eruption of Huaynaputina, Peru," *Nature* 393 (1998): 455–458.

27. M. Sigl et al., "Timing and Climate Forcing of Volcanic Eruptions for the Past 2,500 Years," *Nature* 523 (2015): 543–549. Similar findings can be found in Martin P. Tingley and Peter Huybers, "Recent Temperature Extremes at High Northern Latitudes Unprecedented in the Past 600 Years," *Nature* 496 (2013): 201–205; Raphael Neukom et al., "Inter-Hemispheric Temperature Variability over the Past Millennium," *Nature Climate Change* 4 (2014): 362–367; Raymond S. Bradley and Philip D. Jones, " 'Little Ice Age' Summer Temperature Variations: Their Nature and Relevance to Recent Global Warming Trends," *Holocene* 3 (1993): 367–376; Rosanne D'Arrigo et al., *Dendroclimatic Studies: Tree Growth and Climate Change in Northern Forests* (Washington, DC:

American Geophysical Union, 2014); Lea Schneider et al., "Revising Midlatitude Summer Temperatures back to AD 600 Based on a Wood Density Network," *Geophysical Research Letters* 42 (2015): 4556–4562; Pei Xing et al., "The Extratropical Northern Hemisphere Temperature Reconstruction during the Last Millennium Based on a Novel Method," *PLoS ONE* 11 (2016): e0146776; and Markus Stoffel et al., "Estimates of Volcanic-Induced Cooling in the Northern Hemisphere over the Past 1,500 Years," *Nature Geoscience* 8 (2015): 784–788.

28. For a compilation of weather descriptions in this period, see, e.g., Curt Weikinn, *Quellentexte zur Witterungsgeschichte Europas von der Zeitwende bis zum Jahre 1850* (Berlin: Akademie-Verlag, 1958), 2:374–435. For complete discussions of the climate sources of the Netherlands, Germany, Switzerland, and Czech lands, see J. Buisman and A. F. V. van Engelen, *Duizend jaar weer, wind en water in de Lage Landen* (Franeker: Van Wijnen, 1995); Rüdiger Glaser, *Klimageschichte Mitteleuropas* (Darmstadt: Primus Verlag, 2001), 134–135; Christian Pfister, *Wetternachhersage: 500 Jahre Klimavariationen und Natur Katastrophen (1496–1995)* (Bern: Paul Haupt, 1999); R. Brázdil and O. Kotyza, *History of Weather and Climate in the Czech Lands*, 3 vols. (Zurich: ETH Geographisches Institut, 1995–1999).

29. The first major synthesis of sixteenth-century European historical climatology appeared in a 1999 special issue of *Climatic Change*, including Christian Pfister and Rudolf Brázdil, "Climatic Variability in Sixteenth-Century Europe and Its Social Dimension: A Synthesis," *Climatic Change* 43 (1999): 5–53, and Rudolf Brázdil et al., "Flood Events of Selected European Rivers in the Sixteenth Century," *Climatic Change* 43 (1999): 239–285. For the modeled pressure anomaly, see Pfister, "Weeping in the Snow." Further research has confirmed and extended these results; see Rudolf Brázdil et al., "Historical Climatology in Europe—the State of the Art," *Climatic Change* 70 (2005): 363–430; Rudolf Brázdil et al., "European Climate of the Past 500 Years: New Challenges for Historical Climatology," *Climatic Change* 101 (2010): 7–40; Petr Dobrovolný et al., "Monthly, Seasonal and Annual Temperature Reconstructions for Central Europe Derived from Documentary Evidence and Instrumental Records since AD 1500," *Climatic Change* 101 (2010): 69–107; J. Luterbacher, "Circulation Dynamics and Its Influence on European and Mediterranean January–April Climate over the Past Half Millennium: Results and Insights from Instrumental Data, Documentary Evidence and Coupled Climate Models," *Climatic Change* 101 (2010): 201–234; and Alexis Metzger and Martine Tabeaud, "Reconstruction of the Winter Weather in East Friesland at the Turn of the Sixteenth and Seventeenth Centuries (1594–1612)," *Climatic Change* 141 (2017): 331. See also M. V. Shabalova and A. F. V. van Engelen, "Evaluation of a Reconstruction of Winter and Summer Temperatures in the Low Countries, AD 764–1998," *Climatic Change* 58 (2003): 219–242, and Rüdiger Glaser and Dirk Riemann, "A Thousand-Year Record of Temperature Variations for Germany and Central Europe Based on Documentary Data," *Journal of Quaternary Science* 24 (2009): 437–449.

30. For tree ring and multiproxy studies, see Carlo Casty et al., "Temperature and Precipitation Variability in the European Alps since 1500," *International*

Journal of Climatology 25 (2005): 1855–1880; Ulf Büntgen et al., "Summer Temperature Variations in the European Alps AD 755–2004," *Journal of Climate* 19 (2006): 5606–5623; Mathias Trachsel et al., "Multi-Archive Summer Temperature Reconstruction for the European Alps, AD 1053–1996," *Quaternary Science Reviews* 46 (2012): 66–79; I. Larocque-Tobler et al., "A Last Millennium Temperature Reconstruction Using Chironomids Preserved in Sediments of Anoxic Seebergsee (Switzerland): Consensus at Local, Regional and Central European Scales," *Quaternary Science Reviews* 41 (2012): 49–56. For Alpine glaciers, see H. Holzhauser and H. J. Zumbuhl, "Glacier Fluctuations in the Western Swiss and French Alps in the 16th Century," *Climatic Change* 43 (1999): 223–237. For harvest dates, see O. Wetter and C. Pfister, "Spring-Summer Temperatures Reconstructed for Northern Switzerland and Southwestern Germany from Winter Rye Harvest Dates, 1454–1970," *Climate of the Past* 7 (2011): 1307–1326. Autumn temperatures did not decline as much as winter and summer; see Alex Blass et al., "Decadal-Scale Autumn Temperature Reconstruction back to AD 1580 Inferred from the Varved Sediments of Lake Silvaplana (Southeastern Swiss Alps)," *Quaternary Research* 68 (2007): 184–195.

31. For a regional breakdown based on various proxies, see Joel Guiot and Christophe Corona, "Growing Season Temperatures in Europe and Climate Forcings over the Past 1400 Years," *PLoS ONE* 5 (2010): e9972. Parts of Eastern Europe may form an exception; see Ulf Büntgen et al., "Filling the Eastern European Gap in Millennium-Long Temperature Reconstructions," *Proceedings of the National Academy of Sciences* 110 (2013): 1773–1778.

32. P. Zhang et al., "1200 Years of Warm-Season Temperature Variability in Central Scandinavia Inferred from Tree-Ring Density," *Climate of the Past* 12 (2016): 1297–1312; Hans W. Linderholm et al., "Fennoscandia Revisited: A Spatially Improved Tree-Ring Reconstruction of Summer Temperatures for the Last 900 Years," *Climate Dynamics* 45 (2015): 933–947; Jan Esper et al., "Northern European Summer Temperature Variations over the Common Era from Integrated Tree-Ring Density Records," *Journal of Quaternary Science* 29 (2014): 487–494; Lotta Leijonhufvud et al., "Five Centuries of Stockholm Winter / Spring Temperatures Reconstructed from Documentary Evidence and Instrumental Observations," *Climatic Change* 101 (2010): 109–141.

33. For an overview of events, see Peter Clark, "Introduction," in Clark, ed., *European Crisis*, 3–22. On Scandinavia, see also Utterström, "Climatic Fluctuations."

34. D. Camuffo and S. Enzi, "Reconstructing the Climate of Northern Italy from Archive Sources," in *Climate since A.D. 1500*, ed. R. S. Bradley and P. D. Jones, rev. ed. (London: Routledge, 1995), 143–154. On the freezing of the Venetian lagoon and severe weather in Venice, see Francesco Sansovino, *Venezia città nobilissima et singolare* (Venice: Steffano Curti, 1663), 632, and F. S. Zanon, "Fattori metereologici straordinari in Venezia e nei dintorni ricordati dai cronisti," in *La laguna di Venezia*, ed. G. Magrini (Venice: Ferrari, 1933). For the freezing of the Arno, see *Relazione delle feste fatte in Fiorenza sopra il ghiaccio*

del Fiume d'Arno l'ultimo di Dicembre MCDIV [sic] (Florence: Alessandro Guiducci, 1604). It was apparently the first time it had frozen in sixty years.

35. N. S. Davidson, "Northern Italy in the 1590s," in Clark, ed., *European Crisis,* 158–160; Guido Alfani, "Climate, Population and Famine in Northern Italy: General Tendencies and Malthusian Crisis, ca. 1450–1800," *Annales de Démographie Historique* 120 (2011): 23–53.

36. Leo Noordegraf, "Dearth, Famine and Social Policy in the Dutch Republic at the End of the Sixteenth Century," in Clark, ed., *European Crisis,* 67–83.

37. Parker, *Global Crisis,* xxii.

38. White, *Climate of Rebellion;* Oktay Özel, "The Reign of Violence: The Celalis, c. 1550–1700," in *The Ottoman World,* ed. Christine Woodhead (New York: Routledge, 2011), 184–202; Neil Roberts et al., "Palaeolimnological Evidence for an East–West Climate See-Saw in the Mediterranean since AD 900," *Global and Planetary Change* 84–85 (2012): 23–34.

39. Christopher Dunning, *Russia's First Civil War: The Time of Troubles and the Founding of the Romanov Dynasty* (State College: Pennsylvania State University Press, 2001), 21, 96–104. Proxy evidence of the extreme cold is discussed in Rosanne D'Arrigo et al., "1738 Years of Mongolian Temperature Variability Inferred from a Tree-Ring Width Chronology of Siberian Pine," *Geophysical Research Letters* 28 (2001): 543–546, and R. M. Hantemirov, L. A. Gorlanova, and S. G. Shiyatov, "Extreme Temperature Events in Summer in Northwest Siberia since AD 742 Inferred from Tree Rings," *Palaeogeography, Palaeoclimatology, Palaeoecology* 209 (2004): 155–164.

40. Colin Breen, "Famine and Displacement in Plantation-Period Munster," in *Plantation Ireland: Settlement and Material Culture, c. 1550–c. 1700,* ed. James Lyttleton and Colin Rynne (Dublin: Four Courts Press, 2009), 132–139, quote from 135. Within the context of the dialogue, this is a description of the Munster famine of 1581, but I assume—as does Breen—it was also meant to refer to Ireland's "present state" in 1596, when similar atrocities were being reported. See also R. B. Outhwaite, "Dearth, the English Crown, and the 'Crisis of the 1590s,'" in Clark, ed., *European Crisis,* 23–43, at 30–33. For similar examples in other contemporary conflicts, see Cunningham and Grell, *Four Horsemen,* 227–234.

41. Key works comparing colonial Ireland and Virginia include David B. Quinn, *The Elizabethans and the Irish* (Ithaca, NY: Cornell University Press, 1966), 106–122; James Muldoon, "The Indian as Irishman," *Essex Institute Historical Collections* 111 (1975): 267–289; Jane H. Ohlmeyer, "A Laboratory for Empire? Early Modern Ireland and English Imperialism," in *Ireland and the British Empire,* ed. Kevin Kenny (New York: Oxford University Press, 2006), 26–60; Keith Pluymers, "Taming the Wilderness in Sixteenth- and Seventeenth-Century Ireland and Virginia," *Environmental History* 16 (2011): 610–632; and esp. Audrey J. Horning, *Ireland in the Virginian Sea: Colonialism in the British Atlantic* (Chapel Hill: University of North Carolina Press, 2013), which makes the strongest case that, despite certain similarities and overlap of personalities, these were different kinds of enterprises. For a specific comparison of warfare

in the two contexts, see Wayne E. Lee, *Barbarians and Brothers: Anglo-American Warfare, 1500–1865* (New York: Oxford University Press, 2011).

42. F. S. Rodrigo et al., "A 500-Year Precipitation Record in Southern Spain," *International Journal of Climatology* 19 (1999): 1233–1253. Breaking it down seasonally, it appears the winters and springs were exceptionally rainy but summers were dry, as indicated in F. S. Rodrigo and Mariano Barriendos, "Reconstruction of Seasonal and Annual Rainfall Variability in the Iberian Peninsula (16th–20th Centuries) from Documentary Data," *Global and Planetary Change* 63 (2008): 243–257, and José Creus Novau and Miguel Angel Saz Sánchez, "Las precipitaciones de la época cálida en el sur de la provincia de Alicante desde 1550 a 1915," *Revista de historia moderna: anales de la Universidad de Alicante,* 23 (2005): 35–48. See also lake level and lake sediment data for southern and eastern Spain in Roberts et al., "Palaeolimnological Evidence," and Marc Oliva et al., "Environmental Evolution in Sierra Nevada (South Spain) since the Last Glaciation, Based on Multi-Proxy Records," *Quaternary International* 353 (2014): 195–209.

43. M. B. Vallve and J. Martin-Vide, "Secular Climatic Oscillations as Indicated by Catastrophic Floods in the Spanish Mediterranean Coastal Area (14th–19th Centuries)," *Climatic Change* 38 (1998): 473–491; Rüdiger Glaser et al., "The Variability of European Floods since AD 1500," *Climatic Change* 101 (2010): 235–256; José Miguel Ruiz, Pilar Carmona, and Alejandro Pérez Cueva, "Flood Frequency and Seasonality of the Jucar and Turia Mediterranean Rivers (Spain) during the 'Little Ice Age,'" *Méditerranée* 122 (2015): 121–130. T. Bullón, "Winter Temperatures in the Second Half of the Sixteenth Century in the Central Area of the Iberian Peninsula," *Climate of the Past* 4 (2008): 357–367, creates a temperature index based on written evidence and finds that temperatures of the 1590s were low, but not exceptionally so. This may be because the temperatures of the preceding decades were already unusually low, and so the cold was not especially noted.

44. D. Ruiz-Labourdette et al., "Summer Rainfall Variability in European Mediterranean Mountains from the Sixteenth to the Twentieth Century Reconstructed from Tree Rings," *International Journal of Biometeorology* 58 (2014): 1627–1639; Miguel Angel Saz Sánchez and José Creus Novau, "El clima de la Rioja desde el siglo XV: reconstrucciones dendroclimáticas del observatorio de Haro," *Zubía* 13 (2001): 41–64. There is some confirmation of the drought in rogation (religious intercession for rain) records in central Spain; see Fernando Domínguez Castro et al., "Reconstruction of Drought Episodes for Central Spain from Rogation Ceremonies Recorded at the Toledo Cathedral from 1506 to 1900: A Methodological Approach," *Global and Planetary Change* 63 (2008): 230–242.

45. M. Genova, "Extreme Pointer Years in Tree-Ring Records of Central Spain as Evidence of Climatic Events and the Eruption of the Huaynaputina Volcano (Peru, 1600 AD)," *Climate of the Past* 8 (2012): 751–764; E. Tejedor et al., "Temperature Variability in the Iberian Range since 1602 Inferred from Tree-Ring Records," *Climate of the Past* 13 (2017): 93–105. The extremes are also reported in contemporary sources, e.g., Luis Cabrera de Cordoba, *Relaciones de*

las cosas sucedidas en la Corte de España desde 1599 hasta 1614 (Madrid: J. Martin Alegria, 1857), 57, 166, 205–206 (flooding in Andalusia). See also Inocencio Font Tullot, *Historia del clima de España: cambios climáticos y sus causas* (Madrid: Instituto Nacional de Meteorologia, 1988), 75–82.

46. Act 2, scene 1, lines 109–114. On seasons and seasonal disruption in Shakespeare, see François Laroque, *Shakespeare's Festive World: Elizabethan Seasonal Entertainment and the Professional Stage* (New York: Cambridge University Press, 1991). Examples are found in other plays of the period, including Thomas Nashe, *Summer's Last Will and Testament,* lines 1759–1813.

47. Excellent summaries in Clark, ed., *Crisis of the 1590s:* R. B. Outhwaite, "Dearth, the English Crown, and the 'Crisis of the 1590s,'" 23–43; Peter Clark, "A Crisis Contained? The Condition of English Towns in the 1590s," 44–66; and James Casey, "Spain: A Failed Transition," 209–229. For more comprehensive evidence of population pressure in sixteenth-century Castile, see David Vassberg, *Land and Society in Golden Age Castile* (New York: Cambridge University Press, 1984), and Gonzalo Anes Álvarez, "The Agrarian Depression in Castile in the Seventeenth Century," in *The Castilian Crisis of the Seventeenth Century,* ed. I. A. A. Thompson and Bartolomé Yun Casalilla (Cambridge: Cambridge University Press, 1999), 60–76.

48. Quoted in R. B. Outhwaite, *Dearth, Public Policy, and Social Disturbance in England, 1550–1800* (Cambridge: Cambridge University Press, 1995), 11.

49. Campbell, "Nature as Historical Protagonist."

50. Hugh Plat, *Sundrie New and Artificiall Remedies against Famine Written upon the Occasion of This Present Dearth* (London: Peter Short, 1596), quote from B1r-v.

51. Parker, "History and Climate." Further examples in Fernando Sánchez Rodrigo, "Clima y producción agrícola en Andalucía durante la edad moderna (1587–1729)," in *Naturaleza transformada: estudios de historia ambiental en España,* ed. Manuel Gozález de Molina and Joan Martínez Alier (Barcelona: Icaria, 2001), 161–182.

52. Wrigley and Schofield, *Population History,* 321–324, 666–668; Campbell, "Nature as Historical Protagonist."

53. Quotation from *The Life of Guzmán de Alfarache,* in Geoffrey Parker, *Imprudent King: A New Life of Philip II* (New Haven, CT: Yale University Press, 2014), 348. For histories of the plague, see Bartolomé Bennassar, *Recherches sur les grandes épidémies dans le nord de l'Espagne à la fin du XVIe siècle* (Paris: SEVPEN, 1969); Bernard Vincent, "La peste atlántica de 1596–1602," *Asclepio* 28 (1976): 5–25; Vicente Pérez Moreda, *Las crisis de mortalidad en la España interior (siglos XVI–XIX)* (Madrid: Siglo Veintiuno de España, 1980), 94–96, 120–128, 254–295; and Vicente Pérez Moreda, "The Plague in Castile at the End of the Sixteenth Century," in *The Castilian Crisis of the Seventeenth Century: New Perspectives on the Economic and Social History of Seventeenth-Century Spain,* ed. I. A. A. Thompson and Bartolomé Yun Casalilla (Cambridge: Cambridge University Press, 1994), 32–59.

54. Overview in Ángel García Sanz, "Castile 1580–1680: Economic Crisis and the Policy of 'Reform,'" in *The Castilian Crisis of the Seventeenth Century: New*

Perspectives on the Economic and Social History of Seventeenth-Century Spain, ed. I. A. A. Thompson and Bartolomé Yun Casalilla (Cambridge: Cambridge University Press, 1994), 13–31.

55. On populations affected by famine, see Andrew Appleby, *Famine in Tudor and Stuart England* (Stanford, CA: Stanford University Press, 1978). On the geographic distribution of mortality, see Wrigley and Schofield, *Population History*, 670–673.

56. See Pérez Moreda, *Crisis de mortalidad*.

57. Impacts summarized in García Sanz, "Castile 1580–1680." See also Carlos Álvarez-Nogal, Leandro Prados de la Escosura, and Carlos Santiago-Caballero, "Spanish Agriculture in the Little Divergence," *European Review of Economic History* 20 (2016): 452–477.

58. Campbell, "Nature as Historical Protagonist."

59. E.g., Earl J. Hamilton, "The Decline of Spain," *Economic History Review* 8 (1938): 168–179; J. H. Elliott, "The Decline of Spain," *Past and Present* 20 (1961): 52–75; and Henry Kamen, "The Decline of Spain: A Historical Myth? A Rejoinder," *Past and Present* 91 (1981): 181–185. A historiographical overview is Helen Rawlings, *The Debate on the Decline of Spain* (Manchester: Manchester University Press, 2012).

60. See esp. J. H. Elliott, "Self-Perception and Decline in Early Seventeenth-Century Spain," in *Spain and Its World, 1500–1700* (New Haven, CT: Yale University Press, 1989), 241–261.

61. For a summary of explanations for depopulation, see Annie Molinié-Bertrand, *Au siècle d'or l'Espagne et ses hommes: la population du Royaume de Castille au XVIe siècle* (Paris: Economica, 1985), 377–390. Lope de Deza, *Govierno polytico de agricultura* (Madrid: Alonso Martin de Balboa, 1618), was the exception in emphasizing weather and climate, to the point of advocating the establishment of a sort of weather prediction bureau (see 123r–v). Pedro Fernández Navarrete, *Conservación de monarquías y discursos políticos* [1621], ed. Michael D. Gordon (Madrid: Instituto de Estudios Fiscales, 1982), 326, also blamed "tantas inclemencias de tiempos," while emphasizing social and political factors. Martín González de Cellorigo, *Memorial de la política necesaria y útil restauración a la República de España y estados de ella y del desempeño universal de estos reinos (1600)*, ed. José Luis Pérez de Ayala (Madrid: Instituto de Cooperación Iberoamericana, 1991), likewise mentioned the impact of plague, while dwelling more on other reasons for Spain's crisis. Sancho de Moncada, *Restauración politica de España* [1619] (Madrid: Juan de Zúñiga, 1746), 99–100, 133, 137, mentioned these causes as well but only in passing. On mules, see Juan de Valverde Arrieta, *Diálogos de la fertilidad y abundancia de España* (Madrid: Alonso Gomez, 1578) and Vassberg, *Land and Society*, 161–163.

62. Fernández Navarrete, *Conservación de monarquías*, 27 (quote), 371–379.

63. Moncada, *Restauración*, 98.

64. Livi Bacci, *Population of Europe*, 116–122; Ida Altman, "A New World in the Old: Local Society and Spanish Emigration to the Indies," in *"To Make America": European Emigration in the Early Modern Period*, ed. Ida Altman and James

Horn (Berkeley: University of California Press, 1991), 33. To put the figure in perspective, roughly twice that share of English population would emigrate to the American colonies during the seventeenth century; see Ida Altman and James Horn, "Introduction," in *"To Make America,"* 4.

65. Fernández Navarrete, *Conservación de monarquías*, 75.

66. A close analysis of Spanish emigrants of the 1590s shows clear signs that many were escaping poverty, particularly at the behest of relatives already in the New World. See Peter Boyd-Bowman, "Patterns of Spanish Emigration to the Indies until 1600," *Hispanic American Historical Review* 56 (1976): 584.

67. The negative historiographical sentiment is perhaps best captured in recent popular histories, such as G. J. Meyer, *The Tudors: The Complete Story of England's Most Notorious Dynasty* (New York: Delacorte Press, 2010).

68. Richard Hakluyt, "Discourse on Western Planting," in *The Original Writings and Correspondence of the Two Richard Hakluyts*, ed. E. G. R. Taylor (London: Hakluyt Society, 1935), 2:234–235. See also Mancall, *Hakluyt's Promise*, 145.

69. J. F. Larkin and Paul L. Hughes, eds., *Royal Proclamations of King James I, 1603–1625* (Oxford: Oxford University Press, 1973), 47–48, 51–53.

70. *The Meeting of Gallants at an Ordinary; or, The Walks in Paul's: A Dialogue between War, Famine, and the Pestilence, Blazing Their Several Evils* [1603], lines 91–94, in Gary Taylor and John Lavagnino, eds., *Thomas Middleton: The Collected Works* (New York: Oxford University Press, 2007).

71. Thomas Dekker, *Newes from Graves-End: Sent to Nobody* (London: Thomas Archer, 1604), F4r–v.

72. Robert Gray, *A Good Speed to Virginia* (London: William Welby, 1609), fol. B4. See also Chapter 9.

73. Karen Ordahl Kupperman, "The Beehive as a Model for Colonial Design," in *America in European Consciousness 1493–1750*, ed. Karen Ordahl Kupperman (Chapel Hill: University of North Carolina Press, 1995), 272–292.

74. E.g., Fernández Navarrete, *Conservación de monarquias*, 63; González de Cellorigo, *Memorial*, 11–12; Joannes de Laet, *Hispania, sive, de regis Hispaniae regnis et opibus commentarius* (Amsterdam: Elzevir, 1629), 86–90. See also J. N. Hillgarth, *The Mirror of Spain, 1500–1700* (Ann Arbor: University of Michigan Press, 2000), 503–516.

75. Peter C. Mancall, *Hakluyt's Promise: An Elizabethan's Obsession for an English America* (New Haven, CT: Yale University Press, 2007), 171–172; Waldemar Zacharasiewicz, *Die Klimatheorie in der englischen Literatur und Literaturkritik von der Mitte des 16. bis zum frühen 18. Jahrhundert* (Vienna: Braunmüller, 1977), esp. 105.

76. Letter of Ambassador Velasco to Philip III in Alexander Brown, *The Genesis of the United States* (London: W. Heinemann, 1891), 1:455–457.

77. On vagrancy, see A. L. Beier, *Masterless Men: The Vagrancy Problem in England 1560–1640* (London: Methuen, 1985). On crime and public order in the 1590s, see M. J. Power, "London and the Control of the 'Crisis' of the 1590s," *History* 70 (1985): 371–385. On the poor law, see Paul Slack, *Poverty and Policy in Tudor and Stuart England* (London: Longman, 1988), esp. 39, 101, 126–129. On the demographic effects, see Morgan Kelly and Cormac Ó Gráda, "The Poor Law

of Old England: Institutional Innovation and Demographic Regimes," *Journal of Interdisciplinary History* 41 (2011): 339–366.

78. James Horn, " 'To Parts beyond the Seas': Free Emigration to the Chesapeake in the Seventeenth Century," in *"To Make America": European Emigration in the Early Modern Period*, ed. Ida Altman and James Horn (Berkeley: University of California Press, 1991), 117–118.

79. The original order has still not been located, but it is paraphrased in AGI Mexico, leg. 5, fols. 44–46. See also Hoffman, *Florida's Frontiers*, 97–99, for a summary of correspondence.

80. AGI Santo Domingo 224, r. 6, n. 60; AGI Santo Domingo 224, r. 6, n. 62.

81. AGI Mexico 1065, leg. 5, fols. 44–46. Irene A. Wright, "Spanish Policy toward Virginia, 1606–1612: Jamestown, Ecjia, and John Clark of the *Mayflower*," *American Historical Review* 25 (1920): 449, discusses the *consulta* of the Junta de Guerra de Indias in the context of reports about the Jamestown colony.

82. The following chronology has been pieced together by cataloguing more than eighty documents—including diplomatic dispatches; *consultas* from the councils of state, war, and war in the Indies; and royal *cédulas*—transcribed, translated, or paraphrased in the following publications: Brown, *Genesis;* Wright, "Spanish Policy"; *JVFC;* and William S. Goldman, "Spain and the Founding of Jamestown," *WMQ* 68 (2011): 427–450.

83. Brown, *Genesis*, 1:45–46; Goldman, "Spain," 435 (AG Simancas, E leg. 841, fol. 99); *JVFC*, 1:69–71 (AG Simancas, E leg. 2685, fols. 78–79); *JVFC*, 1:117–120 (AG Simancas, E leg. 2586, fol. 68). See also Brown, *Genesis*, 1:172.

84. Goldman, "Spain," 437 (AGI Ind. Gen., leg. 1867, fol. 114) and 438; *JVFC*, 1:115–116.

85. Goldman, "Spain," 444–445; AGI Mexico 1065, leg. 5, fol. 64.

86. For documents on the expedition, see *JVFC*, 2:23–319, and John H. Hann, "Translation of the Ecija Voyages of 1605 and 1609 and the González Derrotero of 1609," *Florida Archaeology* 2 (1986): 1–80.

87. Brown, *Genesis*, 1:197.

88. *JVFC*, 2:259–260.

89. *JVFC*, 2:286; Brown, *Genesis*, 1:392–393.

90. Brown, *Genesis*, 1:455–457; Goldman, "Spain," 448 (quoting AGI Simancas, E leg. 844, fol. 61). The intelligence on wind patterns is credited to William Monson, who later wrote that "the wind and current sets with that violence and constancy that it is impossible to keep to windward of any port if we keep the sea, or to recover a height if we are put to leeward of it. Therefore the error of our planters in Virginia and the Bermudas shall appear, who were drawn principally into those enterprises, in hopes to annoy the Spaniards' trade in the West Indies, not knowing that the current sets with such force from Cape Florida to the northward that it is impossible to beat up with a tack. There were so ignorant as not to know that if they go from those places to the West Indies, they must first fetch the Canaries for a wind, which is a thousand leagues nearer the West Indies going out from England than out of Virginia." See "An Admonition Addressed to Gentlemen to Beware How They Engage in Sea Voyages, or Give Ear to Projectors That Shall Advise Them to Such

Actions," in *Naval Tracts of Sir William Monson*, ed. Michael Oppenheim (London: Navy Records Society, 1913), 3:323–325.

91. Brown, *Genesis*, 1:523.

92. Wright, "Spanish Policy," 458 (AG Simancas, E leg. 844, fol. 113); Brown, *Genesis*, 2:537–538, quote from 575.

93. Brown, *Genesis*, 2:593–594, 632–634, 656. Cf. John Chamberlain, *The Letters of John Chamberlain*, ed. Norman Egbert McClure (Philadelphia: American Philosophical Society, 1939), 2:366–367 (July 9, 1612): "It is generally looked for that [Pedro de Zúñiga] will expostulate about our planting in Virginia, wherein there will need no great contestation, seeing yt is to be feared that that action will fall to the ground of yt self, by the extreem beastly ydlenes of our nation (notwithstanding any cost or diligence used to support them) will rather die and starve then be brought to any labor or industrie to maintain themselves."

5. We Had Changed Summer with Winter

1. "Orders of the Council of Virginia," *JVFC*, 1:46; Philip L. Barbour, *The Three Worlds of Captain John Smith* (Boston, 1964), 108 (quote), 113; James Horn, *A Land as God Made It: Jamestown and the Birth of America* (New York: Basic Books, 2005), 39–40; Karen Kupperman, *The Jamestown Project* (Cambridge, MA: Harvard University Press, 2007), 184.

2. John Smith, "The Proceedings of the English Colonie in Virginia . . . [1612]," *CWCJS*, 1:204; George Percy, "Observations Gathered out of a Discourse," *JVFC*, 1:129.

3. *CWCJS*, 1:204–205.

4. *JVFC*, 1:132.

5. Laurence C. Smith and Scott R. Stephenson, "New Trans-Arctic Shipping Routes Navigable by Midcentury," *Proceedings of the National Academy of Sciences* 110 (2013): 1191–1195.

6. Dagomar Degroot, "Testing the Limits of Climate History: The Quest for a Northeast Passage during the Little Ice Age, 1594–1597," *Journal of Interdisciplinary History* 45 (2015): 459–484; Dagomar Degroot, "Exploring the North in a Changing Climate: The Little Ice Age and the Journals of Henry Hudson, 1607–1611," *Journal of Northern Studies* 9 (2015): 69–91. For further climate reconstructions and another overview of climate and exploration in Arctic Russia, see Vladimir Klimenko, "Thousand-Year History of Northeastern Europe Exploration in the Context of Climatic Change: Medieval to Early Modern Times," *Holocene* 26 (2016): 365–379.

7. J. Luterbacher et al., "Circulation Dynamics and Its Influence on European and Mediterranean January–April Climate over the Past Half Millennium: Results and Insights from Instrumental Data, Documentary Evidence and Coupled Climate Models," *Climatic Change* 101 (2010): 201–234.

8. See note 10 below.

9. J. Luterbacher and C. Pfister, "The Year without a Summer," *Nature Geoscience* 8 (2015): 246–248; Gillen D'Arcy Wood, *Tambora: The Eruption That Changed the World* (Princeton, NJ: Princeton University Press, 2014).

10. For Arctic temperature reconstructions and models, see Nicholas P. McKay
and Darrell S. Kaufman, "An Extended Arctic Proxy Temperature Database for
the Past 2,000 Years," *Scientific Data* 1 (2014): 140026, and E. Crespin et al.,
"Arctic Climate over the Past Millennium: Annual and Seasonal Responses to
External Forcings," *Holocene* 23 (2013): 321–329. On northern Scandinavian
temperatures, see Danny McCarroll et al., "A 1200-Year Multiproxy Record of
Tree Growth and Summer Temperature at the Northern Pine Forest Limit of
Europe," *Holocene* 23 (2013): 471–484. For an overview of sea ice dynamics and
reconstructions, see Leonid Polyak et al., "History of Sea Ice in the Arctic,"
Quaternary Science Reviews 29 (2010): 1757–1778; updated databases can also be
found at the National Snow and Ice Data Center (http://nsidc.org). For
multi-proxy land-based sea ice reconstructions, see Christophe Kinnard et al.,
"Reconstructed Changes in Arctic Sea Ice over the Past 1,450 Years," *Nature*
479 (2011): 509–512; M. Macias Fauria et al., "Unprecedented Low Twentieth
Century Winter Sea Ice Extent in the Western Nordic Seas since AD 1200,"
Climate Dynamics 34 (2010): 781–795. For Iceland sea ice, see Guillaume
Massé et al., "Abrupt Climate Changes for Iceland during the Last Millen-
nium: Evidence from High Resolution Sea Ice Reconstructions," *Earth
and Planetary Science Letters* 269 (2008): 565–569; Longbin Sha et al.,
"Palaeo-Sea-Ice Changes on the North Icelandic Shelf during the Last
Millennium: Evidence from Diatom Records," *Science China Earth Sciences* 58
(2015): 962–970; Astrid Ogilvie, personal communication. See also Patricia
Cabedo-Sanz et al., "Variability in Drift Ice Export from the Arctic Ocean to the
North Icelandic Shelf over the Last 8000 Years: A Multi-Proxy Evaluation,"
Quaternary Science Reviews 146 (2016): 99–115, for a regime shift at 0.4ka in sea
ice north of Iceland in connection to a strengthened East Greenland Current.
On Svalbard temperatures, see A. Grinsted et al., "Svalbard Summer Melting,
Continentality, and Sea Ice Extent from the Lomonosovfonna Ice Core,"
Journal of Geophysical Research: Atmospheres 111 (2006): D07110; William J.
D'Andrea et al., "Mild Little Ice Age and Unprecedented Recent Warmth in an
1800 Year Lake Sediment Record from Svalbard," *Geology* 40 (2012); E.
Isaksson et al., "Two Ice-Core Delta O-18 Records from Svalbard Illustrating
Climate and Sea-Ice Variability over the Last 400 Years," *Holocene* 15 (2005):
501–509; Dmitry Divine et al., "Thousand Years of Winter Surface Air
Temperature Variations in Svalbard and Northern Norway Reconstructed from
Ice-Core Data," *Polar Research* 30 (2011): 7379. On sea ice cover around
Svalbard, see Patrycja Jernas et al., "Palaeoenvironmental Changes of the Last
Two Millennia on the Western and Northern Svalbard Shelf," *Boreas* 42 (2013):
236–255. On temperatures and currents in the Fram Strait and off Svalbard,
see Sophie Bonnet et al., "Variability of Sea-Surface Temperature and Sea-Ice
Cover in the Fram Strait over the Last Two Millennia," *Marine Micropaleon-
tology* 74 (2010): 59–74; Jernas, "Palaeoenvironmental Change"; Tor Lien Mjell
et al., "Multidecadal Changes in Iceland Scotland Overflow Water Vigor over
the Last 600 Years and Its Relationship to Climate," *Geophysical Research
Letters* 43 (2016): GL068227; J. Pawłowska et al., "Palaeoceanographic Changes
in Hornsund Fjord (Spitsbergen, Svalbard) over the Last Millennium: New

Insights from Ancient DNA," *Climate of the Past* 12 (2016): 1459–1472. On sea ice and temperatures in western Greenland and Baffin Bay, see Nancy S. Grumet et al., "Variability of Sea-Ice Extent in Baffin Bay over the Last Millennium," *Climatic Change* 49 (2001): 129–145; Sofia Ribeiro et al., "Climate Variability in West Greenland during the Past 1500 Years: Evidence from a High-Resolution Marine Palynological Record from Disko Bay," *Boreas* 41 (2012): 68–83; Longbin Sha et al., "A Diatom-Based Sea-Ice Reconstruction for the Vaigat Strait (Disko Bugt, West Greenland) over the Last 5000 Yr," *Palaeogeography, Palaeoclimatology, Palaeoecology* 403 (2014): 66–79. On sea ice in arctic Canada, see Jochen Halfar et al., "Arctic Sea-Ice Decline Archived by Multicentury Annual-Resolution Record from Crustose Coralline Algal Proxy," *Proceedings of the National Academy of Sciences* 110 (2013): 19737–19741. On temperature and sea ice around Novaya Zemlya, see Ivar Murdmaa et al., "Paleoenvironments in Russkaya Gavan' Fjord (NW Novaya Zemlya, Barents Sea) during the Last Millennium," *Palaeogeography, Palaeoclimatology, Palaeoecology* 209 (2004): 141–154. On a shift in the NAO, see C. J. Proctor et al., "A Thousand Year Speleothem Proxy Record of North Atlantic Climate from Scotland," *Climate Dynamics* 16 (2000): 815–820; Jan Esper et al., "Long-Term Drought Severity Variations in Morocco," *Geophysical Research Letters* 34 (2007): L17702; Valérie Trouet et al., "Persistent Positive North Atlantic Oscillation Mode Dominated the Medieval Climate Anomaly," *Science* 324 (2009): 78–80; Pablo Ortega et al., "A Model-Tested North Atlantic Oscillation Reconstruction for the Past Millennium," *Nature* 523 (2015): 71. On eruptions and historical ice cover around the Hudson Strait, see A. J. W. Catchpole, "Hudson's Bay Company Ships' Log-Books as Sources of Sea Ice Data, 1751–1870," in *Climate since A.D. 1500*, ed. R. S. Bradley and P. D. Jones, rev. ed. (London: Routledge, 1995), 17–39. On the Labrador Current, see Alan D. Wanamaker et al., "Coupled North Atlantic Slope Water Forcing on Gulf of Maine Temperatures over the Past Millennium," *Climate Dynamics* 31 (2008): 183–194, and G. W. K. Moore et al., "Amplification of the Atlantic Multidecadal Oscillation Associated with the Onset of the Industrial-Era Warming," *Scientific Reports* 7 (2017): 40861. On Baffin Island temperatures, see Christopher R. Florian et al., "Algal Pigments in Arctic Lake Sediments Record Biogeochemical Changes Due to Holocene Climate Variability and Anthropogenic Global Change," *Journal of Paleolimnology* 54 (2015): 53–69; J. J. Moore et al., "Little Ice Age Recorded in Summer Temperature Reconstruction from Varved Sediments of Donard Lake, Baffin Island, Canada," *Journal of Paleolimnology* 25 (2001): 503–517; Elizabeth K. Thomas and Jason P. Briner, "Climate of the Past Millennium Inferred from Varved Proglacial Lake Sediments on Northeast Baffin Island, Arctic Canada," *Journal of Paleolimnology* 41 (2009): 209–224; Anne Beaudoin et al., "Palaeoenvironmental History of the Last Six Centuries in the Nettilling Lake Area (Baffin Island, Canada): A Multi-Proxy Analysis," *Holocene* 26 (2016): 1835–1846.

11. Eduardo Moreno-Chamarro et al., "An Abrupt Weakening of the Subpolar Gyre as Trigger of Little Ice Age-Type Episodes," *Climate Dynamics* 48 (2017): 727. The study does not assume any particular cause (volcanic or otherwise) for the initial buildup of sea ice.

12. *HGNI*, 2:215, 4:330; Tomás López Medel, "Tratado cuyo título es: De los tres elementos . . . ," in *Tomás López Medel: Trayectoria de un clérigo-oidor ante el Nuevo Mundo,* ed. Berta Ares Queija (Guadalajara: Instituto Provincial de Cultura, 1993), 414. Cf. Robert Thorne, "The Booke . . . Being an Information of the Parts of the World [1527]," in *PN,* 2:178. For a more detailed discussion of attempts to justify a Northern Passage, see John Kirtland Wright, "The Open Polar Sea," in *Human Nature in Geography: Fourteen Papers, 1925–1965* (Cambridge, MA: Harvard University Press, 1966), 89–118, and Sam White, "Unpuzzling American Climate: New World Experience and the Foundations of a New Science," *Isis* 106 (2015): 544–566.

13. See the various accounts in R. A. Skelton and James Alexander Williamson, eds., *The Cabot Voyages and Bristol Discovery under Henry VII* (London: Hakluyt Society, 1962), 229, 231, 267, 270, 274–275 (quote). The exact location of the voyages remains uncertain, although Cape Breton and Newfoundland are commonly proposed.

14. Overviews of early Northwest Passage expeditions are in John L. Allen, "From Cabot to Cartier: The Early Exploration of Eastern North America, 1497–1543," *Annals of the Association of American Geographers* 82 (1992): 500–521, and John L. Allen, "The Indrawing Sea: Imagination and Experience in the Search for the Northwest Passage, 1497–1632," in *American Beginnings: Exploration, Culture, and Cartography in the Land of Norumbega,* ed. Emerson W. Baker et al. (Lincoln: University of Nebraska Press, 1994), 7–36 (quote from 24).

15. The surviving evidence of the voyage is reproduced in *PN,* 2:212–270. For descriptions of the cold (from writings found among the victims), see *PN,* 2:223, 253. On the death of the expedition's members, see Eleanora C. Gordon, "The Fate of Sir Hugh Willoughby and His Companions: A New Conjecture," *Geographical Journal* 152 (1986): 243–247.

16. Steven Burrough, "The Navigation and Discoverie toward the River of Ob . . . in the Yere 1556," *PN,* 2:342.

17. David B. Quinn, *Voyages and Colonising Enterprises of Sir Humphrey Gilbert* (London: Hakluyt Society, 1940), 1:105–117.

18. On Frobisher's background and investors, see James McDermott, " 'A Right Heroicall Heart': Sir Martin Frobisher," in *Meta Incognita: A Discourse of Discovery,* ed. Thomas H. B. Symons (Quebec: Canadian Museum of Civilization, 1999), 1:55–118.

19. Original accounts of the expedition are in Richard Collinson, ed., *The Three Voyages of Martin Frobisher* (London: Hakluyt Society, 1867), quote from 80–81. On Inuit encounters and the fate of the captured sailors, see Peter C. Mancall, "The Raw and the Cold: Five English Sailors in Sixteenth-Century Nunavut," *WMQ* 70 (2013): 3–40.

20. Michael Lok in Collinson, ed., *Three Voyages,* 87.

21. In Collinson, ed., *Three Voyages,* 110.

22. For details on the assay, see Bernard Allaire, "Methods of Assaying Ore and Their Application in the Frobisher Ventures," in Symons, ed., *Meta Incognita,* 2:477–504.

23. "Articles of Graunt" in Collinson, ed., *Three Voyages*, 111–131.

24. Dionyse Settle, "Dionyse Settle's Account of the Second Voyage [1577]," in *The Three Voyages of Martin Frobisher*, ed. Vilhjamur Stefansson (London: Argonaut Press, 1938), 2:13, 14.

25. On the background to the voyage and primary sources, see James McDermott, ed., *The Third Voyage of Martin Frobisher to Baffin Island, 1578* (London: Hakluyt Society, 2001). The cost of the voyage is given in the introduction, 49.

26. George Best, "True Discourse [1578]," in McDermott, ed., *Third Voyage*, 214.

27. Ibid., 221.

28. James Watt, "The Medical Record of the Frobisher Voyages of 1576, 1577 and 1588," in Symons, ed., *Meta Incognita*, 2:607–632.

29. McDermott, ed., *Third Voyage*, 41.

30. James McDermott, "The Construction of the Dartford Furnaces," in Symons, ed., *Meta Incognita*, 2:505–522.

31. Robert Thorne, "Declaration [1527]," in *PN*, 2:162; Wright, "Open Polar Sea," 100.

32. John Davis, "The Worlde's Hydrographical Discription [1595]," in *The Voyages and Works of John Davis the Navigator*, ed. Albert Hastings Markham (London: Hakluyt Society, 1880), 222–223.

33. E.g., George Best, *A True Discourse of the Late Voyages of Discoverie, for the Finding of a Passage to Cathaya, by the Northweast, under the Conduct of Martin Frobisher Generall* (London: Henry Bynnyman, 1578), 11–12; William Bourne, "A Regiment for the Sea (1580)," in *A Regiment for the Sea and Other Writings*, ed. E. G. R. Taylor (London: Hakluyt Society, 1963), 301–312.

34. Colin Coates and Dagomar Degroot, "Les bois engendrent les frimas et les gelées: comprendre le climat en Nouvelle-France," *Revue d'histoire de l'Amérique française* 68 (2015): 197–219. For in-depth discussion of these arguments, see White, "Unpuzzling American Climate."

35. Karen Kupperman, "Fear of Hot Climates in the Anglo-American Colonial Experience," *WMQ* 41 (1984): 213–240; Vladimir Jankovic, *Confronting the Climate: British Airs and the Making of Environmental Medicine* (New York: Palgrave Macmillan, 2010).

36. Anthony Parkhurst, "A Letter Written to M. Richard Hakluyt of the Middle Temple, Conteining a Report of the True State and Commodities of Newfound-land," in *PN*, 8:13. See also Joyce E. Chaplin, *Subject Matter: Technology, the Body, and Science on the Anglo-American Frontier, 1500–1676* (Cambridge, MA: Harvard University Press, 2001), 152–153, and White, "Unpuzzling American Climate." For later writings about Newfoundland and efforts to justify settlement in the climate of northeastern North America, see Mary Fuller, "The Poetics of a Cold Climate," *Terrae Incognitae* 30 (1998): 41–53.

37. Humphrey Gilbert, "A Briefe Relation of Newfound-land, and the Commodities Thereof," in *PN*, 8:57.

38. The principal primary accounts of the expedition can be found in Edward Hayes, "Edward Hayes' Narrative of Sir Humphrey Gilbert's Last Expedition," in *The Voyages and Colonizing Enterprises of Sir Humphrey Gilbert*, ed. David B. Quinn (London: Hakluyt Society, 1940), 385–423; and George Peckham, *A True*

Reporte, of the Late Discoueries, and Possession, Taken in the Right of the Crowne of Englande, of the New-Found Landes (London: John Hinde, 1583).

39. John Davis, "The Worlde's Hydrographical Discription [1595]," in Markham, *Voyages and Works of John Davis,* 205; John Janes, "The First Voyage of Master John Davis," 4, 9 (quote).

40. John Davis, "The Second Voyage Attempted by Master John Davis with Others, for the Discoverie of the Northwest Passage, in Anno 1586," in Markham, *Voyages and Works of John Davis,* 15.

41. "And there it seemed best to our Generall, M. Davis, to divide his fleet, himselfe sailed to the Northwest and to direct the *Sunshine,* wherein I was, and the pinnesse called the *Northstar,* to seeke a passage Northward betweene Groenland and Island, to the latitude of 80 degrees, if land did not let us," according to Henry Morgan, "The Relation of the Sunshine and the Northstarre to Discover the Passage betweene Groenland and Island," in Markham, *Voyages and Works of John Davis,* 33–34. "At this place the chiefe ship whereupon I trusted, called the *Mermayd of Dartmouth,* found many occasions of discontentment, and being unwilling to proceed, shee there forsook me," according to Davis, "Worlde's Hydrographical Discription," 207.

42. Morgan, "Relation," 36, 38 (quote).

43. Davis, "Second Voyage," 15, 24–25 (quote).

44. Davis, "Second Voyage," 26, 30; Davis, "Worlde's Hydrographical Discription," 209; John Davis, "Letter to William Sanderson," in Markham, *Voyages and Works of John Davis,* 32.

45. Davis, "Worlde's Hydrographical Discription," 209, 210 (quote); John Janes, "The Third Voyage Northwestward, Made by John Davis, Gentleman," in Markham, *Voyages and Works of John Davis,* 47; John Davis, "Letter to M. Sanderson," in Markham, *Voyages and Works of John Davis,* 59.

46. Thomas Rundall, ed., *Narratives of Voyages towards the North-West in Search of a Passage to Cathay and India, 1496 to 1631* (London: Hakluyt Society, 1849), 62; George Waymouth, "The Voyage of Captaine George Weymouth, Intended for the Discoverie of the North-West Passage toward China, with Two Fly Boates," in *Hakluytus Posthumus, or Purchas His Pilgrimes,* ed. Samuel Purchas (London: Hakluyt Society, 1905), 14:311, 313, 317–318.

47. Rundall, *Narratives of Voyages towards the North-West,* 72.

48. Clements R. Markham, "The Journal of the Voyage of John Knight," in *The Voyages of Sir James Lancaster* (London: Hakluyt Society, 1877), 287.

49. "The Voyage of Master John Knight . . . for the Discovery of the North-West Passage, Begun the Eighteenth of Aprill 1606," in Purchas, ed., *Hakluytus Posthumus,* 14:358–359.

50. Ibid., 363.

51. See contributions in Karen Kupperman, ed., *America in European Consciousness 1493–1750* (Chapel Hill: University of North Carolina Press, 1995), esp. Christian F. Feest, "Collecting of American Indian Artifacts in Europe, 1493–1750," 324–360. For a recent overview of European exploration and publishing about the New World, see Joyce Oldham Appleby, *Shores of*

Knowledge: New World Discoveries and the Scientific Imagination (New York: W. W. Norton, 2013).

52. Kupperman, *Jamestown*, 109–144; Alison Games, *The Web of Empire: English Cosmopolitans in an Age of Expansion, 1560–1660* (Oxford: Oxford University Press, 2008); Lesley B. Cormack, *Charting an Empire: Geography at the English Universities, 1580–1620* (Chicago: University of Chicago Press, 1997); Peter C. Mancall, *Hakluyt's Promise: An Elizabethan's Obsession for an English America* (New Haven, CT: Yale University Press, 2007).

53. J. H. Elliott, "Learning from the Enemy: Early Modern Britain and Spain," in *Spain, Europe, and the Wider World, 1500–1800* (New Haven, CT: Yale University Press, 2009), 25–51; J. H. Elliott, *Empires of the Atlantic World: Britain and Spain in America, 1492–1830* (New Haven, CT: Yale University Press, 2006).

54. Antonio Barrera-Osorio, *Experiencing Nature: The Spanish American Empire and the Early Scientific Revolution* (Austin: University of Texas Press, 2006); Jorge Cañizares-Esguerra, *Nature, Empire, and Nation: Explorations of the History of Science in the Iberian World* (Stanford, CA: Stanford University Press, 2006); Víctor Navarro Brotons and William Eamon, eds., *Más allá de la leyenda negra: España y la revolución científica* (Valencia: Instituto de Historia de la Ciencia, 2007); Daniela Bleichmar et al., eds., *Science in the Spanish and Portuguese Empires, 1500–1800* (Stanford, CA: Stanford University Press, 2009); María M. Portuondo, *Secret Science: Spanish Cosmography and the New World* (Chicago: University of Chicago Press, 2009).

55. The geographical reports are discussed in Raquel Álvarez Peláez, *La Conquista de la naturaleza americana* (Madrid: Consejo Superior de Investigaciones Científicas, 1993). For modern editions, see René Acuña, ed., *Relaciones geográficas del siglo XVI*, 10 vols. (México: Universidad Nacional Autónoma de México, Instituto de Investigaciones Antropológicas, 1980). On the dissemination of information in the reports, see Portuondo, *Secret Science*, 211–223.

56. Richard Hakluyt, "Discourse on Western Planting," in *The Original Writings and Correspondence of the Two Richard Hakluyts*, ed. E. G. R. Taylor (London: Hakluyt Society, 1935), 254.

57. See, e.g., Richard Hakluyt, "Epistle Dedicatory to Raleigh, 1587" in *Original Writings*, 2:376–377.

58. Richard Hakluyt, "Pamphlet for the Virginia Enterprise, 1585," in *Original Writings*, 2:333. See also ch. 20 of Hakluyt, "Discourse on Western Planting." For more on the English imagination of commodities in America, see James Walter Findley, " 'Went to Build Castles in the Aire:' Colonial Failure in the Anglo-North Atlantic World, 1570–1640," Ph.D. diss., University of North Carolina at Greensboro, 2015, 1–83.

59. "Instructions by Way of Advice," *JVFC*, 1:49–54.

60. Ibid.

61. *JVFC*, 1:133; *CWCJS*, 1:205.

62. For the story of Jamestown's rediscovery, excavation, and finds, see William M. Kelso and Beverly A. Straube, *Jamestown Rediscovery, 1994–2004* (Richmond, VA: Association for the Preservation of Virginia Antiquities,

2004); William M. Kelso, *Jamestown, the Buried Truth* (Charlottesville: University of Virginia Press, 2006). Kelso, *Jamestown*, 12–14, 170–173, describes the site's natural advantages and disadvantages.

63. Gabriel Archer (?), "Description of the River and Country (21 May–21 June 1607)," *JVFC*, 1:99.

64. David B. Quinn, "A List of Books Purchased for the Virginia Company," in *Explorers and Colonies: America 1500–1625* (London: Hambledon Press, 1990), 383–396.

65. David B. Quinn, *England and the Discovery of America: 1481–1620* (New York: Alfred A. Knopf, 1974), 289.

6. Destroyed with Cruel Disease

1. Virginia Council, "Instructions by Way of Advice," in *JVFC*, 1:51.

2. John Smith, "A True Relation of Such Occurences and Accidents of Noate as Hath Happened in Virginia . . . [1608]," in *CWCJS*, 1:29–31; Gabriel Archer, "A Relatyon Written by a Gentleman of the Colony," *JVFC*, 1:88.

3. *JVFC*, 1:94–95; *CWCJS*, 1:31, claims 400 Indians.

4. George Percy, "Observations Gathered out of a Discourse of the Plantation of the Southern Colonie in Virginia by the English, 1606," *JVFC*, 1:139.

5. *CWCJS*, 1:142; Karen Kupperman, *The Jamestown Project* (Cambridge, MA: Harvard University Press, 2007), 221; James Horn, *A Land as God Made It: Jamestown and the Birth of America* (New York: Basic Books, 2005), 53.

6. Edward Maria Wingfield, "Discourse (1608)," *JVFC*, 1:214–215.

7. *JVFC*, 1:143.

8. John Smith, "The Proceedings of the English Colonie in Virginia . . . [1612]," *CWCJS*, 1:209.

9. *JVFC*, 1:143–144; *CWCJS*, 1:35; *JVFC*, 1:214–215.

10. *JVFC*, 1:144.

11. E.g., Wyndham B. Blanton, "Epidemics, Real and Imaginary, and Other Factors Influencing Seventeenth-Century Virginia's Population," *Bulletin of the History of Medicine* 31 (1957): 454–462; Gordon W. Jones, "The First Epidemic in the English America," *Virginia Magazine of History and Biography* 71 (1963): 3–10; Darrett B. Rutman and Anita H. Rutman, "Of Agues and Fevers: Malaria in the Early Chesapeake," *WMQ* 33 (1976): 31–60; Carville Earle, "Environment, Disease, and Mortality in Early Virginia," in *The Chesapeake in the Seventeenth Century: Essays on Anglo-American Society*, ed. Thad W. Tate and David L. Ammerman (Chapel Hill: University of North Carolina Press, 1979), 96–125; Karen Kupperman, "Apathy and Death in Early Jamestown," *Journal of American History* 66 (1979): 24–40.

12. David W. Stahle et al., "The Lost Colony and Jamestown Droughts," *Science* 280 (1998), 564–567. See also the discussion of proxy evidence in Chapter 2.

13. For pollen-based temperature reconstructions, see Debra A. Willard, Thomas M. Cronin, and Stacey Verardo, "Late-Holocene Climate and Eco-system History from Chesapeake Bay Sediment Cores, USA," *Holocene* 13 (2003): 201–214, and D. A. Willard et al., "Impact of Millennial-Scale Holocene

Climate Variability on Eastern North American Terrestrial Ecosystems: Pollen-Based Climatic Reconstruction," *Global and Planetary Change* 47 (2005): 17–35. Studies of $\partial^{18}O$ in benthic foraminifera confirm the cold anomaly around 1600, while Mg / Ca ratios locate cold anomalies ca. 500 and 300 B.P. within a cool LIA. See T. M. Cronin et al., "Medieval Warm Period, Little Ice Age and 20th Century Temperature Variability from Chesapeake Bay," *Global and Planetary Change* 36 (2003): 17–29; T. M. Cronin et al., "Multiproxy Evidence of Holocene Climate Variability from Estuarine Sediments, Eastern North America," *Paleoceanography* 20 (2005): PA4006; and T. M. Cronin et al., "The Medieval Climate Anomaly and Little Ice Age in Chesapeake Bay and the North Atlantic Ocean," *Palaeogeography, Palaeoclimatology, Palaeoecology* 297 (2010): 299–310.

14. Ben Hardt et al., "The Seasonality of East Central North American Precipitation Based on Three Coeval Holocene Speleothems from Southern West Virginia," *Earth and Planetary Science Letters* 295 (2010): 342–348; R. Stockton Maxwell et al., "A Multicentury Reconstruction of May Precipitation for the Mid-Atlantic Region Using *Juniperus virginiana* Tree Rings," *Journal of Climate* 25 (2012): 1045–1056; Rosanne D'Arrigo et al., "Regional Climatic and North Atlantic Oscillation Signatures in West Virginia Red Cedar over the Past Millennium," *Global and Planetary Change* 84–85 (2012): 8–13.

15. Rutman and Rutman, "Of Agues and Fevers." See also J. R. McNeill, *Mosquito Empires: Ecology and War in the Greater Caribbean, 1620–1914* (New York: Cambridge University Press, 2010) for the spread of malaria in North America.

16. *CWCJS*, 1:210; "A Briefe Declaration of the Plantation of Virginia duringe the First Twelve Yeares, When Sir Thomas Smith Was Governor of the Companie, and downe to This Present Tyme. By the Ancient Planters Nowe Remaining Alive in Virginia. 1624," in *Colonial Records of Virginia* (Richmond: State of Virginia, 1874), 69.

17. *CWCJS*, 1:211; William M. Kelso, *Jamestown: The Buried Truth* (Charlottesville: University of Virginia Press, 2006), 178.

18. *CWCJS*, 1:33; Kelso, *Jamestown*, 186.

19. William Strachey, *The Historie of Travell into Virginia Britania* [1612], ed. Lewis B. Wright and Virginia Freund (London: Hakluyt Society, 1953), 37–38; "Briefe Declaration of the Planters," 69.

20. *JVFC*, 1:144–145. Compare John Smith's description in *CWCJS*, 1:33. The equation between starvation and cold can be found in other examples of Elizabethan English, such as William Shakespeare, *Taming of the Shrew*, epilogue, lines 5–6.

21. Earle, "Environment, Disease, and Mortality," 8–9; Kupperman, "Apathy and Death."

22. Jones, "First Epidemic," 5–9. Kelso, *Jamestown*, 33, also notes that the more immunologically exposed colonists from London died just as fast as those from rural England, suggesting this was not merely a common gastrointestinal infection but something more serious. On the famines and epidemics of the 1590s, see Chapter 4.

23. *JVFC*, 1:144.
24. *CWCJS*, 1:263. Smith's version of the well has been backed up by recent archaeology; see Kelso, *Jamestown*, 115–124.
25. Audrey J. Horning, *Ireland in the Virginian Sea: Colonialism in the British Atlantic* (Chapel Hill: University of North Carolina Press, 2013), 152–153.
26. Earle, "Environment, Disease, and Mortality."
27. Craig Muldrew, *Food, Energy and the Creation of Industriousness: Work and Material Culture in Agrarian England, 1550–1780* (New York: Cambridge University Press, 2011), 65–83; Keith Thomas, *Religion and the Decline of Magic* (New York: Scribner, 1971), 19; James McWilliams, *A Revolution in Eating: How the Quest for Food Shaped America* (New York: Columbia University Press, 2005), 246–247.
28. Brown, *Genesis*, 2:659–662.
29. Quoted from George Thorpe, a leader of the Berkeley plantation in Virginia in 1620, in Bernard Bailyn, *The Barbarous Years: The Peopling of British North America: The Conflict of Civilizations, 1600–1675* (New York: Alfred A. Knopf, 2012), 96.
30. Horn, *Land as God Made It*, 58–59; Kupperman, *Jamestown*, 219–220. On the role of Gosnold, see *JVFC*, 1:144.
31. *CWCJS*, 1:33.
32. *JVFC*, 1:219–222. See also Michael A. Lacombe, " 'A Continuall and Dayly Table for Gentlemen of Fashion': Humanism, Food, and Authority at Jamestown, 1607–1609," *American Historical Review* 115 (2010): 669–687.
33. *JVFC*, 1:145; *CWCJS*, 1:35.
34. Discussion of population in Helen C. Rountree, *Pocahontas, Powhatan, Opechancanough: Three Indian Lives Changed by Jamestown* (Charlottesville: University of Virginia Press, 2005), 13–14.
35. Strachey, *Historie*. Examples include the Jamestown colonists' first visit to the Paspaheghs in *JVFC*, 1:136–137, and their first attempt to sail past the fall line in May–June 1607 in *JVFC*, 1:87–88. See also the reaction at Stadacona when Jacques Cartier's expedition departed for Hochelaga, discussed in Chapter 9.
36. On early modern England as a commercial society, see Keith Wrightson, *Earthly Necessities: Economic Lives in Early Modern Britain* (New Haven, CT: Yale University Press, 2000).
37. See especially Karen Ordahl Kupperman, *Indians and English: Facing Off in Early America* (Ithaca, NY: Cornell University Press, 2000).
38. On gift exchange violations and conflict, see especially Martin Quitt, "Trade and Acculturation at Jamestown, 1607–1609: The Limits of Understanding," *WMQ* 52 (1995): 227–258; Daniel K. Richter, "Tsenacommacah and the Atlantic World," in *The Atlantic World and Virginia, 1550–1624*, ed. Peter C. Mancall (Chapel Hill: University of North Carolina Press, 2007), 29–65; and Seth Mallios, *The Deadly Politics of Giving: Exchange and Violence at Ajacan, Roanoke, and Jamestown* (Tuscaloosa: University of Alabama Press, 2006). On the potential threat to political hierarchies posed by the dispersal of English goods, see Stephen R. Potter, "Early English Effects on Virginia Algonquian Exchange and Tribute in the Tidewater Potomac," in *Powhatan's Mantle:*

Indians in the Colonial Southeast, ed. Gregory A. Waselkov, Peter H. Wood, and M. Thomas Hatley, rev. ed. (Lincoln: University of Nebraska Press, 2006), 215–241.

39. "A Briefe Declaration," 69–70.
40. *CWCJS,* 1:35, 211–212. While the various chapters of the "Proceedings" are credited to several authors, they are highly partisan and pro-Smith; they mostly report events directly from Smith's point of view. As a result, Smith felt free to copy them almost verbatim in his *General History.* Therefore, I attribute most of the material in the "Proceedings" to Smith himself.
41. For biographies of Smith and discussions of his work, see Philip L. Barbour, *The Three Worlds of Captain John Smith* (Boston: Houghton Mifflin, 1964); Karen Ordahl Kupperman, *Captain John Smith: A Select Edition of His Writings* (Chapel Hill: University of North Carolina Press, 1988); and David Price, *Love and Hate in Jamestown: John Smith, Pocahontas, and the Heart of a New Nation* (New York: Alfred A. Knopf, 2003).
42. *CWCJS,* 1:43.
43. E.g., *JVFC,* 1:226–227.
44. *CWCJS,* 1:213.
45. *CWCJS,* 1:57.
46. *CWCJS,* 1:146–151.
47. For the defense of Smith, J. A. Leo Lemay, *Did Pocahontas Save Captain John Smith?* (Athens: University of Georgia Press, 1992). On the possibility of an "adoption" ceremony, see Kupperman, *Jamestown,* 228; Horn, *Land as God Made It,* 68; Camilla Townsend, *Pocahontas and the Powhatan Dilemma* (New York: Hill and Wang, 2004), 44–65. For a skeptical argument, see Rountree, *Pocahontas Powhatan Opechancanough,* 76–82.
48. *JVFC,* 1:227. Smith claimed that Archer and his followers plotted to escape on the colony's pinnace, leaving the settlers to starve, but the freezing weather kept them from leaving; see *CWCJS,* 1:213.
49. *CWCJS,* 1:214.
50. *CWCJS,* 1:61. Further details are in Francis Perkins, "Letter (28 March 1608)," in *JVFC,* 1:158–164; 1:228; and *CWCJS,* 1:217. On the construction of the dwellings, see Kelso, *Jamestown,* 80–114. For speculation about how the fire might have spread, see Ivor Noël Hume, *The Virginia Adventure, Roanoke to James Towne: An Archaeological and Historical Odyssey* (New York: Alfred A. Knopf, 1994), 183–189.
51. *CWCJS,* 2:148.
52. *JVFC,* 1:160. The upper James River does occasionally freeze over, but it was extraordinary even during the Little Ice Age for the wide, brackish tidal stretch at Jamestown to freeze. See Tamara Dietrich, "Arctic Blast Not Enough to Freeze James, York Rivers," *Daily Press,* Newport News, VA, February 19, 2015.
53. John Smith, "A Map of Virginia. With a Description of the Countrey, the Commodities, People, Government and Religion [1612]," in *CWCJS,* 1:143–144.
54. *CWCJS,* 1:61, 218 (quote).
55. Details of Newport's voyage and time in London in Barbour, *Three Worlds,* 172–174.

56. Horn, *Land as God Made It*, 75.

57. *CWCJS*, 1:73.

58. *CWCJS*, 1:215–219. Cf. *CWCJS*, 1:71: "Hee seeming to despise the nature of a Merchant, did scorne to sell, but we freely should give him, and he liberally would requite us."

59. Dennis Blanton, "Drought as a Factor in the Jamestown Colony, 1607–1612," *Historical Archaeology* 34 (2000): 74–81.

60. "A Map of Virginia [1612]," in *CWCJS*, 1:162–163.

61. On Powhatan subsistence, see Helen C. Rountree, *The Powhatan Indians of Virginia: Their Traditional Culture* (Norman: University of Oklahoma Press, 1989), 32–57.

62. *CWCJS*, 1:158–159, 163.

63. The 1570 mission is described in Chapter 3. See also Rountree, *Pocahontas Powhatan Opechancanough*, 43, 64, 116. On hunger in early spring, see, e.g., "Letter of the Governor and Council of Virginia to the Virginia Company (7 July 1610)," in Brown, *Genesis*, 1:403: "Indeed at this time of yeare they live poore, their corne being but newly putt into the ground, and their old store spent."

64. For relevant examples of bioarchaeological research, see Barbara A. Burnett and Katherine A. Murray, "Death, Drought, and de Soto: The Bioarchaeology of Depopulation," in *The Expedition of Hernando de Soto West of the Mississippi, 1541–1543*, ed. Gloria A. Young and Michael P. Hoffman (Fayetteville: University of Arkansas Press, 1993), 227–236; Patricia M. Lambert, *Bioarchaeological Studies of Life in the Age of Agriculture: A View from the Southeast* (Tuscaloosa: University of Alabama Press, 2000); Martin Gallivan, "The Archaeology of Native Societies in the Chesapeake: New Investigations and Interpretations," *Journal of Archaeological Research* 19 (2011): 281–325; Clark Spencer Larsen and Christopher B. Ruff, "The Stresses of Conquest in Spanish Florida: Structural Adaptation and Changes Before and After Contact," in *In the Wake of Contact: Biological Responses to Conquest*, ed. Clark Spencer Larsen and George R. Milner (New York: Wiley-Liss, 1994), 21–34; Susan Pfeiffer et al., "Stable Dietary Isotopes and mtDNA from Woodland Period Southern Ontario People: Results from a Tooth Sampling Protocol," *Journal of Archaeological Science* 42 (2014): 334–345; Clark Spencer Larsen, ed., *Bioarchaeology of Spanish Florida: The Impact of Colonialism* (Gainesville: University Press of Florida, 2001); John W. Verano and Douglas H. Ubelaker, *Disease and Demography in the Americas* (Washington, DC: Smithsonian Institution Press, 1992); Douglas H. Ubelaker and Philip D. Curtin, "Human Biology of Populations in the Chesapeake Watershed," in *Discovering the Chesapeake: The History of an Ecosystem*, ed. Philip D. Curtin, Grace S. Brush, and George W. Fisher (Baltimore, MD: Johns Hopkins University Press, 2001), 127–148. For a methodological overview, see Alan H. Goodman and Debra L. Martin, "Reconstructing Health Profiles from Skeletal Remains," in *The Backbone of History: Health and Nutrition in the Western Hemisphere*, ed. Richard H. Steckel and Jerome Carl Rose (New York: Cambridge University Press, 2002), 11–60.

65. For these adaptations and examples among Native Americans, see John M. O'Shea, "The Role of Wild Resources in Small-Scale Agricultural Systems: Tales

from the Lakes and the Plains," in *Bad Year Economics: Cultural Responses to Risk and Uncertainty*, ed. Paul Halstead and John O'Shea (New York: Cambridge University Press, 1989), 57–67; Frances B. King, "Climate, Culture, and Oneota Subsistence in Central Illinois," in *Foraging and Farming in the Eastern Woodlands*, ed. C. Margaret Scarry (Gainesville: University Press of Florida, 1993), 232–254; David Demeritt, "Agriculture, Climate, and Cultural Adaptation in the Prehistoric Northeast," *Archaeology of Eastern North America* 19 (1991): 183–202; Jason Hall, "Maliseet Cultivation and Climatic Resilience on the Welastekw / St. John River During the Little Ice Age," *Acadiensis* 44 (2015): 3–25; B. A. Nicholson et al., "Climatic Challenges and Changes: A Little Ice Age Period Response to Adversity—The Vickers Focus Forager / Horticulturalists Move On," *Plains Anthropologist* 51 (2006): 325–333; Thomas M. Wickman, "Snowshoe Country: Indians, Colonists, and Winter Spaces of Power in the Northeast, 1620–1727," Ph.D. diss., Harvard University, 2012; James Woollett, "Labrador Inuit Subsistence in the Context of Environmental Change: An Initial Landscape History Perspective," *American Anthropologist* 109 (2007): 69–84; Natale A. Zappia, *Traders and Raiders: The Indigenous World of the Colorado Basin, 1540–1859* (Chapel Hill: University of North Carolina Press, 2014), 27–47; Richard White, *The Roots of Dependency: Subsistence, Environment, and Social Change among the Choctaws, Pawnees, and Navajos* (Lincoln: University of Nebraska Press, 1983), esp. 31.

66. Rountree, *Pocahontas Powhatan Opechancanough*, 64.

67. On conflict, see David H. Dye, "Warfare in the Protohistoric Southeast 1500–1700," in *Between Contacts and Colonies: Archaeological Perspectives on the Protohistoric Southeast*, ed. Cameron B. Wesson and Mark A. Rees (Tuscaloosa: University of Alabama Press, 2002), 126–141. On social inequality and political hierarchy, see Stephen R. Potter, *Commoners, Tribute, and Chiefs: The Development of Algonquian Culture in the Potomac Valley* (Charlottesville: University Press of Virginia, 1993), 151–154, 166–168; Kupperman, *Indians and English*, 35–38; Martin D. Gallivan, *James River Chiefdoms: The Rise of Social Inequality in the Chesapeake* (Lincoln: University of Nebraska Press, 2003); and Margaret Holmes Williamson, *Powhatan Lords of Life and Death: Command and Consent in Seventeenth-Century Virginia* (Lincoln: University of Nebraska Press, 2003). On drought and the threat to chiefdoms, see David G. Anderson, David W. Stahle, and Malcolm K. Cleaveland, "Paleoclimate and Potential Food Reserves of Mississippian Societies: A Case Study from the Savannah River Valley," *American Antiquity* 60 (1995): 258–286; David G. Anderson, *The Savannah River Chiefdoms: Political Change in the Late Prehistoric Southeast* (Tuscaloosa: University of Alabama Press, 1994), 101, 285–289.

68. *CWCJS*, 1:219.

69. *CWCJS*, 1:228. For a detailed itinerary of the voyages and discussion of the Chesapeake environment and peoples, see Helen C. Rountree et al., *John Smith's Chesapeake Voyages, 1607–1609* (Charlottesville: University of Virginia Press, 2007).

70. *CWCJS*, 1:229, 233, 2:180.

71. *CWCJS*, 2:184, 189.

72. *CWCJS*, 1:234. The Virginia Company later explained its reasoning for the ceremony in *A True Declaration of the Estate of the Colonie in Virginia, with a Confutation of Such Scandalous Reports as Have Tended to the Disgrace of so Worthy an Enterprise* (London: William Stansby, 1610), 10–11. See also Horn, *Land as God Made It*, 103–104.

73. *CWCJS*, 1:234–237; Horn, *Land as God Made It*, 107–108; Kupperman, *Jamestown*, 228–230.

74. *CWCJS*, 1:238.

75. *CWCJS*, 2:187–190. Kelso, *Jamestown*, 180–186, describes archaeological evidence of crafts and craftsmen at Jamestown.

76. In *JVFC*, 1:150. John Smith also described this episode in similar language in the *Proceedings* and *Map of Virginia*, *CWCJS*, 1:172, 266. In his "Discourse," George Percy noted that "William White (having lived with the Natives) reported to us of their customes"; *JVFC*, 1:145.

77. The following discussion is based on Karen Kupperman, "Environmental Stress and Rainmaking: Cosmic Struggles in Early Colonial Times," *ReVista: Harvard Review of Latin America* (Winter 2007): n.p., and Sam White, "'Shewing the Difference betweene Their Conjuration, and Our Invocation on the Name of God for Rayne': Weather, Prayer, and Magic in Early American Encounters," *WMQ* 72 (2015): 33–56.

78. Cabeza de Vaca, "Relación," in Rolena Adorno and Patrick Charles Pautz, *Álvar Núñez Cabeza de Vaca: His Account, His Life and the Expedition of Pánfilo de Narváez* (Lincoln: University of Nebraska Press, 1999), 1:224–225. Baltasar de Obregón, writing in the 1580s, claimed that while he was on campaign in the region during the 1560s, Querecho Indians confirmed this incident to him; see his *Historia de los descubrimientos de Nueva España*, ed. Eva María Bravo García (Seville: Alfar, 1997), 178–179.

79. Original accounts in Lawrence A. Clayton, Vernon J. Knight, and Edward C. Moore, eds., *The De Soto Chronicles: The Expedition of Hernando de Soto to North America in 1539–1543* (Tuscaloosa: University of Alabama Press, 1993), 1:114–120, 239–240, 299–305, 2:389–406.

80. Original accounts in Juan Carlos Mercado, ed., *Menéndez de Avilés y la Florida: crónicas de sus expediciones* (Lewiston, NY: Edwin Mellen Press, 2006), 149–164, 330–339.

81. AGI Santo Domingo 224, 27 Dec 1583.

82. Thomas Hariot, *Briefe and True Report of the New Found Land of Virginia* (Frankfurt: Theodor de Bry, 1590), 27–28.

83. E.g., Gabriel Sagard, *The Long Journey to the Country of the Hurons* [1632], ed. George M. Wrong (Toronto: Champlain Society, 1939), 78, 178–181, 313; Reuben G. Thwaites, *The Jesuit Relations and Allied Documents: Travels and Explorations of the Jesuit Missionaries of New France, 1610–1791* (Cleveland, OH: Burrows, 1896–1901), 10:43–49.

84. Edward Johnson, *Wonder-Working Providence, 1628–1651*, ed. John Franklin Jameson (New York: Charles Scribner's Sons, 1910), 86–87.

85. William Bradford, *Of Plymouth Plantation*, ed. Samuel Eliot Morison (Philadelphia: Franklin Library, 1983), 125.

86. On literary conventions in colonial writings, see especially Stuart B. Schwartz, ed., *Implicit Understanding: Observing, Reporting, and Reflecting on the Encounters between Europeans and Other Peoples in the Early Modern Era* (New York: Cambridge University Press, 1994). On "white gods," see Camilla Townsend, "Burying the White Gods: New Perspectives on the Conquest of Mexico," *American Historical Review* 108 (2003): 659–687. For examples of weather miracles in contemporary reports of missionaries, see, e.g., *Annuæ Litteræ Societatis Iesu Anni 1606* (Mainz, 1608), 221–222; *Annuæ Litteræ Societatis Iesu Anni 1608* (Mainz, 1608), 101; Pierre du Jarric, *L'histoire des choses les plus memorables* (Bordeaux, 1614), 3:1017.

87. Based on data from the North American Drought Atlas, http://www.ncdc .noaa.gov/paleo/newpdsi.html. On Spanish La Florida, see also Dennis B. Blanton, "The Factors of Climate and Weather in Sixteenth-Century La Florida," in *Native and Spanish New Worlds: Sixteenth-Century Entradas in the American Southwest and Southeast*, ed. Clay Mathers, Jeffrey M. Mitchem, and Charles M. Haecker (Tucson: University of Arizona Press, 2013), 99–121; Anderson, *Savannah River Chiefdoms*, 283–284.

88. On rogation ceremonies and climate reconstruction, see, e.g., Fernando Domínguez Castro et al., "Reconstruction of Drought Episodes for Central Spain from Rogation Ceremonies Recorded at the Toledo Cathedral from 1506 to 1900: A Methodological Approach," *Global and Planetary Change* 63 (2008): 230–242.

89. Jeffrey Snyder-Reinke, *Dry Spells: State Rainmaking and Local Governance in Late Imperial China* (Cambridge, MA: Harvard University Press, 2009); Sam White, *The Climate of Rebellion in the Early Modern Ottoman Empire* (New York: Cambridge University Press, 2011), 140–162; Bernard Rosenberger and Hamid Triki, "Famines et épidémies au Maroc aux XVIe et XVIIe siècles (suite)," *Hesperis Tamuda* 15 (1974): 29–33; Kate Raphael, *Climate and Political Climate: Environmental Disasters in the Medieval Levant* (Leiden: Brill, 2013), 69–70.

90. Thomas, *Religion and the Decline of Magic*, esp. 41.

91. Stuart Clark, *Thinking with Demons: The Idea of Witchcraft in Early Modern Europe* (New York: Oxford University Press, 1997), 161–178, 457–471; Wolfgang Behringer, "Climatic Change and Witch Hunting: The Impact of the Little Ice Age on Mentalities," *Climatic Change* 43 (1999): 335–351; Emily Oster, "Witchcraft, Weather and Economic Growth in Renaissance Europe," *Journal of Economic Perspectives* 18 (2004): 215–228; Christian Pfister, "Climatic Extremes, Recurrent Crises and Witch Hunts: Strategies of European Societies in Coping with Exogenous Shocks in the Late Sixteenth and Early Seventeenth Centuries," *Medieval History Journal* 10 (2007): 33–73.

92. Alexander Whitaker, *Good Newes from Virginia* (London: William Welby, 1613), 26. Other examples are in, e.g., Pierre Biard, "Relation de La Nouvelle France (1616)," in Thwaites, *Jesuit Relations*, 3:130, and Brown, *Genesis*, 1:498–499. See also Jorge Cañizares-Esguerra, "The Devil in the New World: A Transnational Perspective," in *The Atlantic in Global History*, ed. Jorge Cañizares-Esguerra and Erik R. Seeman (Upper Saddle River, NJ: Pearson, 2007), 21–37; Fernando Cervantes, "The Devil's Encounter with America," in *Witchcraft in Early*

Modern Europe: Studies in Belief and Culture, ed. Jonathan Barry, Marianne Hester, and Gareth Roberts (New York: Cambridge University Press, 1996), 119–144; Edward L. Bond, "Source of Knowledge, Source of Power: The Supernatural World of English Virginia, 1607–1624," *Virginia Magazine of History and Biography* 108 (2000): 105–138; Richard Beale Davis, "The Devil in Virginia in the Seventeenth Century," *Virginia Magazine of History and Biography* 65 (1957): 131–149.

93. For examples of shamans and the control or prediction of the weather, see, e.g., Rountree, *Powhatan Indians of Virginia*, 132–133; Kathleen Joan Bragdon, *Native People of Southern New England, 1500–1650* (Norman: University of Oklahoma Press, 1996), 200–216; Elisabeth Tooker, *An Ethnography of the Huron Indians, 1615–49* (Syracuse, NY: Syracuse University Press, 1991), 143–149. Later ethnographies found the persistence of similar beliefs in these regions into modern times; see, e.g., William C. Sturtevant, *The Mikasuki Seminole: Medical Beliefs and Practices* (Ann Arbor: University of Michigan Press, 1955), 368–369; William Newcomb, *The Culture and Acculturation of the Delaware Indians* (Ann Arbor: University of Michigan Press, 1956), 69–70; Harald E. L. Prins, *The Mi'kmaq: Resistance, Accommodation, and Cultural Survival* (Fort Worth, TX: Harcourt Brace, 1996), 36–37; White, *Roots of Dependency*, 29. Canadian missionaries reported Indian attempts to copy European religious rituals as incantations. See, e.g., Sagard, *Long Journey*, 173–174. Quotation from Strachey, *Historie*, 89.

94. Bartolomé Barrientos, *Pedro Menéndez de Avilés, Founder of Florida*, ed. Anthony Kerrigan (Gainesville: University of Florida Press, 1965), 122.

95. *RV*, 1:277. For further examples, see White, "Shewing the Difference."

96. James Axtell, *Natives and Newcomers: The Cultural Origins of North America* (Oxford: Oxford University Press, 2001), 33.

97. *CWCJS*, 1:239, 242.

98. Quitt, "Trade and Acculturation."

99. *CWCJS*, 1:242–243.

100. Ibid., 1:245.

101. Ibid., 1:245–246.

102. Ibid., 1:249.

103. Ibid., 1:254–256. See also Kupperman, *Jamestown*, 223–225.

104. *CWCJS*, 1:258–259.

7. Our Former Hopes Were Frozen to Death

1. See descriptions in Curt Weikinn, *Quellentexte zur Witterungsgeschichte Europas vor der Zeitwende bis zum Jahre 1850* (Berlin: Academie-Verlag, 1958–1963), 3:15–26; Rüdiger Glaser, *Klimageschichte Mitteleuropas: 1000 Jahre Wetter, Klima, Katastrophen* (Darmstadt: Primus Verlag, 2001), 133–134; J. Buisman and A. F. V. van Engelen, *Duizend jaar weer, wind en water in de Lage Landen* (Franeker: Van Wijnen, 1995), 4:232–245.

2. Buisman and van Engelen, 4:245–254, quote from 246. On art, see A. M. J. de Kraker, "The Little Ice Age: Harsh Winters between 1550 and 1650," in

Hendrick Avercamp: Master of the Ice Scene, ed. Pieter Roelofs (Amsterdam: Rijksmuseum, 2009), 23–29.

3. C. Litton Falkiner, "William Farmer's Chronicles of Ireland," *English Historical Review* 22 (1907): 104–130 and 527–552, quote from 536.

4. Buisman and Engelen, *Duizend jaar,* 4:251; Glaser, *Klimageschichte,* 134.

5. ASV Senato—Dispacci—Germania, filza 39, docs. 41, 48, 49.

6. Pierre de L'Estoile, *Mémoires-Journaux de Pierre de l'Estoile,* ed. Gustave Brunet et al. (Paris: Librairie des Bibliophiles, 1881), 9:38–44 (quote from 42); ASV Senato—Dispacci—Francia, filza 38, docs. 19, 23, 24; Jacques-Auguste de Thou, *Histoire universelle* (London, 1732), 15:30; *Le Mercure François* (Paris: Jean Richer, 1612), 1:251v, 289r–v.

7. Inocencio Font Tullot, *Historia del clima de España: Cambios climáticos y sus causas* (Madrid: Instituto Nacional de Meteorologia, 1988), 82; Armando Alberola Romá, *Los Cambios Climáticos: La Pequeña Edad de Hielo en España* (Madrid: Cátedra, 2014), 94.

8. D. Camuffo and S. Enzi, "Reconstructing the Climate of Northern Italy from Archive Sources," in *Climate since A.D. 1500,* ed. R. S. Bradley and P. D. Jones, rev. ed. (London: Routledge, 1995), 143–154; ASV Senato—Dispacci—Firenze, filza 22, docs. 50, 51, 53, 59–61; Milano, filza 31, docs. 62–69; Roma, filza 58, docs. 35, 49, 52, 54, 58.

9. Sam White, *The Climate of Rebellion in the Early Modern Ottoman Empire* (New York: Cambridge University Press, 2011), 182–185.

10. William Andrews, *Famous Frosts and Frost Fairs in Great Britain: Chronicled from the Earliest to the Present Time* (London: G. Redway, 1887), 10–11; Thomas Dekker, *The Great Frost* (London: Henry Gosson, 1608); John Chamberlain, *The Letters of John Chamberlain,* ed. Norman Egbert McClure (Philadelphia: American Philosophical Society, 1939), 1:251–253.

11. Edwin F. Gay, "The Midland Revolt and the Inquisitions of Depopulation of 1607," *Transactions of the Royal Historical Society,* new series, 18 (1904): 195–244. On the particular connection between food and the revolt, see Joan Thirsk, *Food in Early Modern England: Phases, Fads, Fashions 1500–1760* (London: Continuum, 2006), 60–61. On the wider context of riots and rebellions, see Roger B. Manning, *Village Revolts: Social Protest and Popular Disturbances in England, 1509–1640* (New York: Oxford University Press, 1988). For resentment against enclosure, see, e.g., Francis Trigge, *The Humble Petition of Two Sisters, the Church and Common-Wealth* (London: George Bishop, 1604), and Arthur Standish, *The Commons Complaint* (London: William Stansby, 1611).

12. It has been debated whether the episode in *Coriolanus* refers specifically to the Midland Revolt. This appears unlikely, for the reasons presented in Buchanan Sharp, "Shakespeare's Coriolanus and the Crisis of the 1590s," in *Law and Authority in Early Modern England: Essays Presented to Thomas Garden Barnes,* ed. Buchanan Sharp and Mark Charles Fissel (Newark: University of Delaware Press, 2007), 27–63, and Elyssa Y. Cheng, "Moral Economy and the Politics of Food Riots in Coriolanus," *Concentric: Literary and Cultural Studies* 36 (2010): 17–31. However, see also the interpretations in E. C. Pettet, "Coriolanus and the

Midlands Insurrection of 1607," *Shakespeare Survey* 3 (1950): 34–42; Shannon Miller, "Topicality and Subversion in William Shakespeare's Coriolanus," *Studies in English Literature 1500–1900* 32 (1992): 287–310; and Steve Hindle, "Imagining Insurrection in Seventeenth-Century England: Representations of the Midland Rising of 1607," *History Workshop Journal*, 66 (2008): 21–61.

13. "Wee doe observe," he proclaimed, "that there was not so much as any necessitie of famine or dearth of corne, or any other extraordinary accident, that might stirre or provoke them in that maner to offend; but that it may be thought to proceede of a kinde of insolencie and contempt of our milde and gracious Government." J. F. Larkin and Paul L. Hughes, *Stuart Royal Proclamations I: Royal Proclamations of King James I, 1603–1625* (Oxford: Oxford University Press, 1973), 152–202 (quote from 161); *Foure Statutes, Specially Selected and Commanded by His Majestie to Be Carefully Put in Execution by All Justices and Other Officers of the Peace throughout the Realme* (London: Robert Barker, 1609).

14. Geoffrey Parker, *Europe in Crisis, 1598–1648*, 2nd ed. (London: Blackwell, 2001), 7; R. B. Outhwaite, *Dearth: Public Policy, and Social Disturbance in England, 1550–1800* (Cambridge: Cambridge University Press, 1995), 13.

15. Thomas Dekker, *The Ravens Almanacke Foretelling of a Plague, Famine, and Civill Warre* (London: Thomas Archer, 1609), B3v.

16. *Ravens Almanacke*, D2v. Detailed figures on plague mortality can be found in J. F. D. Shrewsbury, *A History of Bubonic Plague in the British Isles* (Cambridge: Cambridge University Press, 1970), 264–347. The Venetian ambassador in London sent regular reports on food scarcity and plague as well as a description of the frost: e.g., ASV Senato—Dispacci—Inghilterra, filza 6, docs. 42, 46, 77, 84; filza 7, docs. 23–25.

17. On Bristol's economic interest and Atlantic commerce, see David H. Sacks, *The Widening Gate: Bristol and the Atlantic Economy, 1450–1700* (Berkeley: University of California Press, 1991).

18. Lawrence C. Wroth, ed., *The Voyages of Giovanni da Verrazzano, 1524–1528* (New Haven, CT: Yale University Press, 1970), 140; David B. Quinn, "The Early Cartography of Maine in the Setting of Early European Exploration of New England and the Maritimes," in *American Beginnings: Exploration, Culture, and Cartography in the Land of Norumbega*, ed. Emerson W. Baker et al. (Lincoln: University of Nebraska Press, 1994), 37–60.

19. Richard D'Abate, "On the Meaning of a Name: 'Norumbega' and the Representation of North America," in Baker et al., eds., *American Beginnings*, 61; Richard D'Abate and Victor A. Konrad, "General Introduction," in Baker et al., eds., *American Beginnings*, xxiii, xxv–xxvi.

20. David B. Quinn, ed., "Examination of David Ingram," and "A True Discourse of the Adventures and Travalles of David Ingram," both in *The Voyages and Colonizing Enterprises of Sir Humphrey Gilbert* (London: Hakluyt Society, 1940), 2:281–283, 292; Colonial State Papers, National Archives, London, BNA SP 12/175, No. 95, http://colonial.chadwyck.com/home.do (accessed July 30, 2015). See also Éric Thierry, *La France de Henri IV en Amérique du Nord: de la création de l'Acadie à la fondation de Québec* (Paris: H. Champion, 2008), 38–39.

21. Bruce J. Bourque and Ruth Holmes Whitehead, "Trade and Alliances in the Contact Period," in Baker et al., eds., *American Beginnings,* 131–148. See Chapter 9 for further discussion of early trade between European fishermen and Native Americans.

22. For documents and narrative of these voyages, see *ENEV.* For further details and historiography, see also Christopher J. Bilodeau, "The Paradox of Saga-dahoc: The Popham Colony, 1607–1608," *Early American Studies* 12 (2014): 1–35.

23. Gabriel Archer, "The Relation of Captain Gosnols Voyage . . . [1602]," *ENEV,* 117.

24. This at least is my reading of the confusing description in ibid., 135, 137.

25. *ENEV,* 212–214; Samuel Purchas, "A Voyage . . . for the Discoverie of the North Part of Virginia in the Yeere 1603," *ENEV,* 215; Robert Salterne, "Brief Narrative of Pring's Voyage," *ENEV,* 229.

26. Purchas, "A Voyage," 216.

27. Ibid., 221. On this incident and other first encounters with European canines, see Joshua Abram Kercsmar, "Wolves at Heart: How Dog Evolution Shaped Whites' Perceptions of Indians in North America," *Environmental History* 21 (2016): 516–540.

28. *ENEV,* 55–58.

29. James Rosier, "A True Relation of the Most Prosperous Voyage Made This Present Yeere 1605, by Captain George Waymouth, in the Discovery of Virginia," *ENEV,* 254, 272–273.

30. Ibid., 283, 289, 293.

31. "Introduction," *ENEV,* 72. On fears of English overpopulation, see Chapter 4. For more on Popham's background and official life, see Douglas Walthew Rice, *The Life and Achievements of Sir John Popham, 1531–1607: Leading to the Establishment of the First English Colony in New England* (Madison, NJ: Fairleigh Dickinson University Press, 2005) and Bilodeau, "Paradox," 12–13. Popham's central role is evident in a dispatch sent by the Spanish ambassador shortly after Popham's death, predicting that the whole Virginia enterprise might be abandoned now that he was no longer there to support it—see *JVFC,* 1:77.

32. Archer, "Relation," 132.

33. John Brereton, "A Briefe and True Relation of the Discoverie of the North Part of Virginia . . . [1602]," *ENEV,* 160–161.

34. *ENEV,* 10–11.

35. E.g., Brereton, "Briefe and True Relation," 151, and Purchas, "A Voyage," 224.

36. Brereton, "Briefe and True Relation," 159; cf. Gosnold in *ENEV,* 209–210.

37. *ENEV,* 300.

38. E.g., Hayes in *ENEV,* 168, 173, 175–176; John Smith, "A Description of New England . . . [1616]," in *CWCJS,* 1:332–334.

39. Sir Ferdinando Gorges, "A Briefe Narration of the Originall Undertakings of the Advancement of Plantations into the Parts of America," in *Sir Ferdinando Gorges and His Province of Maine* (Boston: Prince Society, 1890), 2:9. See also Bilodeau, "Paradox," 19–20.

40. *ENEV,* 209–210.

41. E.g., Brereton, "Briefe and True Relation," 146.

42. K. M. Cuffey and G. D. Clow, "Temperature, Accumulation, and Ice Sheet Elevation in Central Greenland through the Last Deglacial Transition," *Journal of Geophysical Research* 102 (1997): 26383–26396; Elizabeth K. Thomas and Jason P. Briner, "Climate of the Past Millennium Inferred from Varved Proglacial Lake Sediments on Northeast Baffin Island, Arctic Canada," *Journal of Paleolimnology* 41 (2009): 209–224.

43. The finding holds for both ring width and maximum latewood density measurements, although ring-width data appear to "smear" the cooling impact of eruptions; see Jan Esper et al., "Signals and Memory in Tree-Ring Width and Density Data," *Dendrochronologia* 35 (2015): 62–70. Key studies include Fabio Gennaretti et al., "Volcano-Induced Regime Shifts in Millennial Tree-Ring Chronologies from Northeastern North America," *Proceedings of the National Academy of Sciences* 111 (2014): 10077–10082; Rosanne D'Arrigo et al., "Tree Growth and Inferred Temperature Variability at the North American Arctic Treeline," *Global and Planetary Change* 65 (2009): 71–82, and Rosanne D'Arrigo et al., *Dendroclimatic Studies: Tree Growth and Climate Change in Northern Forests* (Washington, DC: American Geophysical Union, 2014). On the likely impact of volcanic eruptions of summer sunlight and plant growth, see Martin P. Tingley, Alexander R. Stine, and Peter Huybers, "Temperature Reconstructions from Tree-Ring Densities Overestimate Volcanic Cooling," *Geophysical Research Letters* 41 (2014): 7838–7845. On the "year without a summer" in America, see William K. Klingaman and Nicholas P. Klingaman, *The Year without Summer: 1816 and the Volcano That Darkened the World and Changed History* (New York: St. Martin's Press, 2013).

44. "The Relation of Daniel Tucker," *ENEV*, 360; see also the statement by John Stoneman, the pilot, in *ENEV*, 364–375.

45. The late timing and purpose of George Popham's appointment are apparent in his letter to Salisbury of May 31, 1607, *ENEV*, 446.

46. The principal sources of information on the outbound voyage and first weeks in the colony are the journal of John Davies, pilot of the *Mary and John*, and William Strachey's "Historie of travaile into Virginia," both reproduced in *ENEV*, 397–441.

47. On the Hunt map and the archaeology of the colony, see Jeffrey P. Brain, Peter Morrison, and Pamela Crane, *Fort St. George: Archaeological Investigation of the 1607–1608 Popham Colony on the Kennebec River in Maine* (Augusta: Maine State Museum, 2007), and Andrew J. Wahll, ed., *Sabino: Popham Colony Reader, 1602–2000* (Bowie, MD: Heritage Books, 2000), 245–280, 298–302, 341–351, 371–423. At the time of research (summer 2015), further information and artifacts from the excavation were on display at the Maine State Museum.

48. Brain, Morrison, and Crane, *Fort St. George*, 14–17; Reuben G. Thwaites, *The Jesuit Relations and Allied Documents: Travels and Explorations of the Jesuit Missionaries of New France, 1610–1791* (Cleveland, OH: Burrows, 1896), 2:35.

49. Strachey, "Narrative," *ENEV*, 407–414; Brain, Morrison, and Crane, *Fort St. George*, 61–62, 70–71, 103. On the *Virginia*, see displays at the Bath Maritime Museum and Maine's First Ship, http://www.mfship.org/Maines_First _Ship/Home.html (accessed March 16, 2017).

50. Strachey, "Narrative," *ENEV*, 414.

51. Reproduced in Brain, Morrison, and Crane, *Fort St. George*, 165–166.

52. Letter to Salisbury, February 7, 1608, *ENEV*, 455.

53. Ferdinando Gorges to Salisbury, December 3, 1607, *ENEV*, 450.

54. On the fire, see Brain, Morrison, and Crane, *Fort St. George*, 74, which corrects Ferdinando Gorges's much later recollection in "A Briefe Narration of the Originall Undertakings of the Advancement of Plantations into the Parts of America," in *Sir Ferdinando Gorges and His Province of Maine*, ed. James Phinney Baxter (Boston: Prince Society, 1890), 2:15.

55. Ferdinando Gorges to Salisbury, March 20, 1608, *ENEV*, 458. We only know the ships arrived from the account in Smith, "General Historie," *CWCJS*, 2:399.

56. *ENEV*, 415.

57. Alexander Brown, ed., *The Genesis of the United States* (London: W. Heineman, 1891), 1:197.

58. Quoted in Henry O. Thayer, *The Sagadahoc Colony* (Portland, ME: Gorges Society, 1892), 89–90.

59. Gorges, "Briefe Narration," 16–17.

60. *CWCJS*, 2:399. The exception is William Strachey, who insisted the decisive factor was the death of John Popham. See William Strachey, *The Historie of Travell into Virginia Britania (1612)*, ed. Lewis B. Wright and Virginia Freund (London: Hakluyt Society, 1953), 35.

61. The last monograph on the colony was Thayer, *Sagadahoc Colony*. For subsequent interpretations, see *ENEV*, 7; Alfred A. Cave, "Why Was the Sagadahoc Colony Abandoned? An Evaluation of the Evidence," *New England Quarterly* 68 (1995): 625–640; Brain, Morrison, and Crane, *Fort St. George*, 13; and Bilodeau, "Paradox." Karen Kupperman, *The Jamestown Project* (Cambridge, MA: Harvard University Press, 2007), 190, also emphasizes the severe winter as a decisive factor in an already tenuous enterprise.

62. The first indication of cold weather is for October 8 (N.S.), *ENEV*, 413: "the weather turned fowle and full of fogg and rayne."

63. See "Court Proceedings," *ENEV*, 461, 462, 463.

64. Letter to Salisbury, February 7, 1608, *ENEV*, 455; Strachey, *Historie of Travell*, 35; Brain, Morrison, and Crane, *Fort St. George*, 103, 149.

65. Letter to Salisbury, February 7, 1608, *ENEV*, 455.

66. See, e.g., William Monson, "The Two Worlds Undiscovered, besides the Four Known," and "A Project in the Days of Queen Elizabeth for the Settling of Subjects in Guinea," both in *Naval Tracts of Sir William Monson*, ed. Michael Oppenheim (London: Navy Records Society, 1913), 4:350–351, 5:67–68.

67. "Being the mayne intended benefitt expected to uphold the Charge of the plantacion," as William Strachey put it, apparently based on the description of Captain Davies of the *Gift of God;* see *ENEV*, 415.

68. See letter of Ferdinando Gorges, December 1, 1607, *ENEV*, 447.

69. Gorges to Salisbury, March 20, 1608, *ENEV*, 458–459.

70. *ENEV*, 426–427, 408–409, 409–412.

71. Neal Salisbury, *Manitou and Providence: Indians, Europeans, and the Making of New England, 1500–1643* (New York: Oxford University Press, 1982), 93;

Kenneth M. Morrison, *The Embattled Northeast: The Elusive Ideal of Alliance in Abenaki-Euramerican Relations* (Berkeley: University of California Press, 1984), 24; Emerson W. Baker, "Trouble to the Eastward: The Failure of Anglo-Indian Relations in Early Maine," Ph.D. diss., College of William and Mary, 1986, 40–42; Cave, "Why Was the Sagadahoc Colony," 637–638; Bilodeau, "Paradox," 22; Thwaites, *Jesuit Relations,* 2:44–47, 3:222–223; Samuel Purchas, "North Virginia Voyages, 1606–1608," *ENEV,* 347–351. Purchas's exact phrase—"Mr. Patterson was slain by the savages of Nanhoc, a river of the Tarentines"—suggests some geographical confusion, as Quinn points out.

72. *ENEV,* 455.
73. *ENEV,* 415.
74. *ENEV,* 351.
75. *ENEV,* 350–351; Thomas M. Wickman, "Snowshoe Country: Indians, Colonists, and Winter Spaces of Power in the Northeast, 1620–1727," Ph.D. diss., Harvard University, 2012; Thomas Wickman, "'Winters Embittered with Hardships': Severe Cold, Wabanaki Power, and English Adjustments, 1690–1710," *WMQ* 72 (2015): 57–98. Purchas's account suggests as well that the winter turned not only cold but also very snowy, at least after a storm in late January.
76. *ENEV,* 463. All of those deposed mentioned the lack of beer.
77. Brain, Morrison, and Crane, *Fort St. George,* 73–74; Peter Morrison, "Architecture of the Popham Colony, 1607–1608: An Archaeological Portrait of English Building Practice at the Moment of Settlement," M.A. thesis, University of Maine, 2002, 88. Later writers suggested that the colonists themselves were criminals who unfairly slandered the colony and leadership—e.g., William Alexander, *Encouragement to Colonies* (London: William Stansby, 1624), 30—but there is no real evidence for this claim.
78. Gorges, "Briefe Narration," 16–17. See also Douglas R. McManis, *European Impressions of the New England Coast, 1497–1620* (Chicago: University of Chicago Press, 1972), 108. The impression is evident as well in proposals that Maryland would be the new temperate median between (hot) Virginia and (cold) New England—e.g., Andrew White, "An Account of the Colony of Lord Baltimore [1633]," in *Narratives of Early Maryland, 1633–1684,* ed. Clayton Colman Hall (New York: Barnes and Noble, 1946), 7–8. The cold climate of New England continued to be a subject of debate well into the eighteenth century; see Anya Zilberstein, *A Temperate Empire: Making Climate Change in Early America* (New York: Oxford University Press, 2016).
79. *ENEV,* 351.
80. Popular and scholarly biographies can be found in Donald S. Johnson, *Charting the Sea of Darkness: The Four Voyages of Henry Hudson* (Camden, ME: International Marine, 1993); Doug Hunter, *Half Moon: Henry Hudson and the Voyage That Redrew the Map of the New World* (New York: Bloomsbury Press, 2009); Peter C. Mancall, *Fatal Journey: The Final Expedition of Henry Hudson* (New York: Basic Books, 2009). Dagomar Degroot, "Exploring the North in a Changing Climate: The Little Ice Age and the Journals of Henry Hudson, 1607–1611," *Journal of Northern Studies* 9 (2015): 69–91.

81. Dagomar Degroot, *The Frigid Golden Age: Climate Change, Crisis, and Opportunity in the Early Modern Dutch Republic* (New York: Cambridge University Press, forthcoming), chapter 12; John Kirtland Wright, "The Open Polar Sea," in *Human Nature in Geography: Fourteen Papers, 1925–1965* (Cambridge, MA: Harvard University Press, 1966), 89–118. The end of the "open polar sea" is recounted in the popular history by Hampton Sides, *In the Kingdom of Ice: The Grand and Terrible Polar Voyage of the USS Jeannette* (New York: Anchor, 2015).

82. John Playse and Henry Hudson, "First Voyage, His Discoverie toward the North Pole," in *Henry Hudson the Navigator*, ed. G. M. Asher (London: Hakluyt Society, 1860), 8, 10, 16, 22 (quote).

83. Henry Hudson, "Second Voyage or Employment of Master Henry Hudson in 1608," in *Henry Hudson the Navigator*, ed. G. M. Asher (London: Hakluyt Society, 1860), 24.

84. Ibid., 35–36, 43–44 (quote). Hudson did mention the possibility of passing the strait south of Novaya Zemlya ("the Waygats") but insisted he didn't have the time or resources to try it.

85. Hunter, *Half Moon*, 31, estimates the cost of the voyage at around 1 percent that of a VOC trading expedition to the East Indies—so it did not need a high probability of success.

86. Robert Juet, "Third Voyage of M. Henry Hudson in 1609," in *Henry Hudson the Navigator*, ed. G. M. Asher (London: Hakluyt Society, 1860), 45.

87. The principal account of Hudson's decision is in Emmanuel van Meteren, "Hudson's Third Voyage (1609)," in *Henry Hudson the Navigator*, ed. G. M. Asher (London: Hakluyt Society, 1860), 148, and it is confirmed by what Hudson evidently told the mayor of Dartmouth when he landed in England. English officials would probably have welcomed his defection from the VOC to England regardless of the reason. See Hunter, *Half Moon*, 246–248.

88. On the Verrazzano Sea, see Chapter 3. On the "Strait of Anian," see Juan Gil, *Mitos y utopías del Descubrimiento* (Madrid: Alianza Editorial, 1989), 2:315–336.

89. Richard Ruggles, "The Cartographic Lure of the Northwest Passage: Its Real and Imaginary Geography," in Thomas H. B. Symons, *Meta Incognita: A Discourse of Discovery* (Quebec: Canadian Museum of Civilization: 1999), 1:179–256. See Marcel Trudel, *Atlas de la Nouvelle-France* (Quebec: Presses de l'Université Laval, 1968), 62–63, and Derek Hayes, *Historical Atlas of Canada* (Seattle: University of Washington Press, 2002), 37, for versions of the John Dee and Michael Lok maps that may have influenced Hudson.

90. Meteren, "Hudson's Third Voyage," 148. Hunter, *Half Moon*, 45–46, argues Hudson was also inspired by hints of a Northwest Passage in unpublished accounts of the Waymouth expedition (published accounts were scrubbed of all information that could aid enemy navigation).

91. Juet, "Third Voyage," 77.

92. Hudson, "Second Voyage," 44.

93. Abacuk Prickett, "A Larger Discourse of the Same Voyage," in *Henry Hudson the Navigator*, ed. G. M. Asher (London: Hakluyt Society, 1860), 98, 101, 110; Henry Hudson, "Fourth Voyage in 1610. An Abstract of the Journal of

M. Henry Hudson," in *Henry Hudson the Navigator,* ed. G. M. Asher (London: Hakluyt Society, 1860), 95.

94. Prickett, "Larger Discourse," 113; Robert Boyle, *New Experiments and Observations Touching Cold* (London: John Crook, 1665), 518–519. See also Mancall, *Fatal Journey,* 110–114.

95. Mancall, *Fatal Journey.*

96. Prickett, "Larger Discourse," 124, 133–134.

8. Winter for Eight Months and Hell for Four

1. Peter C. Mancall, *Hakluyt's Promise: An Elizabethan's Obsession for an English America* (New Haven, CT: Yale University Press, 2007), 172–173, 256–257; Francesco Avanzi, Juan González de Mendoza, and Marín Ignacio de Loyola, *New Mexico. Otherwise, the Voiage of Anthony of Espeio* (London: Thomas Cadman, 1587); Luis de Tribaldo, "A Letter Written from Valladolid by Ludovicos Tribaldus Toletus to Master Richard Hakluyt, Translated out of Latine, Touching Juan de Onate His Discoveries in New Mexico, One Hundred Leagues to the North from the Old Mexico [1605]," in *Hakluytus Posthumus,* ed. Samuel Purchas (Glasgow: Glasgow University Press, 1906), 18:76–80.

2. The phrase comes from Diego Vargas, *Remote beyond Compare: Letters of Don Diego de Vargas to His Family from New Spain and New Mexico, 1675–1706* (Albuquerque: University of New Mexico Press, 1989).

3. For an overview of climate patterns, see Paul Sheppard et al., "The Climate of the US Southwest," *Climate Research* 21 (2002): 219–238.

4. For tree-ring-reconstructed drought in the Pueblo region, see Henri D. Grissino-Mayer et al., *Multi-Century Trends in Past Climate for the Middle Rio Grande Basin* (Albuquerque, NM: U.S. Department of Agriculture, 2002); Ramzi Touchan et al., "Millennial Precipitation Reconstruction for the Jemez Mountains, New Mexico, Reveals Changing Drought Signal," *International Journal of Climatology* 31 (2011): 896–906; Ellis Q. Margolis, David M. Meko, and Ramzi Touchan, "A Tree-Ring Reconstruction of Streamflow in the Santa Fe River, New Mexico," *Journal of Hydrology* 397 (2011): 118–127; Holly L. Faulstich, Connie A. Woodhouse, and Daniel Griffin, "Reconstructed Cool- and Warm-Season Precipitation over the Tribal Lands of Northeastern Arizona," *Climatic Change* 118 (2012): 457–468.

   On the late sixteenth-century megadrought, see David Stahle et al., "Tree-Ring Reconstructed Megadroughts over North America since AD 1300," *Climatic Change* 83 (2007): 133–149; Edward R. Cook et al., "North American Drought: Reconstructions, Causes, and Consequences," *Earth Science Reviews* 81 (2007): 93–134; D. W. Stahle et al., "Cool- and Warm-Season Precipitation Reconstructions over Western New Mexico," *Journal of Climate* 22 (2009): 3729–3750; Daniel Griffin et al., "North American Monsoon Precipitation Reconstructed from Tree-Ring Latewood," *Geophysical Research Letters* 40 (2013): 954–958. The wet years in 1539–1541 are best captured in R. Kyle Bocinsky and Timothy A. Kohler, "A 2,000-Year Reconstruction of the Rain-Fed Maize Agricultural Niche in the US Southwest," *Nature Communications* 5

(2014): 5618 (see online data). For tree-ring-reconstructed temperatures in the western United States, see K. Briffa, P. D. Jones, and F. H. Schweingruber, "Tree-Ring Density Reconstructions of Summer Temperature Patterns across Western North America since 1600," *Journal of Climate* 5 (1992): 735–754; Sheppard, "Climate of the US Southwest"; Matthew W. Salzer et al., "Five Millennia of Paleotemperature from Tree-Rings in the Great Basin, USA," *Climate Dynamics* 42 (2014): 1517–1526. Matthew W. Salzer and Kurt F. Kipfmueller, "Reconstructed Temperature and Precipitation on a Millennial Timescale from Tree-Rings in the Southern Colorado Plateau, U.S.A.," *Climatic Change* 70 (2005): 465–487, also reveals a cold anomaly 1599–1612, but less extreme (Z-score of –0.9 with respect to the past millennium). For speleothem-reconstructed cold and drought, including a late sixteenth-century "superdrought," see Yemane Asmerom et al., "Multidecadal to Multicentury Scale Collapses of Northern Hemisphere Monsoons over the Past Millennium," *Proceedings of the National Academy of Sciences* 110 (2013): 9651–9656. For lake sediment (diatom, ∂^{13}C, and ∂^{18}O) data, see S. E. Metcalfe et al., "Climate Variability over the Last Two Millennia in the North American Monsoon Region, Recorded in Laminated Lake Sediments from Laguna de Juanacatlán, Mexico," *Holocene* 20 (2010): 1195–1206, and Jay Y. S. Hodgson, Amelia K. Ward, and Clifford N. Dahm, "An Independently Corroborated, Diatom-Inferred Record of Long-Term Drought Cycles Occurring over the Last Two Millennia in New Mexico, USA," *Inland Waters* 3 (2013): 459–472. For previous overviews of climate and Spanish *entradas* in the Southwest, see Carla R. Van West et al., "The Role of Climate in Early Spanish-Native American Interactions in the US Southwest," in *Native and Spanish New Worlds: Sixteenth-Century Entradas in the American Southwest and Southeast,* ed. Clay Mathers, Jeffrey M. Mitchem, and Charles M. Haecker (Tucson: University of Arizona Press, 2013), 81–98, and Sam White, "Cold, Drought, and Disaster: The Little Ice Age and the Spanish Conquest of New Mexico," *New Mexico Historical Review* 89 (2014): 425–458.

5. Ann F. Ramenofsky and Jeremy Kulisheck, "Regarding Sixteenth-Century Native Population Change in the Northern Southwest," in Mathers Mitchem, and Haecker, eds., *Native and Spanish New Worlds,* 123–139; Elinore M. Barrett, *Conquest and Catastrophe: Changing Rio Grande Pueblo Settlement Patterns in the Sixteenth and Seventeenth Centuries* (Albuquerque: University of New Mexico Press, 2002), 12–13, 50; Carroll L. Riley, *The Kachina and the Cross: Indians and Spaniards in the Early Southwest* (Salt Lake City: University of Utah Press, 1999), 52, 60; Ann M. Palkovitch, "Historic Population of the Eastern Pueblos: 1540–1910," *Journal of Anthropological Research* 41 (1985): 401–426; Daniel Reff, *Disease, Depopulation, and Cultural Change in Northwestern New Spain* (Salt Lake City: University of Utah Press, 1991), 228–230; Eric Blinman, "2000 Years of Cultural Adaptation to Climate Change in the Southwestern United States," *Ambio* (2008): 489–497.

6. These details appear in various testimonies described below, e.g., Jusepe Brondat in A. Roberta Carlin et al., "The Desertion of the Colonists of New Mexico 1601 (Part 3)," Cibola Project (Research Center for Romance Studies,

UC Berkeley, 2009). On turkeys, see Natalie Munro, "The Role of the Turkey in the Southwest," in *Handbook of North American Indians*, vol. 3, *Environment, Origin, and Population*, ed. Douglas H. Ubelaker (Washington, DC: Smithsonian Institution Press, 2006), 463–470.

7. Elinore M. Barrett, *The Spanish Colonial Settlement Landscapes of New Mexico, 1598–1680* (Albuquerque: University of New Mexico Press, 2012), 10. Barrett points out that Santa Fe's present climate has only 155 frost-free days and that maize requires 80–135 to reach maturity. Bioarchaeology is discussed in Chapter 6.

8. Suzanne L. Eckert, "Zuni Demographic Structure, A.D. 1300–1680: A Case Study on Spanish Contact and Native Population Dynamics," *Kiva* 70 (2005): 207–226; Charles F. Merbs, "Patterns of Health and Sickness in the Precontact Southwest," in *Columbian Consequences*, vol. 1, *Archaeological and Historical Perspectives on the Spanish Borderlands West*, ed. David Hurst Thomas (Washington, DC: Smithsonian Institution Press, 1989), 41–55; David H. Snow, "Tener Comal y Metate: Protohistoric Rio Grande Maize Use and Diet," in *Perspectives on Southwestern Prehistory*, ed. Paul E. Minnis and Charles L. Redman (Boulder, CO: Westview Press, 1990), 289–300; Ann L. W. Stodder et al., "Cultural Longevity and Biological Stress in the American Southwest," in *The Backbone of History: Health and Nutrition in the Western Hemisphere*, ed. Richard H. Steckel and Jerome Carl Rose (New York: Cambridge University Press, 2002), 481–505; Ann L. W. Stodder and Debra L. Martin, "Health and Disease in the Southwest before and after Spanish Contact," in *Disease and Demography in the Americas*, ed. John W. Verano and Douglas H. Ubelaker (Washington, DC: Smithsonian Institution Press, 1992), 55–73; Ann L. W. Stodder, "Paleoepidemiology of Eastern and Western Pueblo Communities in Protohistoric and Early Historic New Mexico," in *Bioarchaeology of Native American Adaptation in the Spanish Borderlands*, ed. Brenda J. Baker and Lisa Kealhofer (Gainesville: University Press of Florida, 1996), 148–176.

9. David R. Wilcox and Jonathan Haas, "The Scream of the Butterfly: Competition and Conflict in the Prehistoric Southwest," in *Themes in Southwest Prehistory*, ed. George J. Gumerman (Santa Fe, NM: School of American Research Press, 1994), 211–238; Steven A. LeBlanc, *Prehistoric Warfare in the American Southwest* (Salt Lake City: University of Utah Press, 1999).

10. Álvar Núñez Cabeza de Vaca, "Relación [1542]," in *Álvar Núñez Cabeza de Vaca: His Account, His Life, and the Expedition of Pánfilo de Narváez*, ed. Rolena Adorno and Patrick Charles Pautz (Lincoln: University of Nebraska Press, 1999), 1:245. On sources and studies of the expedition, see Chapter 2.

11. Cabeza de Vaca, "Relación," 207, 231.

12. *DCE*, 47.

13. Instructions in *DCE*, 65–67, quote from 66.

14. *DCE*, 67–77 (quotes from 71, 75).

15. *DCE*, 98–101.

16. Investments discussed in *DCE*, 118, and various testimonies in *DCE* (e.g., 127). Muster roll of the expedition in *DCE*, 135–363. For more on numbers and expenses, see Richard Flint, *No Settlement, No Conquest: A History of the*

Coronado Entrada (Albuquerque: University of New Mexico Press, 2008), 56–64 (also used as the main source for the following narrative). On the historiography of the Coronado expedition, see Richard Flint and Shirley Cushing Flint, "Catch as Catch Can: The Evolving History of the Contact Period Southwest, 1838-Present," in Mathers, Mitchem, and Haecker, eds., *Native and Spanish New Worlds*, 47–62. For evolving historical and archaeo-logical research on the expedition, see Richard Flint and Shirley Cushing Flint, eds., *The Coronado Expedition to Tierra Nueva: The 1540–1542 Route Across the Southwest* (Niwot: University Press of Colorado, 1997); Richard Flint and Shirley Cushing Flint, eds., *The Coronado Expedition from the Distance of 460 Years* (Albuquerque: University of New Mexico Press, 2003); Richard Flint and Shirley Cushing Flint, eds., *The Latest Word from 1540: People, Places, and Portrayals of the Coronado Expedition* (Albuquerque: University of New Mexico Press, 2011). For the role of Native allies in Spanish expeditions, see Chapter 2.

17. *DCE*, 235–237.
18. According to the later narrative of Castañeda de Nájera in *DCE*, 392, Coronado went "with as many as fifty horsemen and a few footmen, as well as most of the [native] allies."
19. *DCE*, 254, 291. Cristóbal de Escobar's proof of service in *DCE*, 557–558, claims sixty Native allies died.
20. *DCE*, 292, 393, 257.
21. *DCE*, 267.
22. *DCE*, 401.
23. *DCE*, 298: "it snows six months of the year [in Cíbola]"; *DCE*, 403–404: "it snowed so much that for two months nothing could be done."
24. Quotation in *DCE*, 502. The editors of that source note that in nearby Bernalillo, NM, the monthly mean never fell below 1°C (34°F) during the entire period 1940–1970. The Rio Grande has since been channelized, making such a freeze more difficult. Nevertheless, daily temperature records of Albu-querque suggest that during the twentieth century central New Mexico has never experienced the persistent cold needed to create such a long hard freeze, even had the river been left in its natural state. January and February regularly have low temperatures well below freezing, but daytime highs almost always reach above freezing. See data and analysis from the Western Regional Climate Center: http://www.wrcc.dri.edu/cgi-bin/clilcd.pl?nm23050 (accessed March 16, 2017). Castañeda de Nájera later wrote that the river "had been frozen for nearly four months, so that it was regularly crossed on horseback on top of the ice" (*DCE*, 407). However, there is no other evidence to suggest that the freeze lasted that long, and the timing would contradict his description of the Indians who escaped from a besieged pueblo in March 1541 and then sought to cross the Rio Grande "which was flowing rapidly and was extremely cold" (*DCE*, 405). It is possible that the author was only exaggerating, that he referred to a more northerly stretch of the Rio Grande, or even that he meant the upper Pecos River.
25. *DCE*, 390, 395. On their shelters, see Bradley J. Vierra, Martha R. Binford, and David Atlee Phillips, *A Sixteenth-Century Spanish Campsite in the Tiguex Province*

(Santa Fe: Museum of New Mexico, Laboratory of Anthropology, Research Section, 1989); Bradley J. Vierra and Stanley Hordes, "Let the Dust Settle: A Review of the Coronado Campsite in the Tiguex Province," in *The Coronado Expedition to Tierra Nueva*, ed. Richard Flint and Shirley Cushing Flint (Boulder: University Press of Colorado, 1997), 249–261; and Flint, *No Settlement*, 144.

26. Not a *visita*, such as regularly followed a term of office, but a *pesquisa secreta*. Transcriptions and translation in Richard Flint, ed., *Great Cruelties Have Been Reported: The 1544 Investigation of the Coronado Expedition* (Dallas: Southern Methodist University Press, 2002). Specific descriptions of these episodes appear on 64, 80, 131, 171–172, 196, 215–216, and 239.

27. Ibid., 257, 289.

28. Carroll L. Riley, *Rio del Norte: People of the Rio Grande from Earliest Times to the Pueblo Revolt* (Salt Lake City: University of Utah Press, 1995), 178–179; Flint, *No Settlement*, 140–141; Flint, *Great Cruelties*, 171.

29. Flint, *Great Cruelties*, 80, 149, 238, 317, 402. See also Castañeda de Nájera's version in *DCE*, 402–404. Note that in many early modern Spanish documents, witness statements are merely paraphrases of the original questions that were designed to guide their testimony. That is not the case here. Several witnesses volunteered this information even though their questions made no mention of *mantas*. Current anthropological and historical research indicates that the protohistoric pueblos did have some form of political hierarchy, but that the decisions of pueblo leaders (assuming the Spanish identified them correctly) still required the consent of the community. See Tracy L. Brown, *Pueblo Indians and Spanish Colonial Authority in Eighteenth-Century New Mexico* (Tucson: University of Arizona Press, 2013), 21–64.

30. *DCE*, 319–320 (quote), 411.

31. Flint, *Great Cruelties*, 358, 366.

32. *DCE*, 425. For a classic history of typhus, see Hans Zinsser, *Rats, Lice and History* (New York: Bantam, 1935).

33. *DCE*, 321.

34. *DCE*, 425–426.

35. *DCE*, 474. Cf. Antonio Vázquez de Espinosa, *Description of the Indies, c. 1620*, trans. Charles Upson Clark (Washington, DC: Smithsonian Institution Press, 1942), 194: "But since they were suffering great hardships and the country was so cold and poor, and he saw his men were worn out and disheartened, having traveled in this expedition over 1,000 leagues suffering great hardships and much hunger. So he returned to Mexico City."

36. *DCE*, 384–385. See also Flint, *No Settlement*, 230.

37. On the environmental transformation of New Spain, see Elinor Melville, *A Plague of Sheep: Environmental Consequences of the Conquest of Mexico* (Cambridge: Cambridge University Press, 1994).

38. José de Acosta, *Natural and Moral History of the Indies*, ed. Jane E. Mangan, trans. Frances López-Morillas (Durham, NC: Duke University Press, 2002), 172–173.

39. Peter J. Bakewell, *Miners of the Red Mountain: Indian Labor in Potosí, 1545–1650* (Albuquerque: University of New Mexico Press, 1984); John C. Super, *Food, Conquest, and Colonization in Sixteenth-Century Spanish America* (Albuquerque:

University of New Mexico Press, 1988), 19–20; Nicholas A. Robins, *Mercury,*
Mining, and Empire: The Human and Ecological Cost of Colonial Silver Mining in
the Andes (Bloomington: Indiana University Press, 2011).

40. Peter J. Bakewell, *Silver Mining and Society in Colonial Mexico: Zacatecas,*
1546–1700 (Cambridge: Cambridge University Press, 1971); Daviken Studnicki-
Gizbert and David Schecter, "The Environmental Dynamics of a Colonial
Fuel-Rush: Silver Mining and Deforestation in New Spain, 1522 to 1810,"
Environmental History 15 (2010): 94–119.

41. Philip Wayne Powell, *Soldiers, Indians and Silver: North America's First Frontier*
War (Tempe: Center for Latin American Studies, Arizona State University,
1975), 217–220; Philip Wayne Powell, *Mexico's Miguel Caldera: The Taming of*
America's First Frontier, 1548–1597 (Tucson: University of Arizona Press, 1977),
205–222; R. Acuna-Soto et al., "Megadrought and Megadeath in 16th-Century
Mexico," *Emerging Infectious Diseases* 8 (2002): 360–362. On drought in
Mexico's colonial history, see especially Georgina Endfield, *Climate and Society*
in Colonial Mexico (London: Blackwell, 2008).

42. *CDIE*, 16:142–187.

43. George P. Hammond and Agapito Rey, eds., *The Rediscovery of New Mexico,*
1580–1594: The Explorations of Chamuscado, Espejo, Castaño de Sosa, Morlete,
and Leyva de Bonilla and Humaña (Albuquerque: University of New Mexico
Press, 1966), 67–144. The quotation is from the Spanish edition of this document in *CDIE*, 15:150: "Es tierra que toca un poco en fria, aunque no demasiado;
es el temple como el de Castila." At the time of writing, most documents of the
"rediscovery" and Juan de Oñate's *entrada* of New Mexico were being published in new editions—in facsimile, transcription, and translation—as part of
the University of California Berkeley's Cibola Project (https://escholarship.org
/uc/rcrs_ias_ucb_cibola). I have consulted those versions wherever possible,
but have retained citations to the Hammond and Rey versions except where
there are significant differences.

44. See "Diego Pérez de Luxán's Account of the Espejo Expedition into New
Mexico, 1582," in Hammond and Rey, eds., *Rediscovery*, 170, 177, 181, 183–185,
203–204, 206 (quoted).

45. "Castaño de Sosa's 'Memoria,'" in Hammond and Ray, eds., *Rediscovery*,
245–295. Quotations taken from Spanish version in *CDIE*, 15:214, 221,
239–241, 249–250, 260. Locations based on the reconstruction of the
expedition in Albert H. Schroeder and Daniel S. Matson, *A Colony on the*
Move: Gaspar Castaño de Sosa's Journal, 1590–1591 (Santa Fe, NM: School of
American Research, 1965), 63, 75, 112, 147.

46. Charles W. Hackett, *Historical Documents Relating to New Mexico, Nueva*
Vizcaya and Approaches Thereto, to 1773 (Washington, DC: Carnegie Institution,
1923), 1:220–221.

47. Hackett, *Historical Documents*, 1:255–257. On Oñate's background, see
Donald E. Chipman, "The Oñate-Moctezuma-Zaldívar Families of Northern
New Spain," *New Mexico Historical Review* 52 (1977): 297–310.

48. Hackett, *Historical Documents*, 1:225–367; George Peter Hammond and
Agapito Rey, *Don Juan de Oñate* (Albuquerque: University of New Mexico

Press, 1953), 1:94–168; Jerry R. Craddock and Barbara De Marco, "Appointment of Juan de Oñate as Governor and Captain General of the Provinces of New Mexico, 21 October 1595," Cibola Project, 2013.

49. Riley, *Kachina and Cross*, 44.

50. Hammond and Rey, *Don Juan de Oñate*, 1:199–309 (on the inspection), 390–392 (quote).

51. "Fray Juan de Escalona to the Viceroy, October 1, 1601," in Hammond and Rey, *Don Juan de Oñate*, 2:692.

52. Marc Simmons, *The Last Conquistador: Juan de Oñate and the Settling of the Far Southwest* (Norman: University of Oklahoma Press, 1991); Gaspar Pérez de Villagrá, *Historia de la Nueva México, 1610*, ed. Miguel Encinias, Alfred Rodríguez, and Joseph P. Sánchez (Albuquerque: University of New Mexico Press, 1992); Yolanda Leyva, "Monuments of Conformity: Commemorating and Protesting Oñate on the Border," *New Mexico Historical Review* 82 (2007): 343–367; Michael L. Trujillo, "Oñate's Foot: Remembering and Dismembering in Northern New Mexico," *Aztlán: A Journal of Chicano Study* 33 (2008): 91–119.

53. Narrative in Simmons, *Last Conquistador*, 91–123.

54. History and description of the pueblo in Ward Alan Minge, *Ácoma: Pueblo in the Sky*, rev. ed. (Albuquerque: University of New Mexico Press, 2002).

55. Narratives of the Acoma disaster can be found in many histories, including Jack D. Forbes, *Apache, Navaho, and Spaniard* (Norman: University of Oklahoma Press, 1960); Simmons, *Last Conquistador;* and Stan Hoig, *Came Men on Horses: The Conquistador Expeditions of Francisco Vázquez de Coronado and Don Juan de Oñate* (Boulder: University Press of Colorado, 2013). It has even featured in novels such as Lana M. Harrigan, *Ácoma: A Novel of Conquest* (New York: Forge, 1997). On punishments, see Jerry R. Craddock, "Acoma: Teoría y práctica de la guerra justa," *Initum: Revista catalana d'historia del dret* 7 (2002): 331–359; Powell, *Soldiers, Indians and Silver*, 109–110.

56. The complete trial record has been made available for the first time in both transcription and translation in Jerry R. Craddock and John H. R. Polt, *The Trial of the Indians of Acoma 1598–99*, Cibola Project, 2008.

57. Hammond and Rey, *Don Juan de Oñate*, 1:426.

58. Craddock and Polt, *Trial*, 14, 20, 22, 23, 32, 37, 62. Later examples are in, e.g., Jerry R. Craddock and John H. R. Polt, *Fray Juan de Escalona, Comissary of the Franciscan Missions of New Mexico, to King Phillip III Concerning Conditions in the New Colony October 15, 1601*, Cibola Project, 2015, 2.

59. Craddock and Polt, *Trial*, 73; Jerry R. Craddock and Barbara De Marco, *Ytinerario de la Expedición de Juan de Oñate a Nuevo México 1597–1599*, Cibola Project, 2013, 36.

60. From the testimony of Marcelo de Espinosa, in A. Roberta Carlin et al., *The Desertion of the Colonists of New Mexico 1601 (Part 3)*, Cibola Project, 2009, 10.

61. Villagrá, *Historia*, canto 1, lines 51–67.

62. Jerry R. Craddock and John H. R. Polt, *Oñate's Report to the Viceroy March 2, 1599*, Cibola Project, 2009; Hammond and Rey, *Don Juan de Oñate*, 503; Luis Cabrera de Córdoba, *Relaciones de las cosas sucedidas en la corte de España desde*

1599 hasta 1614 (Madrid: J. Martin Alegria, 1857), 64. Monterrey later confided
to his successor that he had doubts about Oñate, but that he regarded the
enterprise as so important he did not want to send along any negative reports
to Spain; see France V. Scholes and Eleanor B. Adams, eds., *Advertimientos
generales que los virreyes dejaron a sus sucesores para el gobierno de Nueva España,
1590–1604* (Mexico: J. Porrúa, 1956), 86.

63. The Valverde inquiry has been published as Jerry R. Craddock and John H. R.
Polt, *Juan de Oñate in Quivira, 1601: The "Relación cierta y verdadera" and the
Valverde Interrogatory*, Cibola Project, 2013. The full record of testimony
relating to the defection of the colonists appears in A. Roberta Carlin et al., *The
Desertion of the Colonists of New Mexico 1601 (Parts 1–3)*, Cibola Project, 2009.

64. Hammond and Rey, *Don Juan de Oñate*, 608–618; Carlin, *Desertion*, 2:4–5, 3:14.

65. Carlin, *Desertion*, 1:2–8, 12–14, 18, 3:6–7, 16, 26. On Indians and famine foods,
see the letter of Fray Escalona in Juan de Torquemada, *Monarquía Indiana*
(Mexico City: UNAM, 1983), 450, and archaeological evidence in Florence
Hawley Ellis, *San Gabriel del Yungue as Seen by an Archaeologist* (Santa Fe, NM:
Sunstone Press, 1989), 79–80.

66. Carlin, *Desertion*, 3:20, 25, 31–32, 36–37 (quote); Hammond and Rey, *Don Juan
de Oñate*, 2:696; Craddock and Polt, *Juan de Oñate in Quivira*, 58, 163. For
zooarchaeology, see Ellis, *San Gabriel*, 80–81, and Stephen C. Lent, "Survey,
Test Excavation Results, and Data Recovery Plan for Cultural Resources near
San Juan Pueblo, Rio Arriba County, New Mexico," Archaeology Notes no. 17
(Santa Fe: Museum of New Mexico, Office of Archaeological Studies, 1991), 12.

67. Carlin, *Desertion*, 1:7–8, 3:25–26; Hammond and Rey, *Don Juan de Oñate*,
2:693.

68. Carlin, *Desertion*, 3:28.

69. Hammond and Rey, *Don Juan de Oñate*, 2:1032 (quote). Hammond and Rey
also reproduce in vol. 2 the long chain of official correspondence after 1601.
However, the *audiencia* and various "discourses" of Viceroy Monterrey
from 1602 onward do not appear to have influenced royal decisions, since by
the time they came up for consideration in Spain, Monterrey was already
being transferred to Peru. Simmons, *Last Conquistador*, 179–187, makes the
case that the return of Oñate's personal friend Luis de Velasco as viceroy
during 1607–1611 delayed his recall and trial.

70. For population estimates, see Barrett, *Spanish Colonial Settlement*, 51–56. On
the complicated evidence about the date of Santa Fe's founding, and the
reasons so many historians still get it wrong, see James Ivey, "The Viceroy's
Order Founding the Villa of Santa Fe: A Reconsideration, 1605–1610," in *All
Trails Lead to Santa Fe*, ed. Marc Simmons (Santa Fe, NM: Sunstone Press,
2010), 97–107.

71. Hammond and Rey, *Don Juan de Oñate*, 2:1001–1005, 1009–1011.

72. Ibid., 2:1109–1124, 1112, 1116 (quote). Villagrá later had his sentence for this
offense commuted; see AGI Indiferente 450, leg. A5, 202v–203.

73. Hammond and Rey, *Don Juan de Oñate*, 2:1056.

74. Henry Raup Wagner, *Spanish Voyages to the Northwest Coast of America in the
Sixteenth Century* (San Francisco: California Historical Society, 1929), 1–10;

Álvaro del Portillo and Diez de Sollano, *Descubrimientos y exploraciones en las costas de California* (Madrid: School of Spanish-American Studies of Madrid, 1947), 139–149; W. Michael Mathes, *Vizcaíno and Spanish Expansion in the Pacific Ocean, 1580–1630* (San Francisco: California Historical Society, 1968), 1–5.

75. Narrative and documentary record in Carlos Lazcano Sahagún, *La Bahía de Santa Cruz: Cortés en California* (Ensenada, Mexico: Fundación Barca, 2006), 113.

76. For sources and summary of the expedition, see Henry R. Wagner, "California Voyages, 1539–1541: The Voyage of Francisco de Ulloa; the Voyage of Hernando de Alarcon; the Voyage of Francisco de Bolaños," *California Historical Society Quarterly* 3 (1924): 307–397; Julio César Montané Martí and Carlos Lazcano Sahagún, *El encuentro de una península: La navegación de Francisco de Ulloa, 1539–1540* (Ensenada, Mexico: Fundación Barca, 2008). Ulloa's original account is reproduced in Wagner, *Spanish Voyages*, 11–50 (quote from 38—original is "seca y esteríl y de tan ruin parecer"). The extract from Francisco López de Gómara comes from Wagner, *Spanish Voyages*, 417. The other quotes come from Pedro de Palencia, "Relatione dello scoprimento," in *Delle navigationi et viaggi*, ed. Giovanni Battista Ramusio (Venice, 1606), 3:288r, 291r, 292v, 293r–294r.

77. Carlos Lazcano Sahagún, *Más allá de la Antigua California: La navegación de Juan Rodríguez Cabrillo, 1542–1543* (Ensenada, Mexico: Fundación Barca, 2007); Harry Kelsey, *Juan Rodríguez Cabrillo* (San Marino, CA: Huntington Library, 1986), 143–163; Wagner, *Spanish Voyages*, 72–93 (quote from 89–90); AGI Patronato 20, n.5, r.13; *HGIO*, decade 7, book 5, chapters 3–4, 4:75. A newer translation and explanation of itinerary can be found in James D. Nauman, ed., *An Account of the Voyage of Juan Rodríguez Cabrillo* (San Diego, CA: Cabrillo National Monument Foundation, 1999). Monterey Bay weather records come from the Western Regional Climate Center at http://www.wrcc .dri.edu/cgi-bin/cliMAIN.pl?ca5795 (accessed March 16, 2017).

78. Wagner, *Spanish Voyages*, 93; *HGIO*, decade 7, book 5, chapters 3–4, 4:77.

79. Briffa, Jones, and Schweingruber, "Tree-Ring Density"; L. J. Graumlich, "A 1000-Year Record of Temperature and Precipitation in the Sierra Nevada," *Quaternary Research* 39 (1993): 249–255; L. A. Scuderi, "A 2000-Year Tree Ring Record of Annual Temperatures in the Sierra Nevada Mountains," *Science* 259 (1993): 1433–1436; Andrea H. Lloyd and Lisa J. Graumlich, "Holocene Dynamics of Treeline Forests in the Sierra Nevada," *Ecology* 78 (1997): 1199–1210; Arndt Schimmelmann et al., "A Large California Flood and Correlative Global Climatic Events 400 Years Ago," *Quaternary Research* 49 (1998): 51–61.

80. The official log of the expedition is reproduced in translation and facsimile in Wagner, *Spanish Voyages*, 114–117.

81. R. García-Herrera et al., "The Use of Spanish Historical Archives to Reconstruct Climate Variability," *Bulletin of the American Meteorological Society* 84 (2003): 1025–1035. The first onset of scurvy usually occurred within four months, just as ships first reached the California coast—see the description by Antonio de la Ascensión in Wagner, *Spanish Voyages*, 245–246, and the

description of the route in the introduction by Edward Baranowski to *Documents from the 1602–1603 Sebastián Vizcaíno Expedition up the California Coast*, Cibola Project, 2011.

82. Francis Fletcher, "The World Encompassed," in *New American World*, ed. David B. Quinn and Alison M. Quinn (London: Arno, 1979), 1:467–469. Spanish intelligence came from the 1584 testimony of Drake's captured cousin John, who testified that "they met with great storms: the whole sky was obscured and covered with clouds," but also described the climate of California as "temperate, cold rather than hot"—see Dora Polk, *The Island of California: A History of the Myth* (Spokane, WA: Arthur H. Clark, 1991), 233–234. The exact location of Drake's landing is a long-standing issue of historical contention; see Raymond Aker, V. Aubrey Neasham, and Robert H. Power, "The Debate: Point Reyes Peninsula / Drakes Estero; Bolinas Bay / Bolinas Lagoon; San Francisco Bay / San Quentin Cove," *California Historical Quarterly* 53 (1974): 203–292.

83. Apocryphal voyage of Maldonado in W. Michael Mathes, ed., *Californiana: Documentos para la historia de la demarcación comercial de California 1583–1632* (Madrid: J. Porrúa Turanzas, 1965), 1:38–60. Maps and geography are discussed in Polk, *Island of California*.

84. Mathes, ed., *Californiana*, 1:18–37, quotes from 22, 34; Wagner, *Spanish Voyages*, 139–153.

85. Mathes, ed., *Californiana*, 1:128–137, 163–178; Wagner, *Spanish Voyages*, 154–167.

86. Vizcaíno's life and early career in Mathes, *Vizcaíno*, 25–32, and Wagner, *Spanish Exploration*, 168–179.

87. Mathes, ed., *Californiana*, 1:262–279, quotes from 265, 278.

88. Ibid., 1:282.

89. Mathes, *Vizcaíno*, 51–53.

90. Instructions in Mathes, ed., *Californiana*, 1:353–364 (quote from 362: "yreis considerando la calidad de la tierra y templanza della y la gente").

91. The basic route of the voyage is best described in the rutter of Gerónimo Martín Palacios in Mathes, ed., *Californiana*, 1:471–565, and the narrative of Antonio de la Ascensión, translated in Wagner, *Spanish Exploration*, 180–272.

92. Letter of December 1602 in Mathes, ed., *Californiana*, 1:379. See also the later account of Antonio de la Ascensión, in Wagner, *Spanish Exploration*, 247: "It is in the same region and parallel of latitude as Seville, and is of almost the same climate. The Spaniards could settle here as an assistance to those sailing from China as it is of the climate and quality of Spain."

93. Martín Fernández de Navarrete, ed., *Colección de diarios y relaciones para la historia de los viajes y descubrimientos* (Madrid: Instituto histórico de marina, 1943), 4:62–64. Translation in Rose Marie Beebe and Robert M. Senkewicz, eds., *Lands of Promise and Despair: Chronicles of Early California, 1535–1846* (Santa Clara, CA: Heyday Books, 2001), 43.

94. Mathes, *Vizcaíno*, 98; Mathes, ed., *Californiana*, 1:380–427, quote from 423; Wagner, *Spanish Exploration*, 253.

95. Real Cédula of August 1606 in Mathes, ed., *Californiana*, 2:672–684, quote from 674.

96. Viceroy's letters to Philip III in Mathes, ed., *Californiana*, 2:693–696, 703–705. On negotiations surrounding the projected colony, see Wagner, *Spanish Exploration*, 273–280, and Mathes, *Vizcaíno*, 108–120. Quotation from Miguel Venegas, *Noticia de la California* (Madrid: Manuel Fernández, 1757), 1:192.

97. Bakewell, *Silver Mining and Society*, 156–157; Bakewell, *Miners of the Red Mountain*, 28–29.

98. For ENSO, see documentary evidence in Luc Ortlieb, "The Documented Historical Record of El Niño Events in Peru: An Update of the Quinn Record (Sixteenth through Nineteenth Centuries)," in *El Niño and the Southern Oscillation: Multiscale Variability and Global and Regional Impacts*, ed. Henry F. Diaz and Vera Markgraf (Cambridge: Cambridge University Press, 2000), 207–295, and the multiproxy reconstruction in Joëlle Gergis and Anthony Fowler, "A History of ENSO Events since A.D. 1525: Implications for Future Climate Change," *Climatic Change* 92 (2009): 343–387. Additional support for ENSO variability and a strong 1607 El Niño comes from R. Dunbar et al., "Eastern Pacific Sea Surface Temperature since 1600 AD: The $\partial^{18}O$ Record of Climate Variability in Galapagos Corals," *Paleoceanography* 9 (1994): 291–316, and S. Fleury et al., "Pervasive Multidecadal Variations in Productivity within the Peruvian Upwelling System over the Last Millennium," *Quaternary Science Reviews* 125 (2015): 78–90. On Andean droughts, see Alain Gioda, M. Rosario Prieto, and Ana Forenza, "Archival Climate History Survey in the Central Andes (Potosí, 16th–17th Centuries)," *Prace Geograficzne* 107 (2000): 107–112, and Bartolomé Arzáns de Orsúa y Vela, *Historia de la villa imperial de Potosí*, ed. Lewis Hanke and Gunnar L. Mendoza (Providence, RI: Brown University Press, 1965), 262–264. Description of Mexico flood and drainage in Fernando de Cepeda, Fernando Alfonso Carillo, and Juan de Álvares Serrano, *Relación universal legítima y verdadera . . . [1637]*, ed. José Manuel Serrano Alvarez and Luis E. Bracamontes (Mexico City: Secretaría de Obras Públicas, 1976), quoted 57–58. For more on the *desagüe*, see, e.g., Vera S. Candiani, *Dreaming of Dry Land: Environmental Transformation in Colonial Mexico City* (Stanford, CA: Stanford University Press, 2014).

99. On revenues from Peru, see Kenneth Andrien, *Crisis and Decline: The Viceroyalty of Peru in the Seventeenth Century* (Albuquerque: University of New Mexico Press, 1985). On silver isotopes, see Anne-Marie Desaulty et al., "Isotopic Ag-Cu-Pb Record of Silver Circulation through 16th–19th Century Spain," *Proceedings of the National Academy of Sciences* 108 (2011): 9002–9007. On money as a source of national anxiety, see Elvira Vilches, *New World Gold: Cultural Anxiety and Monetary Disorder in Early Modern Spain* (Chicago: University of Chicago Press, 2010).

100. Hammond and Rey, *Don Juan de Oñate*, 2:1067–1068. Decrees about the role of soldiers and missionaries in the colony are in ibid., 2:1076–1079. For an overview of New Mexico's seventeenth-century history see, e.g., John L. Kessell, *Pueblos, Spaniards, and the Kingdom of New Mexico* (Norman: University of Oklahoma Press, 2008).

101. For instance, Spain's official chronicler of the time, Antonio de Herrera, thought the winters were as cold as those of Castile. See Montané Martí and Lazcano Sahagún, *Encuentro*, 154.

102. See the proceedings of the Council of State, December 18, 1607, in Emma Helen Blair and James Alexander Robertson, eds., *The Philippine Islands, 1493–1898* (Cleveland, OH: Arthur H. Clark, 1904), 14:216: "In this state of affairs it has seemed best to him to advise your Majesty that it ought to be carefully considered whether it is expedient that each year there should be carried to Eastern India a million eight-real pieces for articles of so little importance as are those which are brought thence; and what plan could be made to obviate this drain of silver, as we are in such need of here." Fray Antonio de la Ascensión, meanwhile, argued for a chain of missions starting in southern California, as would eventually be built during the 1700s; see his letter in Mathes, ed., *Californiana*, 2:715–720.

103. The *cédula* is found in Mathes, ed., *Californiana*, 2:740–753. See also Wagner, *Spanish Exploration*, 280–282. For more on the myth of Rico de Oro and Rico de Plata, see Juan Gil, *Mitos y utopías del Descubrimiento* (Madrid: Alianza Editorial, 1989), 2:126–147.

104. Mathes, ed., *Californiana*, 1:6–10 (quote from 9), 2:739.

105. Mathes, *Vizcaíno*, 134–146; Donald C. Cutter, "Plans for the Occupation of Upper California: A New Look at the 'Dark Age' from 1602 to 1769," *Journal of San Diego History* 24 (1978): n.p. On Spain's first contacts with Japan, see Michael Cooper, *An Unscheduled Visit: Rodrigo de Vivero in Japan, 1609–10* (Tokyo: Asiatic Society of Japan, 2008).

9. Death Follows Us Everywhere

1. Norman Clermont, "A-t-on vécu les hivers d'un Petit Âge Glaciaire en Nouvelle-France?," *Géographie Physique et Quaternaire* 50 (1996): 395–398.

2. V. C. Slonosky, "The Meteorological Observations of Jean-Francois Gaultier, Quebec, Canada: 1742–56," *Journal of Climate* 16 (2003): 2232–2247; V. C. Slonosky, "Daily Minimum and Maximum Temperature in the St-Lawrence Valley, Quebec: Two Centuries of Climatic Observations from Canada," *International Journal of Climatology* 35 (2015): 1662–1681; and V. C. Slonosky, personal communication. Daniel Houle, Jean-David Moore, and Jean Povencher, "Ice Bridges on the St. Lawrence River as an Index of Winter Severity from 1620 to 1910," *Journal of Climate* 20 (2007): 757–764, indicates a greater frequency of severe winters during the nineteenth century than in the twentieth century, but a paucity of historical materials renders the study unreliable as a guide to winter severity before the mid-1800s.

3. On the tree-ring climate relationship in Quebec, see Jianguo Huang et al., "Radial Growth Response of Four Dominant Boreal Tree Species to Climate along a Latitudinal Gradient in the Eastern Canadian Boreal Forest," *Global Change Biology* 16 (2010): 711–731; A. Nicault et al., "Spatial Analysis of Black Spruce (*Picea mariana* (Mill.) BSP) Radial Growth Response to Climate in

Northern Quebec–Labrador Peninsula, Canada," *Canadian Journal of Forest Research* 45 (2015): 343–352. For tree-ring-based summer temperature reconstructions of Quebec, see Serge Payette et al., "Secular Climate Change in Old-Growth Tree-Line Vegetation of Northern Quebec," *Nature* 315 (1985): 135–138; Claude Lavoie and Serge Payette, "Black Spruce Growth Forms as a Record of a Changing Winter Environment at Treeline, Quebec, Canada," *Arctic and Alpine Research* 24 (1992): 40–49; S. Archambault and Y. Bergeron, "An 802-Year Tree-Ring Chronology from the Quebec Boreal Forest," *Canadian Journal of Forest Research* 22 (1992): 674–682; P. E. Kelly, E. R. Cook, and D. W. Larson, "A 1397-Year Tree-Ring Chronology of *Thuja occidentalis* from Cliff Faces of the Niagara Escarpment, Southern Ontario, Canada," *Canadian Journal of Forest Research* 24 (1994): 1049–1057; Lily Wang, Serge Payette, and Yves Bégin, "1300-Year Tree-Ring Width and Density Series Based on Living, Dead and Subfossil Black Spruce at Tree-Line in Subarctic Quebec, Canada," *Holocene* 11 (2001): 333–341; Rosanne D'Arrigo, Rob Wilson, and Gordon Jacoby, "On the Long-Term Context for Late Twentieth Century Warming," *Journal of Geophysical Research: Atmospheres* 111 (2006): D03103. Two recent studies of the same north Quebec tree sample, one measuring ring width and one using $\partial^{18}O$, have both found cooling in the late 1500s–early 1600s, but with very different absolute values. At the time of writing the discrepancy had not been resolved. See Fabio Gennaretti et al., "Volcano-Induced Regime Shifts in Millennial Tree-Ring Chronologies from Northeastern North America," *Proceedings of the National Academy of Sciences* 111 (2014): 10077–10082; M. Naulier et al., "A Millennial Summer Temperature Reconstruction for Northeastern Canada Using Oxygen Isotopes in Subfossil Trees," *Climate of the Past* 11 (2015): 1153–1164; and Fabio Gennaretti, personal communication.

4. For low-resolution temperature proxies, see, e.g., Hugo Beltrami and Jean-Claude Mareschal, "Ground Temperature Histories for Central and Eastern Canada from Geothermal Measurements: Little Ice Age Signature," *Geophysical Research Letters* 19 (1992): 689–692; Hafida El Bilali, R. Timothy Patterson, and Andreas Prokoph, "A Holocene Paleoclimate Reconstruction for Eastern Canada Based on $\delta^{18}O$ Cellulose of Sphagnum Mosses from Mer Bleue Bog," *Holocene* 23 (2013): 1260–1271. For high-resolution lake sediment studies, see Eugene R. Wahl, Henry F. Diaz, and Christian Ohlwein, "A Pollen-Based Reconstruction of Summer Temperature in Central North America and Implications for Circulation Patterns during Medieval Times," *Global and Planetary Change* 85 (2012): 66–74; Daniel Houle et al., "Compositional Vegetation Changes and Increased Red Spruce Abundance during the Little Ice Age in a Sugar Maple Forest of North-Eastern North America," *Plant Ecology* 213 (2012): 1027–1035; Nathalie Paquette and Konrad Gajewski, "Climatic Change Causes Abrupt Changes in Forest Composition, Inferred from a High-Resolution Pollen Record, Southwestern Quebec, Canada," *Quaternary Science Reviews* 75 (2013): 169–180; Karelle Lafontaine-Boyer and Konrad Gajewski, "Vegetation Dynamics in Relation to Late Holocene Climate Variability and Disturbance, Outaouais, Québec, Canada," *Holocene* 24 (2014):

1515–1526; Peter S. Keizer, Konrad Gajewski, and Robert McLeman, "Forest Dynamics in Relation to Multi-Decadal Late-Holocene Climatic Variability, Eastern Ontario, Canada," *Review of Palaeobotany and Palynology* 219 (2015): 106–115. According to Jeannine-Marie St. Jacques et al., "The Bias and Signal Attenuation Present in Conventional Pollen-Based Climate Reconstructions as Assessed by Early Climate Data from Minnesota, USA," *PLoS ONE* 10 (2015): e0113806, pollen-based temperature reconstructions have actually underestimated Little Ice Age cooling in the Great Lakes region. On eruptions, dimming, and tree-ring studies, see Martin P. Tingley, Alexander R. Stine, and Peter Huybers, "Temperature Reconstructions from Tree-Ring Densities Overestimate Volcanic Cooling," *Geophysical Research Letters* 41 (2014): 7838–7845.

5. On the winter cold as part of French Canadian historiography, see, e.g., Christian Morissonneau, "Le Nouveau Monde: les perceptions et représentations de Champlain," in *Le Nouveau Monde et Champlain,* ed. Guy Martinière and Didier Poton (Paris: Les Indes Savantes, 2008), 47–48, and Pierre Carle and Jean Louis Minel, eds., *L'homme et l'hiver en Nouvelle-France* (Montréal: Hurtubise HMH, 1972).

6. R. A. Skelton and James Alexander Williamson, eds., *The Cabot Voyages and Bristol Discovery under Henry VII* (London: Hakluyt Society, 1962), esp. 229; Clements R. Markham, ed., *Journal of Christopher Columbus and Documents Relating to the Voyages of John Cabot and Gaspar Corte Real* (London: Hakluyt Society, 1893), 232–238; Lawrence C. Wroth, ed., *The Voyages of Giovanni da Verrazzano, 1524–1528* (New Haven, CT: Yale University Press, 1970), 19–25.

7. Marcel Trudel, *Les vaines tentatives,* Histoire de la Nouvelle France 1 (Montréal: Fides, 1963), 27–28; Samuel Eliot Morison, *The European Discovery of America: The Northern Voyages* (New York: Oxford University Press, 1971), 228–233; G. Patterson, "The Portuguese on the North-East Coast of America, and the First European Attempt at Colonization There," *Transactions of the Royal Society of Canada* 2 (1890): 127–173; Henry Percival Biggar, *The Precursors of Jacques Cartier: 1497–1534, a Collection of Documents Relating to the Early History of the Dominion of Canada* (Ottawa: Government Printing Bureau, 1911), 195–197 ("e por acharem a terra muito fria"); Jean Alfonse, *La cosmographie avec l'espère et régime du soleil et du nord,* ed. Georges Musset (Paris: Leroux, 1904), 478; Samuel de Champlain, *The Works of Samuel de Champlain,* ed. Henry Percival Biggar (Toronto: Champlain Society, 1922), 1:468.

8. Trudel, *Vaines tentatives,* 69–70; Henry Percival Biggar, ed., *A Collection of Documents Relating to Jacques Cartier and the Sieur de Roberval* (Ottawa: Public Archives of Canada, 1930), 42. Trudel points out there was no priest on the expedition, suggesting no interest in settling among the First Nations.

9. Biggar, *Collection of Documents,* 43–44.

10. For all narratives of the Cartier expedition, I have relied on the notes and text of the critical French edition, Jacques Cartier, *Relations,* ed. Michel Bideaux (Montreal: University of Montreal Press, 1986) (quotes here from 96–97, 101; hereafter "Bideaux, *Relations*"), as well as the English translation, *The Voyages of Jacques Cartier,* ed. Henry Percival Biggar and Ramsay Cook (Toronto: University of Toronto Press, 1993) (hereafter "Cook, *Voyages*"). I follow Bideaux

(*Relations*, 52–54, 61–67) in identifying Cartier as the author of the narratives of the first and second voyages.

11. Bideaux, *Relations*, 113.
12. Bideaux, *Relations*, 116–117.
13. Trudel, *Vaines tentatives*, 86–88; Biggar, *Collection of Documents*, 43–45. On Saguenay, see also J. E. King, "The Glorious Kingdom of Saguenay," *Canadian Historical Review* 31 (1950): 390–400.
14. Bideaux, *Relations*, 358.
15. Ibid., 145.
16. Trudel, *Vaines tentatives*, 102.
17. Bideaux, *Relations*, 162 ("villains de leurs vivres").
18. Cook, *Voyages*, 79.
19. V. C. Slonosky, personal communication.
20. On French shipboard provisions, see Philippe Bonnichon, *Des cannibales aux castors: les découvertes françaises de l'Amérique, 1503–1788* (Paris: France-Empire, 1994), 60. On scurvy and early sixteenth-century expeditions, see Kenneth J. Carpenter, *The History of Scurvy and Vitamin C* (New York: Cambridge University Press, 1987), 3–12.
21. Robert A. Barakat, ed., *The Willoughby Papers: An Historical Record of New-foundland's First English Colony, 1610–c. 1631* (Hopkinton, MA: R. A. Barakat, 1995), 81, 84, 199–202.
22. Cook, *Voyages*, 77.
23. Ibid., 76–77.
24. Bideaux, *Relations*, 169.
25. Cook, *Voyages*, 76–77. Cartier gave the following figures: "In the middle of February, of the 110 men forming our company, there were not ten in good health . . . [a]nd not only were eight men dead already but there were more than fifty whose case seemed hopeless."
26. Ibid., 78–80.
27. Ibid., 80.
28. Jacques Rousseau, "L'annedda et l'arbre de vie," *Revue d'histoire de l'Amérique française* 8 (1954): 171–212; Jacques Mathieu, *L'annedda: l'arbre de vie* (Quebec: Septentrion, 2009).
29. Bideaux, *Relations*, 398–399.
30. Cook, *Voyages*, 81.
31. Ibid., 82.
32. Bideaux, *Relations*, 392–393.
33. Biggar, *Collection of Documents*, 75–81.
34. Bernard Allaire, *La rumeur dorée: Roberval et l'Amérique* (Montreal: Edition La Presse, 2013), 53–57.
35. Spanish intelligence of early 1541 in Biggar, *Collection of Documents*, 275–287. Nature of preparations discussed in Trudel, *Vaines tentatives*, 121–122, and Allaire, *Rumeur*, 94–97.
36. Estimate of expenses in Bideaux, *Relations*, 422; Allaire, *Rumeur*, 146, esti-mates that the entire venture cost some 40–50,000 livres altogether. English intelligence in Biggar, *Collection of Documents*, 188–189 (spelling modernized).

On the use of convicts, see Trudel, *Vaines tentatives*, 125–126, 139, and Allaire, *Rumeur*, 65–71.

37. Cartier's commission, dated October 1540, appears in Biggar, *Collection of Documents*, 128–131; Roberval's, dated January 1541, appears in ibid., 178–185. On Roberval's background, see Allaire, *Rumeur*, 1–52.

38. Biggar, *Collection of Documents*, 377–385. On numbers in the expeditions, see Richard Fiset and Gilles Samson, "Charlesbourg-Royal and France-Roy (1541–43): France's First Colonization Attempt in the Americas," *Post-Medieval Archaeology* 43 (2009): 62.

39. "The Third Voyage of Discovery," in Cook, *Voyages*, 96–106, quotes from 98, 100–101.

40. "The Voyage of John Francis de la Roche," in Cook, *Voyages*, 107–113, quotes from 108–109.

41. Allaire, *Rumeur*, 78, 85–87. Spanish intelligence in Biggar, *Collection of Documents*, 463.

42. Allaire, *Rumeur*, 105, 119–124.

43. "Voyage of John Francis de la Roche," 109, 110–111. Jean Alphonse quoted in Bideaux, *Relations*, 219.

44. Julie-Anne Bouchard-Perron and Allison Bain, "From Myth to Reality: Archaeobotany at the Cartier-Roberval Upper Fort Site," *Post-Medieval Archaeology* 43 (2009): 87–105; Hélène Côté, "The Archaeological Collection from the Cartier-Roberval Site (1541–43): A Remarkable Testimony to French Colonization Efforts in the Americas," *Post-Medieval Archaeology* 43 (2009): 71–86; Fiset and Samson, "Charlesbourg-Royal and France-Roy."

45. Trudel, *Vaines tentatives*, 162–164; Allaire, *Rumeur*, 128, 145.

46. Important diplomatic dispatches in Biggar, *Collection of Documents*, esp. 104–107, 135–137, 206, 279–287, 320–325 (quote).

47. Biggar, *Collection of Documents*, 561–564, 403–405, 447–467 (quote from 462), 430–435.

48. André Thevet, *André Thevet's North America: A Sixteenth-Century View*, ed. Roger Schlesinger and Arthur Stabler (Kingston, ON: McGill-Queen's University Press, 1986), 18. In the same passage, Thevet also reported Europeans' first taste of Canadian maple syrup—a treasure worth as much as any diamond mine, but not apparently lucrative enough to attract more voyages to Quebec.

49. Ibid., 9–10, 17–18, 22, 45, 54.

50. Bideaux, *Relations*, 125–128. It is not clear whether Cartier himself wrote that introduction.

51. Ibid., 220. See also Colin Coates and Dagomar Degroot, " 'Les bois engendrent les frimas et les gelées': comprendre le climat en Nouvelle-France," *Revue d'histoire de l'Amérique française* 68 (2015): 197–219.

52. Unlike many English, few French appear to have bought into the climatic and geographical ideas that underlay Arctic exploration and the search for a Northwest or Northeast Passage. See Bernard Allaire, "French Reactions to the Northwest Voyages and Assays of Geoffroy Le Brumen of the Frobisher Ore (1576–1584)," in *Meta Incognita: A Discourse of Discovery*, ed. Thomas H. B. Symons (Quebec: Canadian Museum of Civilization, 1999), 2:589–606.

53. V. Dickenson, "Cartier, Champlain, and the Fruits of the New World: Botanical Exchange in the 16th and 17th Centuries," *Scientia Canadensis* 31 (2008): 27–47; Charles André Julien, *Les voyages de découverte et les premiers établissements (XVe–XVIe siècles)* (Paris: Presses Universitaires de France, 1948), 306–367; François Marc Gagnon and Denise Petel, *Hommes effarables et bestes sauvages: images du Nouveau-Monde d'après les voyages de Jacques Cartier* (Montreal: Boréal, 1986); Morissonneau, "Le Nouveau Monde," 43–52.

54. On the Desceliers map, see Derek Hayes, *Historical Atlas of Canada* (Seattle: University of Washington Press, 2002), 29. The Bibliothèque Nationale de France has now made many originals available online, such as Guillaume le Testu's *Cosmographie universelle* (1555) (http://gallica.bnf.fr/ark: /12148/btv1b8447838j).

55. Marc Lescarbot, *The History of New France*, ed. W. L. Grant (Toronto: Champlain Society, 1907–1914), 2:190.

56. Laurier Turgeon, "The Cartier Voyages to Canada (1534–42) and the Beginnings of French Colonialism in North America," in *Charting Change in France around 1540*, ed. Marian Rothstein (Selinsgrove, PA: Susquehanna University Press, 2006), 97–118.

57. The history of Atlantic fisheries is the subject of a large specialist literature, much of it summarized for a popular audience in Brian M. Fagan, *Fish on Friday: Feasting, Fasting, and the Discovery of the New World* (New York: Basic Books, 2006). On medieval inland fishing and aquaculture, see Richard C. Hoffmann, "Economic Development and Aquatic Ecosystems in Medieval Europe," *American Historical Review* 101 (1996): 631–669. For zooarchaeology: J. H. Barrett, A. M. Locker, and C. M. Roberts, "The Origins of Intensive Marine Fishing in Medieval Europe: The English Evidence," *Proceedings of the Royal Society B—Biological Sciences* 271 (2004): 2417–2421.

58. On the reported abundance of New World fisheries, see W. Jeffrey Bolster, *The Mortal Sea: Fishing the Atlantic in the Age of Sail* (Cambridge, MA: Harvard University Press, 2012), 34–45. On early French fishermen in Newfoundland, see Bernard Allaire, "Les voyages de Jacques Cartier dans le contexte des navigations nord-américaines au XVIe siècle," *Revue d'histoire maritime* 4 (2005): 93–116.

59. Jürgen Alheit and Eberhard Hagen, "Long-Term Climate Forcing of European Herring and Sardine Populations," *Fisheries Oceanography* 6 (1997): 130–139; Jürgen Alheit, "Klimawandel und Fischbestände. Hering, Sardine und Sardelle," *Biologie in unserer Zeit* 38 (2008): 30–38; Georg H. Engelhard, David A. Righton, and John K. Pinnegar, "Climate Change and Fishing: A Century of Shifting Distribution in North Sea Cod," *Global Change Biology* 20 (2014): 2473–2483; Guðbjörg Ásta Ólafsdóttir et al., "Historical DNA Reveals the Demographic History of Atlantic Cod (*Gadus morhua*) in Medieval and Early Modern Iceland," *Proceedings of the Royal Society of London B: Biological Sciences* 281 (2014): 20132976; Bo Poulsen, "The Variability of Fisheries and Fish Populations prior to Industrialized Fishing: An Appraisal of the Historical Evidence," *Journal of Marine Systems* 79 (2010): 327–332; Geir Ottersen et al.,

"Major Pathways by Which Climate May Force Marine Fish Populations," *Journal of Marine Systems* 79 (2010): 343–360. On catches in the 1590s, see Bolster, *Mortal Sea*, 47. On changes in Arctic climate and North Atlantic currents, see Chapter 5.

60. Bolster, *Mortal Sea*, 39.

61. W. Jeffrey Bolster, "Putting the Ocean in Atlantic History: Maritime Communities and Marine Ecology in the Northwest Atlantic, 1500–1800," *American Historical Review* 113 (2008): 28. See also Richard C. Hoffmann, "A Long Voyage to the Banks of Newfoundland: How Medieval European Fisheries Went Overseas," in *Governing the Sea in the Early Modern Era: Festschrift for Roy Ritchie*, ed. Carole Shammas and Peter C. Mancall (San Marino, CA: Huntington Library, 2015), 15–38.

62. Richard C. Hoffmann, "Frontier Food for Late Medieval Consumers: Culture, Economy, Ecology," *Environment and History* 7 (2001): 131–167; Jan de Vries and Adriaan van der Woude, *The First Modern Economy: Success, Failure, and Perseverance of the Dutch Economy, 1500–1815* (New York: Cambridge University Press, 1997); Michael Khodarkovsky, *Russia's Steppe Frontier: The Making of a Colonial Empire, 1500–1800* (Bloomington: Indiana University Press, 2002); Sam White, *The Climate of Rebellion in the Early Modern Ottoman Empire* (New York: Cambridge University Press, 2011); Brett L. Walker, *The Conquest of Ainu Lands: Ecology and Culture in Japanese Expansion, 1590–1800* (Berkeley: University of California Press, 2001); Richard Eaton, *The Rise of Islam and the Bengal Frontier, 1204–1760* (Berkeley: University of California Press, 1993); Robert B. Marks, *Tigers, Rice, Silt, and Silk: Environment and Economy in Late Imperial South China* (New York: Cambridge University Press, 1998); John Richards, *The Unending Frontier: Environmental History of the Early Modern World* (Berkeley: University of California Press, 2003).

63. Lucien Campeau, "L'origine du commerce des fourrures en Amérique du Nord," in *Colloque Jacques Cartier: Histoire, textes, images* (Montreal: Montreal Historical Society, 1986), 84–99; Bernard Allaire, "L'arrivée des fourrures d'origine canadienne à Paris," in *La France-Amérique (XVIe–XVIIIe siècles)*, ed. Frank Lestringant (Paris: Honoré Champion, 1998), 239–258; Laurier Turgeon, "French Fishers, Fur Traders, and Amerindians during the Sixteenth Century: History and Archeology," *WMQ* 55 (1998): 585–610; Bernard Allaire, *Pelleteries, manchons et chapeaux de castor: les fourrures nord-américaines à Paris, 1500–1632* (Sillery, Marne: Septentrion, 1999).

64. Turgeon, "French Fishers"; Marcel Moussette, "A Universe under Strain: Amerindian Nations in North-Eastern North America in the 16th Century," *Post-Medieval Archaeology* 43 (2009): 30–47; Andrew Lipman, *The Saltwater Frontier: Indians and the Contest for the American Coast* (New Haven, CT: Yale University Press, 2015), 54–84; Matthew Bahar, "People of the Dawn, People of the Door: Indian Pirates and the Violent Theft of an Atlantic World," *Journal of American History* 101 (2014): 401–426.

65. Moussette, "Universe under Strain," 39; Trudel, *Vaines tentatives*, 247–251; Trudel, *Atlas*, 72–73.

66. See Chapter 1.

67. This synthesis is drawn from a now considerable range of evidence and interpretation on the disappearance of the St. Lawrence Iroquois. See Bruce G. Trigger, "Trade and Tribal Warfare on the St. Lawrence in the Sixteenth Century," _Ethnohistory_ 9 (1962): 240–256; Bruce G. Trigger, _Natives and Newcomers: Canada's "Heroic Age" Reconsidered_ (Montreal: McGill-Queen's University Press, 1985), 98, 147; Norman Clermont, "Le Developpement préhistorique des Iroquoiens du St-Laurent," in _Colloque Jacques Cartier: Histoire, Textes, Images_ (Montreal: Montreal Historical Society, 1986), 177–198; J. B. Jamieson, "Trade and Warfare: The Disappearance of the Saint Lawrence Iroquoians," _Man in the Northeast_ 39 (1990): 79–86; Daniel K. Richter, _The Ordeal of the Longhouse: The Peoples of the Iroquois League in the Era of European Colonization_ (Chapel Hill: University of North Carolina Press, 1992), 53; William R. Fitzgerald, "Contact, Neutral Iroquoian Transformation, and the Little Ice Age," in _Societies in Eclipse: Archaeology of the Eastern Woodland Indians, A.D. 1400–1700_, ed. David S. Brose, C. Wesley Cowan, and Robert C. Mainfort (Washington, DC: Smithsonian Institution Press, 2001), 37–47; Roland Tremblay, _The St. Lawrence Iroquoians: Corn People_ (Montreal: Montreal Museum of Archaeology and History, 2006), 119–122; James F. Pendergast, "More on When and Why the St. Lawrence Iroquois Disappeared," in _Essays in St. Lawrence Iroquoian Archaeology_, ed. James F. Pendergast and Claude Chapdelaine, 2nd ed. (St. John's, NL: Copetown Press, 2012), 7–48; James D. Rice, _Nature and History in the Potomac Country: From Hunter-Gatherers to the Age of Jefferson_ (Baltimore: Johns Hopkins University Press, 2009), 26–46. I am also indebted to the thoughtful discussion of Thomas Wickman, "Climate and Indigenous Peoples in History," in _The Palgrave Handbook of Climate History_, ed. S. White, C. Pfister, and F. Mauelshagen (London: Palgrave, forthcoming).
68. David B. Quinn, "The Voyage of Étienne Bellenger to the Maritimes in 1583: A New Document," _Canadian Historical Review_ 43 (1962): 328–343; Éric Thierry, _La France de Henri IV en Amérique du Nord: de la création de l'Acadie à la fondation de Québec_ (Paris: H. Champion, 2008), 36–41. Thierry's monograph, incorporating much new research in French archival sources, has superseded other standard histories of early New France, and it serves as the basic reference for the narrative that follows.
69. David B. Quinn, _England and the Discovery of America, 1481–1620_ (New York: Alfred A. Knopf, 1974), 313–336; "The Voyage of M. Charles Leigh," _PN_, 8:166–182.
70. Thierry, _France_, 41–47; Éric Thierry, "La Paix de Vervins et les ambitions françaises en Amérique," in _Le Traité de Vervins_, ed. Jean-François Labourdette, Jean-Pierre Poussou, and Marie-Catherine Vignal (Paris: Presses de l'Université de Paris-Sorbonne, 2000), 373–390; Juan López de Velasco, _Geografía y descripción universal de las Indias_ (Madrid: Atlas, 1971), 92; Antonio Martí Alanís, ed., _Canada en la correspondencia diplomática de los embajadores de España en Londres, 1534–1813_ (Madrid: Cultura Hispánica del Instituto de Cooperación Iberoamericana, 1980), 13–15, quote from 15.
71. Marc de Villiers and Sheila Hirtle, _Sable Island: The Strange Origins and Curious History of a Dune Adrift in the Atlantic_ (New York: Bloomsbury, 2009)

discusses the island's environment and geography. Parks Canada, which now owns the island, estimates more than 350 ships have wrecked there, earning it the nickname "The Graveyard of the Atlantic." See Parks Canada's website for the Sable Island National Park Reserve, http://www.pc.gc.ca/eng/pn-np/ns /sable/index.aspx (accessed May 20, 2016).

72. Lescarbot, *History*, 2:284–285.

73. E.g., Robert Le Blant and René Baudry, eds., *Nouveaux documents sur Champlain et son époque* (Ottawa: National Archives of Canada, 1967), 26–27.

74. Samuel de Champlain, "The Voyages of the Sieur de Champlain [1613]: Book I," in *The Works of Samuel de Champlain*, ed. Henry Percival Biggar, trans. William Francis Ganong (Toronto: Champlain Society, 1922), 1:228–229; Lescarbot, *History*, 2:194–195; Le Blant and Baudry, eds., *Nouveaux documents*, 78–79 (quote), 109–112; Joseph de Ber, "Un document inédit sur l'île de Sable et le Marquis de la Roche," *Revue d'histoire de l'Amerique française* 2 (1948): 199–213; Thierry, *Histoire*, 49–62, 119–121.

75. Champlain, "Voyages [1632]," in *Works*, 3:19. The "court of King Petaud" refers to a mythical court where no one was in charge—i.e., bedlam. As Thierry, *Histoire*, 74, points out, the usual figure of eleven deaths comes from the ambiguity of Champlain's difficult French: "les unze moururent miserable-ment, les autres patissans fort attendans le retour des vaisseaux." Here "les unze" could just as well be *les ons* (some) as *les onze* (the eleven).

76. Thierry, *Histoire*, 75–87, quote from 87.

77. Lescarbot, *History*, 2:20–21. Cf. Champlain, "Voyages [1632]," in *Works*, 3:264–265.

78. Champlain, "Voyages [1632]," in *Works*, 3:268, 292–293; Lescarbot, *History*, 2:278 (quote), 284–285.

79. Champlain, *Works*, 3:294–295 (I have altered the translation considerably to better match the original French). Cf. Lescarbot, *History*, 2:179, which criticizes Cartier for naming Chaleur Bay.

80. Lescarbot, *History*, 2:302, 306–307, 344–345, 346.

81. On Champlain's background and state of knowledge in 1603, see the introductory essay to Conrad E. Heidenreich and K. Janet Ritch, eds., *Samuel de Champlain before 1604: Des Sauvages and Other Documents Related to the Period* (Toronto: Champlain Society, 2010). On his cartographic skills, see Conrad E. Heidenreich, *Explorations and Mapping of Samuel de Champlain, 1603–1632* (Toronto: B. V. Gutsell, 1976). The authenticity of Champlain's voyage to the Caribbean and his authorship of the *Brief Discours* are discussed in François Marc Gagnon, "Is the *Brief Discours* by Champlain?," in *Champlain: The Birth of French America*, ed. Raymonde Litalien and Denis Vaugeois, trans. Käthe Roth (Montreal: Septentrion, 2004), 83–92. In support of that conclusion: (1) a subsequent document confirms that Champlain had previously served as a quartermaster-sergeant; (2) foreigners did occasionally bribe their way onto Spanish ships for travel to the Indies; (3) Champlain did not implausibly claim to actually captain a ship, only to look after one for his uncle; (4) his itinerary could be reasonably accommodated by some detours and delays; (5) subsequent research has discovered that there really was a "Captain Provençal" who

could have been Champlain's uncle; (6) the use of some Saintonge terms; (7) if we accept that Champlain was willing to borrow from the usual conventions of travel writing, his descriptions of places he probably didn't visit and the presence of some incredible natural wonders in his account can be excused. On Champlain as official *géographe du roi*, see, e.g., David Hackett Fischer, *Champlain's Dream* (New York: Simon and Schuster, 2008), 108–118. Other historians—e.g., Trudel, *Vaines tentatives*, 258—have also objected to this inference, which is drawn from a single line in Lescarbot's history.

82. On Champlain's traditional role as the visionary founding father of French Canada, see Mathieu d'Avignon, *Champlain et les fondateurs oubliés: les figures du père et le mythe de la fondation* (Québec: Presses de l'Université Laval, 2008); Raymonde Litalien, "Samuel Champlain, fondateur du Canada, sa présence dans la mémoire," in *Le Nouveau Monde et Champlain*, ed. Guy Martinière and Didier Poton (Paris: Les Indes Savantes, 2008), 17–26.

83. Heidenreich and Ritch, *Des Sauvages*, 328–329; Thierry, *Histoire*, 91–114.

84. Thierry, *Histoire*, 121–140. Although not a major figure in popular histories of Canada, de Monts has been the subject of several biographies: William Inglis Morse and Walter Edwards Houghton, eds., *Pierre du Gua, Sieur de Monts: Records: Colonial and "Saintongeois"* (London: Bernard Quaritch, 1939) (see 6–8 for his commission); Jean Liebel, *Pierre Dugua, sieur de Mons, fondateur de Québec* (Paris: Le Croît Vif, 1999) (see 83–84 for merchant resistance to his monopoly); Jean-Yves Grenon, *Pierre Dugua de Mons: Founder of Acadie (1604–5), Co-Founder of Quebec (1608)* (Annapolis Royal, NS: Peninsular Press, 2000); Guy Binot, *Pierre Dugua de Mons: gentilhomme royannais, premier colonisateur du Canada, lieutenant général de la Nouvelle-France de 1603 à 1612* (Vaux-sur-Mer, Charente-Maritime: Bonne Anse, 2004).

85. Samuel de Champlain, *Aux origines du Québec*, ed. Éric Thierry (Paris: Cosmopole, 2010), 36; Thierry, *Histoire*, 147–148.

86. Steven R. Pendery and H. W. Borns, eds., *Saint Croix Island, Maine: History, Archaeology, and Interpretation* (Augusta: Maine Historic Preservation Commission and the Maine Archaeological Society, 2012). Description based on the author's visit to both sites in June 2014 and the U.S. National Parks Service site https://www.nps.gov/sacr/index.htm (last visited May 16, 2017).

87. Champlain, *Aux origines*, 41–47, quote from 47.

88. Éric Thierry, "French Settlement, 1604–1613," in Pendery and Borns, eds., *St. Croix Island*, 21–22. On the evolution of French colonial architecture, see also André Robitaille, *Habiter en Nouvelle-France, 1534–1648* (Beauport, QC: MNH, 1996).

89. Steven R. Pendery, "The Cultural Landscape," in Pendery and Borns, eds., *Saint Croix Island*, 67–69.

90. Champlain, *Aux origines*, 49–50; Lescarbot, *History*, 257–258.

91. Steven R. Pendery, Stéphanie Noël, and Arthur Spiess, "Diet and Nutrition," in Pendery and Borns, eds., *Saint Croix Island*, 171–184; Lescarbot, *History*, 2:257–258; Champlain, *Aux origines*, 49.

92. Carpenter, *History of Scurvy*, 12–27; Hugh Plat, *Certaine Philosophical Preparations of Foode and Beverage for Sea-Men, in Their Long Voyages: With Some Necessary*,

Approoved, and Hermeticall Medicines and Antidotes, Fit to Be Had in Readinesse at Sea, for Prevention or Cure of Divers Diseases (London: H. Lownes, 1607); Lescarbot, *History,* 2:265–268; Marc Lescarbot, *Une lettre inédite de Lescarbot,* ed. Marcel Gabriel (Paris: C. Delagrave, 1885), 7; Champlain, *Aux origines,* 51.

93. Lescarbot, *History,* 2:257–258.

94. Steven R. Pendery, "History of Archaeology," in Pendery and Borns, eds., *Saint Croix Island,* 55; Thomas A. Crist et al., "The Skeletons," in Pendery and Borns, eds., *Saint Croix Island,* 185–222. The interpretation is based on the presence of dental lesions, absence of antemortem trauma, absence of enamel defects, and no specific evidence of any infectious disease. Numbers and timing of deaths are in Lescarbot, *History,* 257–258, and Champlain, *Aux origines,* 50. One death, that of René Noel, is precisely dated to March 31 by a document in Le Blant and Baudry, *Nouveaux documents,* 107–108.

95. Champlain, *Aux origines,* 66.

96. Ibid., 66, 71–72.

97. Ibid., 73.

98. Ibid., 73–75; Thierry, "French Settlement," 24–25; Robitaille, *Habiter.*

99. Champlain, *Aux origines,* 76; Thierry, *Histoire,* 170–172.

100. Lescarbot, *History,* 2:285–302; Lescarbot, *Lettre inédite;* Thierry, *Histoire,* 174–178. As with Pierre Dugua, Poutrincourt has not been a major figure in popular history but has been the subject of some biographies: e.g., Adrien Huguet, *Jean de Poutrincourt, fondateur de Port-Royal en Acadie, vice-roi du Canada, 1557–1615: campagnes, voyages et aventures d'un colonisateur sous Henri IV* (Paris: A. Picard, 1932).

101. Champlain, *Aux origines,* 85–104, quote from 103.

102. Lescarbot, *History,* 2:317–319, 324.

103. Champlain, *Aux origines,* 107–108; Lescarbot, *History,* 344–346, quote from 344. Lescarbot added that "the leaves do not appear on the trees until toward the end of the month of May."

104. Champlain, *Aux origines,* 108; Lescarbot, *History,* 2:343–344; Michael Salter, "L'Ordre de Bon Temps: A Functional Analysis," *Journal of Sport History* 3 (1976): 111–119; Éric Thierry, "A Creation of Champlain's: The Order of Good Cheer," in *Champlain: The Birth of French America,* ed. Raymonde Litalien and Denis Vaugeois, trans. Käthe Roth (Montreal: Septentrion, 2004), 135–142.

105. Binot, *Dugua,* 110–127; Le Blant and Baudry, *Nouveaux documents,* 102–106; Thierry, 204–212, 255–256.

106. Le Blant and Beaudry, *Nouveaux documents,* 162; Thierry, *Histoire,* 255–256; T. J. Kupp, "Quelques aspects de la dissolution de la compagnie de M. de Monts, 1607: brève étude de l'influence du commerce hollandais sur les premiers efforts concertés au XVIIe siècle en vue de coloniser le Canada," *Revue d'histoire de l'Amérique française* 24 (1970): 357–374.

107. Le Blant and Baudry, *Nouveaux documents,* 162; Thierry, *Histoire,* 137–138.

108. Lescarbot, *History,* 2:351.

109. Lescarbot, *History,* 2:346.

110. David B. Quinn, "The Preliminaries to New France: Site Selection for the Fur Trade by the French, 1604–1608," in *European Approaches to North America,*

1450–1640 (Aldershot, Hampshire: Ashgate, 1998), 255–271; Thierry, *Histoire,* 264.

111. Champlain, *Aux origines,* 127–128.

112. Thierry, *Histoire,* 290–291.

113. Thierry, *Histoire,* 267–268; Fischer, *Champlain's Dream,* 238–239, 465–478; Le Blant and Baudry, *Nouveaux documents,* 154–161; Leslie Choquette, *Frenchmen into Peasants: Modernity and Tradition in the Peopling of French Canada* (Cambridge, MA: Harvard University Press, 1997).

114. Thierry, *Histoire,* 293–295; Robitaille, *Habiter.*

115. Champlain, *Aux origines,* 145.

116. Ibid., 142–143.

117. Ibid., 145–147.

118. Carpenter, *Scurvy,* 231–232.

119. Champlain, *Aux origines,* 149–151; Lescarbot, *History,* 3:7–8.

120. Françoise Niellon, "Québec in the Time of Champlain," *Post-Medieval Archaeology* 43 (2009): 198–212; Anne Meachem Rick, "Études zooarchéologique du site de l'habitation de Champlain, périodes 1608–1632 et 1675–1700," in Françoise Niellon and Marcel Moussette, *Le site de l'habitation de Champlain a Québec: étude de la collection archéologique* (Quebec: Quebec Ministry of Cultural Affairs, 1985), 309–356.

121. Champlain, *Aux origines,* 147.

122. Ibid., 154.

123. Ibid., 165; Lesley-Ann L. Dupigny-Giroux, "Backward Seasons, Droughts and Other Bioclimatic Indicators," in *Historical Climate Variability and Impacts in North America,* ed. Lesley-Ann Dupigny-Giroux and Cary J. Mock (Berlin: Springer, 2009), 231–250. Past editors of this account and of Marc Lescarbot's retelling of it have doubted whether Champlain could have been telling the truth. However, June snowfalls did occur in the region during the 1810s and 1840s, and the snows on the mountaintops could have taken several weeks to melt. The author thanks Andy Nash at NOAA and Lesley-Ann Dupigny-Giroux, the Vermont state climatologist, for their assistance with this question.

124. Champlain, *Aux origines,* 167–168.

125. Historians have long been divided over the wisdom and humanity of Champlain's dealings with Indians, and indeed over the whole manner of coexistence between Natives and newcomers that would emerge in New France. The debate has pitted some ardent admirers of Champlain against some fierce detractors. On this historiography, see, e.g., Bruce G. Trigger, "Champlain Judged by His Indian Policy: A Different View of Early Canadian History," *Anthropologica* 13 (1971): 85–114; Gilles Havard and Cécile Vidal, *Histoire de l'Amérique française* (Paris: Flammarion, 2003), esp. 172, 208.

10. Such Wonders of Afflictions

1. Karen Kupperman, *The Jamestown Project* (Cambridge, MA: Harvard University Press, 2007), 241–243, quote from 241; James Horn, *A Land as God Made It: Jamestown and the Birth of America* (New York: Basic Books, 2005), 132–134.

2. Alexander Brown, ed., *The Genesis of the United States* (London: W. Heinemann, 1891), 1:206–237. The charter wasn't granted until May 23 (old style) but the letters patent were signed on February 17. The letter to Popham colony investors is in George E. Ellis et al., "March Meeting, 1886. The Parkman Manuscripts; Winslow Papers; Letter from the Virginia Company; 'Exploded' Coats of Arms; Flag of Fort McHenry," *Proceedings of the Massachusetts Historical Society* 2 (1885): 225–257.

3. Brown, *Genesis*, 1:235; see also Lord De La Warr's 1610 commission in ibid., 1:376.

4. William Strachey, *For the Colony in Virginea Britannia: Lawes Divine, Morall and Martiall* (London: Walter Burre, 1612), 7.

5. Virginia Council, "Instructions Orders and Constitutions to Sir Thomas Gates Knight Governor of Virginia (1609)," in *The Records of the Virginia Company of London*, ed. Susan M. Kingsbury (Washington, DC: Government Printing Office, 1906), 3:12–24, quotes from 12–13, 14, 22. On Argall's instructions, see also Emmanuel van Meteren, "Commentarien," in *JVFC*, 2:275–276, and Virginia Council, *A True and Sincere Declaration of the Purpose and Ends of the Plantation Begun in Virginia* (London: J. Stepney, 1610), 8–9.

6. Letter in Brown, *Genesis*, 1:317.

7. Letter in Brown, *Genesis*, 1:250–253.

8. Kupperman, *Jamestown*, 242–243; Theodore K. Rabb, *Enterprise and Empire: Merchant and Gentry Investment in the Expansion of England, 1575–1630* (Cambridge, MA: Harvard University Press, 1967), 72–73, 84.

9. Peter C. Mancall, *Hakluyt's Promise: An Elizabethan's Obsession for an English America* (New Haven, CT: Yale University Press, 2007), 270–272; Richard Hakluyt, trans., *Virginia Richly Valued, by the Description of the Maine Land of Florida, Her Next Neighbour: Out of the Foure Yeeres Continuall Travell and Discoverie . . . of Don Ferdinando De Soto* (London: Felix Kingston, 1609); Marc Lescarbot, *Nova Francia, or The Description of That Part of New France Which Is One Continent with Virginia*, trans. P. Erondelle (London: George Bishop, 1609), 2.

10. Virginia Council, *True and Sincere Declaration*, 2, 3, 4.

11. On these sermons, see Edward L. Bond, *Damned Souls in a Tobacco Colony: Religion in Seventeenth-Century Virginia* (Macon, GA: Mercer University Press, 2000), 1–36; Horn, *Land as God Made It*, 139–140; Kupperman, *Jamestown*, 243–247.

12. Robert Gray, *A Good Speed to Virginia* (London: William Welby, 1609), fol. B3; Robert Johnson, *Nova Britannia. Offering Most Excellent Fruites by Planting in Virginia. Exciting All Such as Be Well Affected to Further the Same* (London: Samuel Macham, 1609), fol. D1. See also William Symonds, *Virginia: A Sermon Preached at White-Chappel, in the Presence of Many, Honourable and Worshipfull, the Adventurers and Planters for Virginia 25 April 1609* (London: Eleazar Edgar, 1609), 19–22.

13. Gray, *Good Speed*, fol. D1. See also Robert Tynley, *Two Learned Sermons* (London: Thomas Adams, 1609), 67–68.

14. Johnson, *Nova Britannia*, B4v, C2v–C3r; Daniel Price, *Sauls Prohibition Staide* (London: Matthew Law, 1609), fol. F2v; Richard Crakanthorpe, *A Sermon at the*

Solemnizing of the Happie Inauguration (London: Thomas Adams, 1609), fol. D2r.

15. William Crashaw, *A Sermon Preached in London before the Right Honorable the Lord Lawarre . . . February 21, 1609* (London: William Welby, 1610), fols. E1–E2, F4.

16. William Strachey, "A True Repertory of the Wracke, and Redemption of Sir Thomas Gates," in *Hakluytus Posthumus,* ed. Samuel Purchas (Glasgow: James MacLehose and Sons, 1906), 19:6. The other eyewitness accounts compiled for this description are: Testimony of William Box, *CWCJS,* 2:219–220; Gabriel Archer, "A Letter of M. Gabriel Archar, Touching the Voyage of the Fleet of Ships, Which Arrived at Virginia, without Sir Tho. Gates, and Sir George Summers, 1609," *JVFC,* 2:279–283; George Somers, "Letter to Salisbury (15 June 1610)," in Brown, *Genesis,* 1:400–402; and Silvester Jourdain, *A Plaine Description of the Barmudas* (London: William Welby, 1613). The one element missing from these descriptions is an accurate location. According to Archer, they were "about one hundred and fiftie leagues distant from the West Indies," and according to Somers, "about some 200 leagues from the Bermooda [Barbuda?]."

17. Jourdain, *Plaine Description;* Box, *CWCJS,* 2:219–220; Archer, "Letter"; Somers, "Letter."

18. Strachey, "True Repertory"; Archer, "Letter."

19. Jourdain, *Plaine Description;* Strachey, "True Repertory."

20. See Chapter 7 for descriptions of the winter.

21. On weather prodigies and providential interpretations, see Vladimir Jankovic, *Reading the Skies: A Cultural History of English Weather* (Chicago: University of Chicago Press, 2000), 33–44, 56–59, and John Emrys Morgan, "Understanding Flooding in Early Modern England," *Journal of Historical Geography* 50 (2015): 37–50. On the causes of the flood, see E. Bryant and S. Haslett, "Was the AD 1607 Coastal Flooding Event in the Severn Estuary and Bristol Channel (UK) Due to a Tsunami?," *Archaeology in the Severn Estuary* 13 (2002): 163–167; S. Haslett and E. Bryant, "The AD 1607 Coastal Food in the Bristol Channel and Severn Estuary: Historical Records from Devon and Cornwall (UK)," *Archaeology in the Severn Estuary* 15 (2004): 81–89; Kevin Horsburgh and Matt Horritt, "The Bristol Channel Floods of 1607: Reconstruction and Analysis," *Weather* 61 (2007): 272–277. Original descriptions of the flood from *1607 Lamentable Newes out of Monmouthshire in Wales. Contayning, The Wonderfull and Most Fearefull Accidents of the Great Overflowing of Waters in the Saide Countye, Drowning Infinite Numbers of Cattell of All Kinds, as Sheepe, Oxen, Kine and Horses, with Others: Together with the Losse of Many Men, Women and Children, and the Subversion of XXVI Parishes in January Last 1607* (London: William Welby, 1607), B4v; Todd Gray, ed., *The Lost Chronicle of Barnstaple 1586–1611* (Exeter, Devon: Devonshire Association, 1998), 94–96; and William Adams, *Adams's Chronicle of Bristol,* ed. Francis F. Fox (Bristol: J. W. Arrowsmith, 1910), 183. Pamphlets on the flood include: *1607. A True Report of Certaine Wonderfull Overflowings of Waters* (London: Edward White, 1607), A3r; William Jones, *Gods Warning to His People of England, by the Great Overflowing*

of the Waters (London: W. Barley and J. Bayly, 1607); *Miracle upon Miracle or A True Relation of the Great Floods* (London: Nathanael Fosbrook and John Wright, 1607); *More Strange Newes: Of Wonderfull Accidents Hapning by the Late Overflowings of Waters* (London: Edward White, 1607); *Een Warachtich Verhael van de Schrickelicke Springh-Vloedt in het Landtschap van Summerset* (Amsterdam: C. Claesz., 1607); *Discours veritable et tres-piteux, de l'inondation* (Paris: Fleury Bourriquant, 1607). On 1607 and the Pilgrims, see Jeremy Dupertuis Bangs, *Strangers and Pilgrims, Travellers and Sojourners: Leiden and the Foundations of Plymouth Plantation* (Plymouth, MA: General Society of Mayflower Descendants, 2009), 1–8, 48–49.

22. *De Orbe Novo*, 2:309. See also Stuart B. Schwartz, *Sea of Storms: A History of Hurricanes in the Greater Caribbean from Columbus to Katrina* (Princeton, NJ: Princeton University Press, 2015), 14.

23. Schwartz, *Sea of Storms*, 17–32, quote from 31. See also Sherry Johnson, *Climate and Catastrophe in Cuba and the Atlantic World in the Age of Revolution* (Chapel Hill: University of North Carolina Press, 2011), 1–20.

24. E.g., Aslak Grinsted, John C. Moore, and Svetlana Jevrejeva, "Homogeneous Record of Atlantic Hurricane Surge Threat since 1923," *Proceedings of the National Academy of Sciences* 109 (2012): 19601–19605; Aslak Grinsted, John C. Moore, and Svetlana Jevrejeva, "Projected Atlantic Hurricane Surge Threat from Rising Temperatures," *Proceedings of the National Academy of Sciences* 110 (2013): 5369–5373. For the state of the field at the time of writing on research associating climatic change and hurricane strength, see Adam H. Sobel et al., "Human Influence on Tropical Cyclone Intensity," *Science* 353 (2016): 242–246.

25. The two recent studies are Valerie Trouet, Grant L. Harley, and Marta Domínguez-Delmás, "Shipwreck Rates Reveal Caribbean Tropical Cyclone Response to Past Radiative Forcing," *Proceedings of the National Academy of Sciences* 113 (2016): 3169–3174, and Michael J. Burn and Suzanne E. Palmer, "Atlantic Hurricane Activity during the Last Millennium," *Scientific Reports* 5 (2015): 12838. For concurring information from Spanish archives, see Ricardo García-Herrera et al., "New Records of Atlantic Hurricanes from Spanish Documentary Sources," *Journal of Geophysical Research: Atmospheres* 110 (2005): D03109. For further documentary sources on past hurricane frequency, see, e.g., Michael Chenoweth and Dmitry Divine, "A Document-Based 318-Year Record of Tropical Cyclones in the Lesser Antilles, 1690–2007," *Geochemistry, Geophysics, Geosystems* 9 (2008): Q08013, and Michael Chenoweth and Dmitry Divine, "Tropical Cyclones in the Lesser Antilles: Descriptive Statistics and Historical Variability in Cyclone Energy, 1638–2009," *Climatic Change* 113 (2012): 583–598. For sedimentary records from Bermuda and the Bahamas, see Peter J. van Hengstum et al., "Low-Frequency Storminess Signal at Bermuda Linked to Cooling Events in the North Atlantic Region," *Paleoceanography* 30 (2015): 52–76, and Peter J. van Hengstum et al., "Heightened Hurricane Activity on the Little Bahama Bank from 1350 to 1650 AD," *Continental Shelf Research* 86 (2014): 103–115.

26. Schwartz, *Sea of Storms*, 54. See also Matthew Mulcahy, *Hurricanes and Society in the British Greater Caribbean, 1624–1783* (Baltimore: Johns Hopkins University

Press, 2006), 33–64. "For some—probably a significant majority during the first several decades of the seventeenth century—the storms came directly from the hand of God. They interpreted hurricanes as 'wondrous events,' divine judgments for human sins" (33).

27. Hobson Woodward, *A Brave Vessel: The True Tale of the Castaways Who Rescued Jamestown and Inspired Shakespeare's* The Tempest (New York: Viking, 2009). For textual evidence that Strachey's "True Repertory" was almost certainly the original text consulted by Shakespeare, see also Tom Reedy, "Dating William Strachey's 'A True Repertory of the Wracke and Redemption of Sir Thomas Gates': A Comparative Textual Study," *Review of English Studies* 61 (2010): 529–552.

28. Amy Mitchell-Cook and William N. Still, *A Sea of Misadventures: Shipwreck and Survival in Early America* (Columbia: University of South Carolina Press, 2013), 51–71; Sarah Parsons, "The 'Wonders in the Deep' and the 'Mighty Tempest of the Sea': Nature, Providence and English Seafarers' Piety, c. 1580–1640," in *God's Bounty? The Churches and the Natural World*, ed. Peter Clarke and Tony Claydon (Woodbridge, Suffolk: Boydell Press, 2010), 194–204. See also the comparison to Jonah and other biblical references in the Virginia Council's *True Declaration*, 21–26.

29. Jourdain, *Plaine Description*, 9–10. Similar description in Strachey, "True Repertory," 13.

30. Strachey, "True Repertory," 13.

31. The discovery and excavation of the *Sea Venture* appears to confirm this account. See Jonathan Adams, "Sea Venture: A Second Interim Report, Part 1," *International Journal of Nautical Archaeology and Underwater Exploration* 14 (1985): 275–299.

32. *Plaine Description*, 10–11. See also Somers, "Letter," 401: "Wee saved all our lives, and afterwardes saved much of our goodes, but all our bread was wet and lost."

33. Strachey, "True Repertory," 13–14; Jourdain, *Plaine Description*, 11.

34. J. H. Lefroy, *Memorials of the Discovery and Early Settlement of the Bermudas or Somers Islands, 1515–1685* (London: Longmans, Green, 1877), 1–9; *HGIO*, decade 4, book II, chapter vi, 2:31.

35. Samuel de Champlain, "Brief Discours," in *The Works of Samuel de Champlain*, ed. Henry Percival Biggar, trans. Hugh Hornby Langton (Toronto: Champlain Society, 1922), 1:76–77.

36. Jourdain, *Plaine Description*, 11–13; Strachey, "True Repertory," 17–24.

37. Strachey, "True Repertory," 16.

38. Jourdain, *Plaine Description*, 11. The last reported freeze occurred on Christmas Eve, 1840; see William S. Zuill, *The Story of Bermuda and Her People*, 3rd ed. (London: Macmillan Education, 1999), 45.

39. *CWCJS*, 1:257–258; Neal Salisbury, "The Indians' Old World: Native Americans and the Coming of Europeans," *WMQ* 53 (1996): 435–458; Martin D. Gallivan, *The Powhatan Landscape* (Gainesville: University Press of Florida, 2016), 89–90; William Denevan, "The Pristine Myth: The Landscape of the Americas in 1492," *Annals of the Association of American Geographers* 82 (1992): 369–385; Shephard Krech, *The Ecological Indian: Myth and History* (New York: Norton, 1999).

40. Strachey, "True Repertory," 28–38.
41. Richard Norwood, "Discovery of the Bermuda Islands (1622)," in *John Pory's Lost Description of Plymouth Colony,* ed. Champlin Burrage (Boston: Houghton Mifflin, 1918), 3–34, quote from 7–8; Nathaniel Butler, *The Historye of the Bermudaes or Summer Islands,* ed. J. H. Lefroy (London: Hakluyt Society, 1882), 34 (quote), 41–42 (quote), 202, 233.
42. *CWCJS,* 2:390–391. Compare similar stories in Richard Grove, *Green Imperialism: Colonial Expansion, Tropical Island Edens, and the Origins of Environmentalism 1600–1800* (New York: Cambridge University Press, 1995), and John R. McNeill, "Of Rats and Men: A Synoptic Environmental History of the Island Pacific," *Journal of World History* 5 (1994): 299–349.
43. *CWCJS,* 1:259–263.
44. Ibid., 1:263–264.
45. "A Briefe Declaration of the Plantation of Virginia duringe the First Twelve Yeares, When Sir Thomas Smith Was Governor of the Companie, and downe to This Present Tyme. By the Ancient Planters Nowe Remaining Alive in Virginia. 1624," in *Colonial Records of Virginia* (Richmond: State of Virginia, 1874), 70 ("flead" in the original).
46. Emmanuel van Meteren, "Commentarien," in *JVFC,* 2:275–276; Virginia Council, *True and Sincere Declaration,* 8–9.
47. *CWCJS,* 1:267; "Briefe Declaration of the Plantation," 70.
48. Testimony of William Box, in *CWCJS,* 2:219–220; Archer, "Letter."
49. "Briefe Declaration of the Plantation," 70.
50. *CWCJS,* 1:176.
51. George Percy, "A Trewe Relacyon," 1625, Elkins Collection, Free Library of Philadelphia, n.p. Transcription by Mark Nicholls available at https://www.history.org/Foundation/journal/Winter07/jamestownDiary.cfm (accessed July 15, 2016). Given the especially difficult orthography, I have modernized Percy's spelling and punctuation throughout.
52. Percy, "Trewe Relacyon." Cf. *CWCJS,* 1:270; "Briefe Declaration of the Plantation," 70–71.
53. *CWCJS,* 1:270–271; Percy, "Trewe Relacyon"; Henry Spelman, *Relation of Virginia* [c. 1613], ed. James Frothingham Hunnewell (London: Chiswick Press, 1872), 16–17. For analysis of whether Smith actually purchased the village or colluded in the assault, see Horn, *Land as God Made It,* 168–169.
54. Percy, "Trewe Relacyon."
55. *CWCJS,* 1:275.
56. For an attempt to reconstruct the progress of relations from Wahunsenacawh's point of view, see Helen C. Rountree, *Pocahontas Powhatan Opechancanough: Three Indian Lives Changed by Jamestown* (Charlottesville: University of Virginia Press, 2006), 134–147. On devalued goods, see esp. Strachey, "True Repertory," 50.
57. *JVFC,* 2:281–282; *CWCJS,* 1:263–264; Strachey, "True Repertory," 45, 49–50, 61; Somers, "Letter," 401. See also Kupperman, *Jamestown,* 251–253.
58. Percy, "Trewe Relacyon"; Spelman, *Relation,* 23–24; *CWCJS,* 2:232. Jeffrey Shortridge was the only survivor among those sent from Jamestown. Henry Spelman was already living among the Powhatans.

59. Percy, "Trewe Relacyon."
60. Strachey, "True Repertory," 44–45.
61. Calculations in J. Frederick Fausz, "An 'Abundance of Blood Shed on Both Sides': England's First Indian War, 1609–1614," *Virginia Magazine of History and Biography* 98 (1990): 56; William M. Kelso, *Jamestown: The Buried Truth* (Charlottesville: University of Virginia Press, 2006), 90.
62. Kelso, *Jamestown*, 154–165; Chris Cesare, "Historic Human Burials Identified," *Nature News*, July 28, 2015, doi:10.1038 / nature.2015.18079.
63. Strachey, "True Repertory," 58–59; Kelso, *Jamestown*, 119–124. Compare the discussion of Jamestown mortality during 1607 in Chapter 6.
64. On animals in the colony, see *CWCJS*, 1:263, 273; Strachey, "True Repertory," 61. On the kind of swine taken to North America and their early proliferation, see Sam White, "From Globalized Pig Breeds to Capitalist Pigs: A Study in Animal Cultures and Evolutionary History," *Environmental History* 16 (2011): 94–120, and Virginia Anderson, *Creatures of Empire: How Domestic Animals Transformed Early America* (New York: Oxford University Press, 2004).
65. Percy, "Trewe Relacyon"; Somers, "Letter," 401; "Briefe Declaration of the Plantation," 71; Kelso, *Jamestown*, 92–93.
66. Percy, "Trewe Relacyon"; *CWCJS*, 1:276; Rountree, *Pocahontas*, 146.
67. "Briefe Declaration of the Plantation," 71.
68. "Briefe Declaration of the Plantation," 71; Percy, "Trewe Relacyon."
69. For a skeptical reading of the evidence, see Rachel B. Herrmann, "The 'Tragicall Historie': Cannibalism and Abundance in Colonial Jamestown," *WMQ* 68 (2011): 47–74. On famine and cannibalism, see Cormac Ó Gráda, *Eating People Is Wrong, and Other Essays on Famine, Its Past, and Its Future* (Princeton, NJ: Princeton University Press, 2015), 11–37.
70. James Horn et al., *Jane: Starvation, Cannibalism, and Endurance at Jamestown* (Williamsburg, VA: Colonial Williamsburg Foundation, 2013).
71. See Horn, *Land as God Made It*, 177. No explicit timing or reason appears in the original sources, but this appears to be the best inference.
72. Percy, "Trewe Relacyon."
73. Strachey, "True Repertory," 44–45; Somers, "Letter," 401; Percy, "Trewe Relacyon."
74. Strachey, "True Repertory," 45. "Briefe Declaration of the Plantation," 71, claims "many soon after died." Percy, "Trewe Relacyon," indicates that at least two more were killed by Indians. However, Somers, "Letter," 401, claims they managed to save all but two or three of the colonists.
75. On sturgeon size, growth, and distribution, see Beverly A. Straube, *The Archaearium: Rediscovering Jamestown, 1607–1699* (Williamsburg: APVA Preservation Virginia, 2007), 63, and Matthew T. Balazik et al., "Changes in Age Composition and Growth Characteristics of Atlantic Sturgeon (*Acipenser oxyrinchus oxyrinchus*) over 400 Years," *Biology Letters* 6 (2010): 708–710. The quotation from Smith comes from *CWCJS*, 1:264. See also *CWCJS*, 1:146–147: "In somer no place affordeth more plentie of Sturgeon." On Argall's voyage, see Emmanuel van Meteren, "Commentarien," *JVFC*, 2:276, and Virginia Council, *True and Sincere Declaration*, 9–10. Strachey's quotation comes from "True Repertory," 52. Concurring accounts are found in "Letter of the

Governor and Council of Virginia to the Virginia Company (7 July 1610)," in Brown, *Genesis*, 1:408, and Thomas West (Lord De La Warr), "Letter to Salisbury [1610]," in Brown, *Genesis*, 1:415.

76. This narrative merges details from Percy, "Trewe Relacyon"; Jourdain, *Plaine Description*, 14–15; Strachey, "True Repertory," 43–54, quote from 53; "Briefe Declaration of the Plantation," 71–72; *CWCJS*, 1:276–277, 2:233–234; and documents in Brown, *Genesis*, 1:401–415. The narratives include conflicting details that are impossible to reconcile. Somers mentions the loading of four ships at Jamestown, and the letter of the governor and council mentions "the aforesaid pinnas [the *Virginia*] which was some 4 or 5 days come away before, to prepare those at Pointe Comforte." However, Percy describes their departure from Jamestown in only three ships—Gates in the *Deliverance*; Somers and Percy in the *Patience*; and Capt. Davis in the *Virginia*—implying that the *Discovery* had already been sent to Point Comfort. Regarding their destination, Strachey, Jourdain, and the letter of the governor and council all specify Newfoundland; Percy and Smith claim it was England; and Somers does not specify. Newfoundland seems by far the most likely option, given its proximity and its role as a haven for previous failed voyages, such as the Knight expedition described in Chapter 5. De La Warr's letter describes Gates "comminge downe the river haveinge shipped the hole companie and Colonie in two small pinnasses with a determination to staie some tenn Daies at Cape Comfort to expect our Commine." No other source mentions any expectation that De La Warr would arrive in time to save them. Moreover, De La Warr wrote that Gates had thirty days' provisions left, while no other source gives him more than sixteen. I suspect one of three possibilities: (1) De La Warr was simply misinformed. After all, De La Warr had even mistaken the number of ships carrying the men, so he could have gotten this detail wrong as well. (2) Gates later told De La Warr that they intended to wait for him at Point Comfort, to help excuse their departure from Jamestown and to emphasize their loyalty to the Virginia Company. (3) Gates did intend a brief stopover at Point Comfort in order to refresh his passengers and perhaps load up on provisions already found at that settlement. Gates or De La Warr simply exaggerated those intentions into plans to wait ten days, until supplies would have been desperately low, even for the shorter voyage to Newfoundland. This could be what Strachey meant by their intention to depart "with all speede convenient" (a phrase copied in the letter of the governor and council).

Acknowledgments

I came to Atlantic history as an outsider, having just written a very different kind of book about the Ottoman Empire. I am grateful to the many people and institutions that welcomed me into the field and made this project possible. Funding for research was provided by fellowships from the John Carter Brown Library, the Huntington Library, Oberlin College, and the Ohio State University. I'd like to start by thanking those institutions, their librarians, and all my fellow fellows who shared afternoon tea and discussion at the Huntington and evening rum and conversation at Fiering House. Many other people and institutions helped make my research and travel an enjoyable and sometimes eye-opening experience. I would especially like to thank those who answered my questions at Acoma Sky City Cultural Center; various historical societies and museums of New Mexico, Maine, Virginia, Quebec, and Bermuda; the National Parks Service and Parks Canada; and Maine's First Ship.

I am indebted to the many colleagues who gave me the chance to test out my work along the way in seminars, conference panels, and invited talks, including Franz Mauelshagen, Ken Pomeranz, Brad Skopyk, Mark Cane, Dagomar Degroot, Colin Coates, Anya Zilberstein, Jim Scott and the Agrarian Studies seminar at Yale, the National Socio-Environmental Synthesis Center, and the PAGES Volcanoes in Climate and Society working group. This project benefited tremendously from the work of editors and reviewers for the several articles and chapters written in preparation for this book—including Durwood Ball, Charles Cutter, Richard Flint, Shirley Cushing Flint, and Joshua Piker—as well as the anonymous reviewers for the book manuscript. Thanks also to assistants Zeb Larson and Sara Halpern. Many experts kindly shared unpublished data with me and patiently helped with what must have been exasperating questions on everything from the geometry of tree growth to the migration of fish: my thanks to Karl Kreutz,

Julie Richey, Andy Nash, Lesley-Ann Dupigny-Giroux, Fabio Gennaretti, Francis Ludlow, Eduardo Moreno-Chamarro, and Matthew Balazik, among many others I have surely forgotten to mention. I am also grateful to those who reviewed particular passages of the book for accuracy, including Richard Hoffmann, Christian Pfister, and Davide Zanchettin. Above all, I am humbled by the generosity of colleagues who read drafts of the entire manuscript and offered invaluable feedback: Tom Wickman, Vicky Slonosky, John Brooke, Geoffrey Parker, and last but not least, John McNeill, without whose unfailing advice and support I never would have made it where I am today.

It has been wonderful to work with Harvard University Press and its production team. Thank you to Andrew Kinney, Olivia Woods, Angela Piliouras, Sue Warga, and especially to cartographer and illustrator Bill Keegan, who has worked so patiently with me getting the maps just right. In the meantime, I have enjoyed the support and good humor of family and friends—and on some happy occasions, my cat, Madeline, who has condescended to come upstairs and keep me company during long hours of writing and rewriting. Thank you, Emily and Violette, for putting up with all my weeks and months away, and all the times I've been here in the flesh but with my mind somewhere in the sixteenth century.

And if after all that help—and more—there are still mistakes to be found in this book, then the reader may rest assured, they're my own damn fault.

Index